Springer Optimization and Its Applications

Volume 184

Aims and Scope

Optimization has continued to expand in all directions at an astonishing rate. New algorithmic and theoretical techniques are continually developing and the diffusion into other disciplines is proceeding at a rapid pace, with a spot light on machine learning, artificial intelligence, and quantum computing. Our knowledge of all aspects of the field has grown even more profound. At the same time, one of the most striking trends in optimization is the constantly increasing emphasis on the interdisciplinary nature of the field. Optimization has been a basic tool in areas not limited to applied mathematics, engineering, medicine, economics, computer science, operations research, and other sciences.

The series **Springer Optimization and Its Applications (SOIA)** aims to publish state-of-the-art expository works (monographs, contributed volumes, textbooks, handbooks) that focus on theory, methods, and applications of optimization. Topics covered include, but are not limited to, nonlinear optimization, combinatorial optimization, continuous optimization, stochastic optimization, Bayesian optimization, optimal control, discrete optimization, multi-objective optimization, and more. New to the series portfolio include Works at the intersection of optimization and machine learning, artificial intelligence, and quantum computing.

Volumes from this series are indexed by Web of Science, zbMATH, Mathematical Reviews, and SCOPUS.

More information about this series at http://www.springer.com/series/7393

Dionysis D. Bochtis • Claus Grøn Sørensen
Spyros Fountas • Vasileios Moysiadis
Panos M. Pardalos
Editors

Information and Communication Technologies for Agriculture—Theme III: Decision

Springer

Editors
Dionysis D. Bochtis
Institute for Bio-Economy and Agri-Technology (iBO)
Centre for Research and Technology Hellas (CERTH)
Thessaloniki, Greece

Claus Grøn Sørensen
Department of Electrical and Computer Engineering
University of Aarhus
Aarhus N, Denmark

Spyros Fountas
Department of Resources Management and Agricultural Engineering
Agricultural University of Athens
Athens, Greece

Vasileios Moysiadis
Institute for Bio-Economy and Agri-Technology (iBO)
Centre for Research and Technology Hellas (CERTH)
Thessaloniki, Greece

Panos M. Pardalos
Department of Industrial and Systems Engineering
University of Florida
Gainesville, FL, USA

ISSN 1931-6828 ISSN 1931-6836 (electronic)
Springer Optimization and Its Applications
ISBN 978-3-030-84151-5 ISBN 978-3-030-84152-2 (eBook)
https://doi.org/10.1007/978-3-030-84152-2

Mathematics Subject Classification: 68U35; 90B50

This Springer imprint is published by the registered company Springer Nature Switzerland AG
The registered company address is: Gewerbestrasse 11, 6330 Cham, Switzerland

Preface

Every action is based on a decision, while every decision requires acquisition and processing of the information available. The first book of the series on ICT for Agriculture (Theme I: Sensors) focuses on the data acquisition with the use of sensors, while the second one (Theme II: Data) focuses on data processing and utilization. This book here, the third of the series (Theme III: Decision) focuses on the transformation of the collected information into valuable decisions. The successful transition to the new digitized era of agriculture requires the implementation of elaborated decision-making for farmers, processors, and distribution channels of agricultural products. The focus is how to better use digital technologies to reduce cost, inputs, and time, and be more efficient and transparent. The book consists of 14 chapters relevant, and complementary to each other, to agricultural production and related products' distribution. Contributions are grouped in three distinct sections, namely: Value Chain, Primary Production, and Environment, based on the thematic area of each individual chapter, as well as the subject of the decision at hand.

The first section of the book is dedicated to decisions in the value chain of agricultural products. Value chain is a sequence of processes that add value to a product or commodity. The processes in the value chain are interconnected and cover the entire production cycle, including transportation, storage, processing, as well as promotion and marketing of products. Each process involves several stakeholders and products rendering a value chain a complex system that requires specialized organizing to make proper decisions. The application of ICT finds suitable grounds in agricultural value chain, improving the effectiveness and sustainability of production by the utilization of innovative techniques that can be integrated in traditional production processes. The key to the integrated processing of the value chain lays to the abundance of the available data. The proper manipulation of different types of data offers a promising potential for the improvement of production processes through custom in-field applications. Farm Information Management Systems are the means through which the data collected across the value chain can be processed for the extraction of useful and applicable information.

The success in the development of smart farming applications is a challenging task, and it is usually impeded by the lack of interoperability between the involved systems due to different technological architectures, standards, and communication protocols. The data collected in the agricultural sector derive from a variety of heterogenous sources. Additionally, there are not widely accepted standards for data collection and processing while there is not a wide application of interoperability mechanisms that allow for the connection of existing models. This fact highlights the lack of connectivity for the exchange and integration of the collected data, which is a matter of utmost importance in the development of smart farming applications. Towards that direction the chapter entitled "**Agricultural Information Model**" addresses these issues through the presentation of the respective model that was developed by the H2020 DEMETER project. The chapter elaborates on the design of an information model adhering to a layered and modular approach to develop a suite of ontologies that are in accordance with best practices by exploiting existing standards and well-scoped models. Aim is to establish alignments between these models to achieve interoperability and integration of existing data. The model is designed in an expandable manner, meaning that new concepts can be incorporated to the models, depending on the emerging needs. Therefore, the model is adjustable to farmers and stakeholders' preferences, through the realization of smart farming solutions that connect different systems and platforms of the agri-food sector. More specifically, the model acts as a mediator that supports the decision-making process by facilitating different systems in data exchange and providing access to different sources of data. The model is designed in three layers (meta-model layer, cross-domain layer, and domain-specific layer) while it is easy to expand and maintain. The chapter presents in detail the information model development process including the collected requirements and a description of the implementation details and the aspects of semantic interoperability that were investigated. Examples of the use of the model are also presented for the support of developers and users.

The increase in the consumption of protein products, along with the adverse environmental impacts that their production is connected to, calls for immediate actions with respect to the respective value chains. With respect to the value chain of beef cattle, Precision Livestock Farming (PLF) as well as Management Information Systems (MIF) are used to increase the efficiency of operations through improved decision-making. Moreover, the use of such systems enables the successful compliance to food quality and sustainability standards adding value to the final product, increasing at the same time consumers' trust. The integration of Internet of Things (IoT) technologies ensures the interoperability between the different points in the agri-food value chains. In the case of beef production, these points include all the stages of production (breeding, growing, finishing, slaughtering, meat processing, transportation, logistics, etc.) creating a value chain that covers the entire agroecosystem. Within this value chain, a vast number and forms of data can be gathered, analyzed, and stored, creating the need for systems for the assessment of the collected information.

Additionally, an increasing demand for improved traceability within the beef value chain is observed, leading to the eventual adoption of innovative technologies, towards a direction that creates a new data flow that needs to be handled properly. The proper processing of the collected data can offer a variety of benefits including the improvement of the livestock welfare, the increase in quality of the products, the reduction of waste, and the satisfaction of consumers' demands. Addressing the above issues, the chapter "**Development of a Framework for Implementing IOT-A on the Beef Cattle Value Chain**" presents the implementation of the IoT paradigm on the beef cattle value chain. The corresponding value chain is thoroughly described considering its main stages, stakeholders, processes, and the informational flow. In sequence, the requirements and services needed for implementing IoT were presented. A model in the form of IoT-A is presented. The presented model and methodology could be utilized to other agri-food sectors as it fulfills the identified requirements, considering also basic interoperability and security aspects.

Closing this section, the chapter "**Food Business Information Systems and Software in Western Greece**" attempts to examine the adoption of new information technologies by traditional food enterprises, having as a case study of the region of Wester Greece. The chapter, after providing an extensive review of the last 30 years' literature, presents the mapping and analysis of the software and food business information systems used by food businesses in Western Greece with the purpose to improve the management of their functioning. The outcome was that such systems are referred to human resources, in accounting and finance, in marketing and sales, in production and in operational functions. Most enterprises use Enterprise Resource Management (ERP) systems. Moreover, a few businesses use tailor-made packages mainly to manage production, human resources, and operational functions, while there is a growth prospective in operational, human recourses, and production functions to apply sophisticated software packages in the short-term future.

The Primary Production section elaborates on the decision-making for the improvement of the processes taking place within the farm, under the implementation of Information and Communication Technologies. The first chapter, entitled "**From Precision Agriculture to Agriculture 4.0: Integrating ICT in Farming**" focuses on the integration of Information and Communication Technologies in farming and the transformation of conventional agriculture to meet the need of increased effectiveness of agricultural practices. The chapter offers an overview of the basic technologies and provides examples of application in the agricultural sector alongside to the challenges of the Agriculture 4.0 era. Finally, the issue of the adoption of innovative technologies by the farmers is elaborated since it is highly linked to their successful penetration in the agricultural production.

The monitoring of the cultivation is an integral part in any agricultural Decision Support System (DSS) for the collection of the required data. The evolution of aerial monitoring technology has made unmanned aerial vehicles (UAVs or drones) a suitable solution for crop observation compared to satellite imagery. UAVs offer full coverage of fields upon request while they deliver on-the-spot images for further processing. UAVs are more flexible to user preferences; however, their limitations call for effective planning during their operation. Such limitations include their low autonomy as they are mostly battery-operated devices along with the need for long distance or heavy duty traveling for the execution of the required task (as for example mapping or spraying). The above constitute the routing process very important for a successful and efficient completion of the required missions. Providing further insight on the subject the chapter "**On the Routing of Unmanned Aerial Vehicles (UAVs) in Precision Farming Sampling Missions**" focuses on the use of UAVs within smart farming applications. The chapter presents the main applications of UAVs in the sector. Moreover, a profound examination of the methods of safe and optimal drones routing are presented along with the most widespread algorithms developed for that purpose. Lastly, representative applications of these algorithms are elaborated, drawing conclusions for further interventions and improvement.

Contributing to the monitoring and documentation in precision agriculture management systems, depth cameras have gained popularity during the last years. They are mostly utilized for the three-dimensional (3D) reconstruction of objects both in indoor and outdoor settings. As agricultural environments comprise of complex elements (e.g., trees or plants), the use of depth cameras, for example, to create models required for simulation purposes, is a significant challenge. Especially in outdoor environments that involve different object structures and uncertain conditions, the depth information collected may vary significantly. The chapter "**3D Scenery Construction of Agricultural Environments for Robotics Awareness**" examines the various technologies used by depth cameras and demonstrates indicative applications both in indoor and outdoor agricultural environments. A 3D orchard reconstruction is presented deriving from the processing of point clouds from Red Green Blue Depth (RGB-D) images collected in real fields. The orchard environment is simulated in Gazebo using the point cloud samples of trees collected by an UGV. This approach is considered significantly useful for the simulation and evaluation of the navigation of robotic systems even though tree characteristics' information (such as volume or height of tree canopies) can be extracted as well. The implementation of the approach shows a promising potential on environment simulation which can be utilized in several robotic applications despite any limitations such as the camera's limitations.

Aiming at controlling specific varieties of weeds, the chapter "**A Weed Control Unmanned Ground Vehicle Prototype for Precision Farming Activities: The Case of Red Rice**" attempts to address the damage of the spread of the red rice (*Oryza sativa f. spontanea*), which is a wild variety of rice, on the production of commercial rice. Red rice, with faster growing rate and better resilience to weather conditions compared to the ordinary variety, is characterized as a weed. Its population shows an increase on an annual basis, resulting in significant yield losses for farmers. What constitutes this species harmful for production, even though it is edible, is the morphology of its seeds. Being thin, covering entirely the plant, their collection is impossible while they tend to scatter on the field during harvesting getting mixed with the regular crops.

Rice is characterized as one of the most widely produced crops and constitutes a basic food source worldwide. In this manner, its effective production is of utmost importance. Its infestation with red rice is a considerable challenge since the commonly used conventional and chemical-based solutions cannot tackle the weed as it is genetically similar with commercial rice. Considering the above, the implementation of a mechanical solution is only accepted for this case. Towards that direction, the chapter presents the development of a prototype robot that handles the weed in field by applying herbicides "from above" at the stage of production when red rice is much higher than the conventional rice. The robot (an unmanned ground vehicle—UGV) carries a specially designed rod mechanism for the application of the herbicide only to the upper part of the weed. For that purpose, the rod carries a sponge that is drenched with herbicides while a sensor-based mechanism ensures the application of the agrochemical to the right plant elements. The study highlights the importance of innovative technologies in precision farming, that increase productivity, adding value to conventional farming processes while reducing the adverse impacts.

Monitoring of agricultural practices involves the collection of a variety of input data in multiple forms. The introduction of agricultural Decision Support Systems (DSSs) focuses on the organization and utilization of the collected information aiming at the optimization of the farmer's decision-making process. Additionally, DSSs can serve towards the sustainable transformation of modern agriculture by setting standards for environmentally improved processes and effective application of materials. One of the most crucial threats of agricultural production are weeds. Weeds compete with crops leading to reduced yield and poor quality, thus increasing production cost since further processing is required. Attempting to address that challenge, the chapter entitled "**Decision-Making and Decision Support System for a Successful Weed Management**," elaborates on the elements that affect decision-making and should be considered in the development of a DSS for weed management. The first step towards an effective weed control is to examine the environmental factors and the agricultural processes that favor the emergence of weed. Aim is to discover development patterns and predict weed growth for timely intervention suggestions from a DSS. Moreover, for the optimization of decision-making, the biological characteristics of weeds variates must be investigated. A thorough knowledge of the weed biology is considered as a prerequisite for the implementation of any weed control strategy. The chapter highlights the importance of the dissemination of such systems to the end-users/farmers. The trust of farmers in these systems is necessary to adopt them.

The chapter entitled "**Zephyrus: Grain Aeration Strategy Based on the Prediction of Temperature and Moisture Fronts**" presents a new aeration control strategy (called Zephyrus) based on the prediction of air velocity changes and changes in temperature and moisture contours, while air is passed through grain bulks. The presented control approach can be used with different aeration system designs and automatically adjusting its set points according to the geographic region and particular season.

The use of smart livestock technologies for decision-making has proved to benefit farmers by increasing the productivity and profitability of farming processes. The real-time farm scenarios that are examined allow for accurate and timely interventions providing several benefits for the animals and the farmer. Nevertheless, the monitored parameters are usually inaccurate and incomplete, while conflicted data may also be generated, constituting classical logic inadequate for their assessment. In that case, the use of non-classical logic in data processing may prove beneficial as it facilitates non-intrusive assessment, without the disturbance of the animals. The evolution of ICT tools has introduced the utilization of online mathematical identification techniques in smart farming. Such methods, can provide online estimations for the unknown parameters, offering models that can adapt to most of the biological processes, though the biological understanding of the causal mechanisms is not feasible due to their complexity and non-linearity. To deal with the emerging uncertainties, the employment of non-classic logic may allow for solutions for the managerial processes and the biological response models in smart farming. To obtain a complete picture of the state of the animals, a real-time continuous monitoring scheme is preferable to the view-in-time assessment that is used traditionally. In this case, farmers and animals are benefited by alerts that facilitate timely and targeted interventions.

The chapter entitled "**Decision-Making Applications on Smart Livestock Farming**" presents the fundamentals of smart livestock farming, introducing the managerial processes that apply non-classic logic and data mining through the demonstration of models implemented on farm actions. Additionally, tools for creating decision-making applications in smart livestock farming are presenting as, for example, paraconsistent logic applications and applications that use machine learning. The introduction of smart livestock farming leads to the automation of many procedures on the spot. The collection of information can be achieved from a variety of sources and different sensors resulting in powerful decision-making tools since farmers can use the information in conjunction with their observations and expertise. Such tools will continue to evolve along with the improvement of the respective technological elements.

The need for increase in food production and the use of resources and energy, in a sustainable manner, has driven the utilization of the recent innovations in ICT for the efficient assessment of the locally available resources (e.g., water, land, solar, and energy). Also considering the need to manage the environmental impact of agricultural production, the development of "climate-smart" solutions is becoming more important. Climate-smart systems that utilize ICT technologies bridge the gap between the fragile agricultural ecosystems and the non-agricultural processes that are used for their assessment. Nevertheless, such agri-environmental systems are characterized by high complexity and uncertainty, thus the establishment of new methodologies is challenging. The aim is to accomplish the sustainable cooperation of human-operated technologies and environment through the modeling of the diversified set of multidisciplinary processes that are involved. The last section of this book is dedicated to the development of innovative decision applications that also consider the protection of environment, recognizing its importance in the preservation and considerate use of resources, along with the mitigation of adverse impact that are related to agricultural production.

The emergence of the new generation of ICT technologies has promoted the development of tools for the support of decisions with respect to the planning and operation of complicated agricultural systems. However, the holistic management of these multifactorial systems requires the evaluation of the complex interactions that arise, as their statistical or structural analysis is not sufficient. To address the relationships that are created between the sub-processes that coexist, the development of predictive models is becoming very important. The amount of data collected increases exponentially, widening the gap with the potential of large-scale and long-term decisions. To bridge that gap, process model engineering of agri-environmental systems works towards the integration of the well-established and innovative frameworks. The complexity of the agricultural processes, that involves a variety of different disciplines, calls for the development of integrated models and methodologies. For the study of a variety of complex agri-environmental processes, the method of Programmable Process Structures (PPS) has been used in the past. The introduction of PPS has been driven by the need for the dynamic simulation of agricultural environmental processes that require the use of easily modifiable, extensible, and connectable models that represent the functional and structural characteristics in a unified manner. In the chapter entitled "**Programmable Process Structures of Unified Elements for Model-Based Planning and Operation of Complex Agri-environmental Processes**", the application of the PPS methodology is presented with the demonstration of three examples of increasing complexity, showing how PPS handles the structure and functionalities of agri-environmental processes. First, an overview of the existing functional and structural approaches is presented, while the PPS method is described along with its innovative characteristics. Moreover, through the demonstrated examples, the most important features of PPS are presented and more specifically its wide applicability in different process systems that facilitates the integrated assessment of dynamics of the interacting systems and the environment.

Planning and modeling the assessment of agricultural practices can provide farmers with valuable information prior to the execution of a task, thus the prediction potential is one of the most important domains to focus on the penetration of ICT technologies in agriculture. However, the assessment of already applied practices is equally important before the implementation of state-of-the-art ICT applications, especially when their evaluation is imminent. For example, sugarcane burning is a usual practice in South American countries. The use of fire in agriculture has been connected to several adverse impacts related to the increase in temperature and the decrease in natural soil moisture which leads to reduced soil fertility due to increased soil compaction, loss of porosity, and erosion. Moreover, the consequences of sugarcane burning are more severe on human health as indicated by many studies. The objective of the chapter titled "**Monitoring and Estimation of Sugarcane Burning in Brazil, Using Linear Mixed Models**" is to evaluate the spatial and temporal distribution of fire incidences in Paranapanema region in Brazil. For this purpose, images from the Landsat satellites, numerical data (regarding area and fire incidences), and categorical data (terrain slope) was also used. A statistical model was used to evaluate data, making it possible to identify a decrease in fires in smooth undulating terrains, corresponding to 99.9% per year, being characterized by the increase in agricultural machinery in these areas. The results were quite positive as the proposed model could forecast for the next 6 years, in which timeframe, considering causes/effects, there would be a high decrease. It is interesting to note that the direct change in land use can be assessed with the use of ICT technologies (such as remote sensing images). In this manner, the areas that were converted to sugarcane plantations, even though were destined to other uses, can be easily monitored demonstrating the applicability of such systems.

Considering the assessment of agricultural practices, the evaluation of on-farm efficiency is gaining the attention with the increasing concern caused by the rising energy cost and environmental impacts due to machinery and input use. The intensification of agriculture has led to an increase in the consumption of non-renewable resources with the well-known environmental, economic, and societal consequences that are related to conventional agricultural practices and threaten energy security and autonomy. The only way to address the imminent impacts is to move to more sustainable agricultural practices. The detailed supervision of agricultural operations by applying innovative ICT technologies is an essential step towards the realization of sustainable agriculture.

To achieve optimum management of agricultural operations, several tools and simulation methodologies have been developed. In these tools, system optimization is attempted at various levels and with different assessment targets such as the minimization of financial or environmental cost, the reduction of greenhouse gas emissions, or the increase of productivity. The common ground in all the tools developed is the large number of input parameters and processes that characterizes each agricultural system. Therefore, performance assessment tools mostly aim at identifying the weak spots of agricultural supply chain. Among the various agricultural in-field operations, fertilization (in its various forms such as organic or chemical fertilization) has proved to be one of the highest energy consuming. The chapter "**A Decision Support System for Green Crop Fertilization Planning**" presents an evaluation of chemical fertilization, using two distinct case studies that include an annual food crop (industrial tomato) and a perennial energy crop (Arundo donax) to highlight the differences in the energy cost on an annual basis. In this study, the energy performance of the two crops is compared with the use of a decision support system providing decisions with respect to the optimal crop allocation considering a set of available fields. The results demonstrate a high variance in energy consumption among various crop, highlighting the need for custom assessment with the employment of operation management tools. Such tools can be developed in the context of integrated Farming Management Information Systems (FMISs), creating comprehensive tools and services for stakeholders. Within an FMIS, innovative ICT technologies can be utilized towards facilitating decision-making for increasing operation efficiency, improving, as an aftermath, the sustainability of agricultural processes through the minimization of input materials.

It is worth noting that increasing the innovation penetration is a common policy in complex and progressive environments such as agricultural systems. Nonetheless, the tendency for development is also driven by the emergence of environmental, economic, and social problems. Agricultural stakeholders must adapt to the new challenges to safeguard a sustainable and viable future. Towards that direction the agroecological movement aims to a more sustainable agriculture by attempting to increase the efficiency of agricultural processes in a sustainable and ethical manner. However, the transition from conventional to sustainable agricultural processes is not always easy, thus appropriate tools, that facilitate the design and analysis of biodiversity friendly production systems and practices, are required. Along with the relevant tools, a change in the farmers' perception, with respect to what constitutes a process as sustainable, is required. More specifically, farmers should understand the functions of an agroecosystem, while also improving their managerial skills. In this way, they will be able to make beneficial decisions, reaching their production goals while abiding by the standards and principles of agroecology. The above highlight the benefits of the farmers' education and training, with respect to the consequences of an action, its suitability with short- or long-term goals, its compatibility with other actions, etc. The use of ICT technologies can facilitate the learning process through the involvement of the farmer to the assessment of the various components and processes of an agroecosystem.

The chapter "**Knowledge Elicitation and Modeling of Agroecological Management Strategies**" attempts to examine, the management strategies used by farmers, at a level that allows for the simulation of the various operational management processes. The concept developed borrows from the Belief-Desire-Intention (BDI) theory that conceptualizes the sequential decision-making behavior of rational agents. Even though a variety of simulation approaches have been proposed for the assessment of agricultural systems, only a few succeed on working up to farm scale. The aim of the chapter is to facilitate the holistic examination of farm management aspects such as the planning of tasks, the goal-based adaptation to circumstances, and the proper allocation of resources in the context of the implementation of a production strategy. The authors' objective is to establish the guidelines for the management of agroecological systems to test and disseminate the knowledge.

Concluding, appropriate decisions are necessary in the entire value chain of agricultural products and concern the planning as well as the assessment of the consequences of the processes, as highlighted through the chapters of this book. Thus, decisions are required for the proper and effective planning, monitoring and execution of agricultural tasks, for the efficient minimization of inputs as well as the evaluation of the processes. The new era of digital agriculture calls for elaborate decisions that optimize the consequent actions. This book leads the way towards the final volume of the series, under the theme **Action** that focuses on the implementation of cutting-edge technologies on real-world applications.

Thessaloniki, Greece — Dionysis D. Bochtis
Aarhus N, Denmark — Claus Grøn Sørensen
Athens, Greece — Spyros Fountas
Thessaloniki, Greece — Vasileios Moysiadis
Gainesville, FL, USA — Panos M. Pardalos

Contents

Part I
Value Chain

Agricultural Information Model

Raul Palma, Ioanna Roussaki, Till Döhmen, Rob Atkinson, Soumya Brahma, Christoph Lange, George Routis, Marcin Plociennik, and Szymon Mueller

1 Introduction

Smart farming or precision agriculture refers to the adoption of various digital technologies in the agriculture domain aiming at automation and improved efficiency of various farming operations and processes considering several aspects (e.g., increased yield quality and quantity, reduced cost and necessary resources, reduced environmental footprint) [1, 2].

There is a worldwide trend towards the smart farming paradigm and particularly in Europe, where EU Member States have joined forces and signed a declaration of cooperation on "*A smart and sustainable digital future for European agriculture and rural areas*"[1] in April 2019, which acknowledges the potential of employing digital

[1] The declaration has been signed by 25 Member States. For details please refer to https://ec.europa.eu/digital-single-market/en/news/eu-member-states-join-forces-digitalisationeuropean-agriculture-and-rural-areas

R. Palma (✉) · S. Brahma · M. Plociennik · S. Mueller
Poznan Supercomputing and Networking Center, Poznań, Poland
e-mail: rpalma@man.poznan.pl; sbrahma@man.poznan.pl; marcinp@man.poznan.pl; mael@man.poznan.pl

I. Roussaki · G. Routis
Institute of Communication and Computer Systems, Athens, Greece
e-mail: ioanna.rousski@cn.ntua.gr; georgios.routis@cn.ntua.gr

T. Döhmen · C. Lange
Fraunhofer Institute for Applied Information Technology FIT, Sankt Augustin, Germany
e-mail: till.doehmen@fit.fraunhofer.de; christoph.lange-bever@fit.fraunhofer.de

R. Atkinson
Open Geospatial Consortium Europe Technologielaan, Leuven, Belgium
e-mail: ratkinson@ogc.org

D. D. Bochtis et al. (eds.), *Information and Communication Technologies for Agriculture—Theme III: Decision*, Springer Optimization and Its Applications 184, https://doi.org/10.1007/978-3-030-84152-2_1

technologies in the agricultural sector and rural areas and supports the setting up of common data spaces. In this framework, the EC has issued the European Strategy for Data[2] in February 2020, which foresees the roll-out of common European data spaces in nine strategic sectors, including agriculture. This strategy considers data to be "*. . . one key element to enhance the sustainability performance and competitiveness of the agricultural sector. Processing and analyzing production data, especially in combination with other data on the supply chain and other types of data, such as earth observation or meteorological data, allows for precise and tailored application of production approaches at farm level.*" The production data are collected by Farm Management Systems (FMS) or similar applications during farm operations that will be coupled with open data (such as satellite images, weather data, soil maps that are for public use) to maximize the potential value of knowledge generated and introduce new opportunities for monitoring and optimizing the consumption of natural resources.

As mentioned in the concept note[3] issued by the EC in the framework of the Expert workshop on a Common European Agricultural Data Space that took place in September 2020, EC aims to "*set up, populate and operate a secure and trusted dataspace in order to enable the agriculture sector to access data relevant for agricultural production.*" However, before this common data space can be established, there are several technical challenges that need to be addressed jointly. Two major ones among these include the establishment of a common agriculture data model able to represent the wealth of information generated and used in the agri-food sector and the delivery of the necessary interoperability mechanisms. These are very difficult to address, given the wide heterogeneity of data models and semantics used to represent data in the agri-food domain, as well as the lack of related standards to dominate this space, that make full-scale interoperability almost impossible to achieve. The situation is further complicated by the growing number of stakeholder groups involved in the agricultural activities, and the increasing volume of data being collected by different devices and produced by a wealth of different systems and platforms. This chapter aims to present in detail the DEMETER Agricultural Information Model (AIM) that has been delivered by the H2020 DEMETER project (857202) supported by the European Union. AIM builds upon a thorough analysis of both the related state of the art and the state of the practice. It is driven by the elicited stakeholder requirements and, of course, by the DEMETER vision for the creation of a common agricultural data space, where semantic interoperability is ensured. AIM aims to establish the basis of a common agricultural data space and enable the interoperation of different systems, potentially from different vendors, and the

[2]"A European strategy for data," Communication from the Commission to the European Parliament, the Council, the European Economic and Social Committee and the Committee of the Regions, COM/2020/66 final, URL: https://eur-lex.europa.eu/legalcontent/EN/TXT/?uri=COM:2020:66:FIN

[3]http://ec.europa.eu/newsroom/dae/document.cfm?doc_id=68838

analysis of data produced by those systems in an integrated manner in order to make economically and environmentally sound decisions.

More specifically, the rest of the chapter is structured as follows: Section 2 provides an analysis of the state of the art (and state of the practice) on related data models and semantic interoperability mechanisms that exist in the domain of Smart Agrifood. Initially, general data models that are widely used are presented, then specific ontologies are elaborated upon and, finally, mechanisms that allow for semantic interoperability are discussed. This analysis, together with the technical requirements (presented in the following section), drives the design and development of AIM.

Section 3 gives an overview of the technical requirements extracted by the project, based on input received from the stakeholders involved and driven by the DEMETER vision. This is a list of specific technical requirements that AIM needs to address, as well as requirements regarding the interoperability with existing systems and ontologies, including the mapping of these data models to AIM. Section 4 presents the design of AIM:

- Section 4.1 presents the core meta-model, which is based on the NGSI-LD information model [3].
- Section 4.2 presents the cross-domain ontology, i.e., the set of generic models that aim at providing common definitions that are not agri-food-specific and that need to be handled by AIM, avoiding conflicting or redundant definitions of the same classes at the domain-specific layer.
- Section 4.3 presents the domain-specific ontologies developed for AIM, which are mainly applicable for the agri-food domain and model information such as crops, animals, agricultural products as well as farms and farmers.

Section 5 elaborates on the approach enabling semantic interoperability between AIM and several existing ontologies and dominant agri-food data models detailing the semantic mapping of these to AIM.

Section 6 presents the implementation of the AIM. More specifically, it presents the implementations for the different AIM layers, highlighting the critical design and implementation choices made, the mappings implemented, and the various tools used during the implementation process.

Section 7 presents the methodology of profiling and demonstrates its usefulness on building AIM. Section 8 completes the circle elaborating on use cases where the AIM has been adopted in operational environments engaged in the pilots of the DEMETER project. Finally, Section 9 concludes the document presenting also future work towards the final version of AIM.

2 Related Work

The DEMETER approach to creating Agriculture Integrated Model can be related to advancements in several areas of scientific work: general data models, specific ontologies, and vocabularies as well as established mechanisms for semantic interoperability. On a broader scale, general data management solutions, such as FIWARE NGSI/NGSI-LD (Next Generation Sensors Initiative—Linked Data), are proposed for use in cross-domain use cases. NGSI is a protocol developed by OMA (Open Model Alliance Ltd.) to manage Context Information about entities: their characteristics and relationships with other entities. Context Information Manager (CIM) is responsible for providing services for exchanging context information (e.g., registration, discovery, notification) between interested parties. FIWARE NGSI[4] is a binding of OMA NGSI-9 and NGSI-10 abstract interfaces for CIM. Updates to the protocol provide further benefits: NGSIv2[5] was designed with RESTful (Representational State Transfer) principles in mind and provide support for JSON (Java Script Object Notation) data format. The extension of the NGSIv2 format to incorporate advances of Linked Data approach is ETSI's (European Telecommunications Standards Institute) NGSI-LD format [4].

NGSI is a protocol that was developed in order to manage Context Information. A Context Information Model (CIM) [5] is a platform or a system—also known as Context Broker that provides the following services: context information registry, discovery, publication, mediation, modification, or notification. An entity consists of a set of characteristics that describe it, including its dynamic state. It also consists of other entities with which it has defined relationships, and the nature of those relationships.

NGSI-LD [6, 7], is an OMA NGSI information model, known as the evolution of NGSI, in order to better support linked data (entity's relationship), property graphs and semantics, so that it can benefit the capabilities offered by JSON-LD (Java Script Object Notation—Linked Data). The NGSI-LD consists of two levels; the first one is the foundation classes, which correspond to the core meta-model, and cross-domain ontology and the second is the domain-specific ontologies or vocabularies.

FIWARE [8, 9] Agrifood data models enable data portability for different applications in the domains of Smart Cities and Smart Agrifood. They are used with FIWARE NGSI 2 and NGSI-LD. FIWARE Agrifood data model represents a standard format which provides support to develop FIWARE solutions in the Smart Agrifood domain. By adopting a standard model, there is improvement in the standardization of information coming from IoT Networks that includes IoT (Internet of Things) sensors, wearables, GPS (Global Positioning System) services, UAVs (Unmanned Aerial Vehicles), robots, and farm machines. It increases the uniformity and interoperability of the data in the FIWARE technologies and ecosystem's applications. FIWARE Data Models that exist in the Agrifood field are the

[4]https://fiware-orion.readthedocs.io/en/1.8.0/user/walkthrough_apiv1/index.html

[5]http://fiware.github.io/specifications/ngsiv2/stable/

following: AgriApp, AgriCrop, AgriFarm, AgriGreenhouse, AgriParcel, AgriParcelOperation, AgriParcelRecord, AgriPest, AgriProductType, AgriSoil, Animal, WeatherObserved, WeatherForecast.

Saref4agri [10] is an OWL-DL (Web Ontology Language Description Logic) ontology, which extends SAREF (Smart Appliances REFerence) for the Smart Agriculture and Food Chain. ETSI's specialist task force STF 534[6] was established in order to extend the SAREF ontology for the domains of Smart Cities Industry and Manufacturing and Smart Agrifood. As it is mentioned in the associated Saref4agri requirements document ETSI TR 103 511, the intention of Saref4agri is to connect SAREF with ontologies that exist, such as W3C (World Wide Web Consortium), SSN (Semantic Sensor Network), W3C SOSA (Sensor, Observation, Sample, and Actuator), and GeoSPARQL (Geographic Simple Protocol and RDF Query Language).

The existence of different brands of farm equipment and software currently collects and uses data in range of proprietary file formats. This is the normal consequence of the industry's growth; however, it makes it hard for the end-users to "construct the full image" and output value from the data. It is clear that there exists the absence of interoperability in agricultural field operations, something that is not only a problem of a deficiency of common data formats or syntax. There has also been a lack of a shared understanding, or semantics in the field operations among the different industry actors, that are involved in field operations. So, there is use of multiple terms or codes to refer to the same concept or use of the same term to refer to multiple concepts. An outcome of the situation described previously, meaning the non-existence of common semantics, is the lack of common Reference Data. For instance, a usual set of code lists, unique identifiers, and controlled vocabularies which could be used in order to identify crop inputs, farm machine, implement and sensor models, so that every part involved can understand. AgGateway,[7] which represents a non-profit consortium of over 200 companies occupied with the implementation of standards aiming in advancing digital agriculture, realized its Precision Agriculture Council in 2011 so that it can tackle the previously described problems. As a result, the Agricultural Data Application Programming Toolkit (ADAPT) [11] was created targeting to realize common object model operations as well as a set of format conversion tools. ADAPT is an object model that represents some part of the world that is of interest. Object models are consisted of classes that represent groups of related data; an object-oriented version of the standard concept of a data type, and their relationships.

Another model is the IDS Information model [12–14], which is an RDF/OWL (Resource Description Framework/Web Ontology Language)—ontology and covers fundamental concepts of the International Data Spaces (IDS) software architecture for data exchange, such as types of digital contents that are exchanged by IDS participants via IDS base components. IDS addresses challenges such as

[6]https://portal.etsi.org/STF/stfs/STFHomePages/STF534

[7]http://www.aggateway.org

interoperability and security, which are related to DEMETER. The IDS Information Model is an extension of DCAT (Data Catalog Vocabulary) [15], DQV (Data Quality Vocabulary) [16], ODRL (Open Digital Rights Language), [17] and related standards for the description of data resources. The rationale behind this model is based on an exchange of ideas with the working group that specified DCAT. From an NGSI-LD point of view, metadata transfer high-level information about content of datasets, while the IDS Information Model takes contextual properties of datasets, and more generally, digital resources. The IDS Information Model supports extension points in order to transfer information about the domain-specific structure and semantics of the content of datasets. More specifically, it reuses the W3C standards, VoID, for referring to domain ontologies and their terms that are used in a dataset, Data Cube in order to describe the structure of tabular or matrix-like datasets, and SHACL (Shapes Constraint Language) to describe general graph-shaped datasets' structures [18].

The European data portals[8] provide a way to expose provided datasets, both metadata and data to reusers. Generally speaking, the federated architectures of International Data Spaces, that was described previously [13], and the more recent GAIA-X, which explicitly aims at delivering a technical solution for the data spaces envisaged by the European Data Strategy, have been created with the rationale to accommodate a wide range of legacy systems and platforms. It has been proven that a federation of existing platforms is possible when following these approaches. As far as DEMETER is concerned, it is committed to building on previous work like IDS. Concerning data exchange, DEMETER will require both service and API brokerage, which is what GAIA-X targets at [19].

GS1 [20], which is a global model, is an identification standard that provides the means to identify entities of real world, so that they may be the subject of electronic information. This information is stored and/or communicated by end-users. GS1 includes unique identifiers that can be used by an information system to refer unambiguously to a real-world entity such as a trade item, logistics unit, physical location, document, service relationship, or any other entity.

FOODON [21] project targets on building an easily accessible and comprehensive global farm-to-fork ontology concerning food, in a way that it accurately and consistently describes foods commonly known in cultures from around the world. It covers terms for the origin of food sources, the processing and cooking that took place in a product (such as heat processing), covering how it is packaged (meaning the container or the wrapping of this product).

FOODIE ontology [22] represents an application vocabulary including different categories of information dealt by typical farm management tools/apps for their representation in semantic format, and in accordance with existing standards and best practices such as INSPIRE, ISO/OGC (International Organization for Standardization/Open Geospatial Consortium) standards. This ontology aims to enable the representation of farm-related data in a semantic format, and also to enable the

[8] https://data.europa.eu/

advantage of semantic technologies for different tasks, such as transformation of semi-structured data to semantic format, ontology-based data access, data integration using linked data as a federation layer, knowledge discovery through inference, interlinking data with established vocabularies and relative datasets in Linked Open Data cloud.

3 Technical Requirements

This section presents the technical requirements extracted, which drive the design of a common Agricultural Information Model. These have been divided into two main classes, the first focuses on the core data modeling aspects and the second on aspects related to the support for semantic interoperability with existing systems and ontologies. These two requirement classes comprise functional requirements and are presented in the next two subsections.

3.1 Core Data Modeling Requirements

DM1. Common data view for heterogeneous models (Mandatory): AIM needs to define a common data model that specifies a common view on all heterogeneous entities connected and all the data involved in the pilots. This common data model shall be used for all data exchanged between software components. Therefore, it needs to support the translation of the obtained data streams to a common data model.

DM2. Representation of crop farms data (Mandatory): AIM needs to enable the common representation of agronomic data (e.g., crops, sensor data from the field, thermal/multispectral imagery from UAVs, production data, geolocation data, planting data, irrigation logs, fertilization logs, spraying logs) including: (1) Farm and economics modeling: agricultural type and economic size, production volumes and types, calculations according to results, etc.; (2) Field data modeling: location and geometry of the field, planting date, planting distance, detailed yield information; (3) Field status modeling, for example, water- or nitrogen-stressed fields, appropriate evaluation indices (e.g., Normalized Difference Moisture Index (NDMI)), need for fertilizing; (4) Soil data modeling: soil temperature and moisture, soil physical and chemical analysis; (5) Crops, treatment and fertilization modeling: crop type, crop developing stages, crop cultivar or variety, crop health status and pests, pesticides, nitrogen levels, information from counting devices used for the control of insects or plagues; (6) Traceability information of crops (production, transport, retail) to be used in the product passport information, (7) Water management modeling: water and energy consumption, water quality (e.g., salinity levels).

DM3. Earth Observation Data Representation (Mandatory): AIM needs to enable the representation of current earth observation (EO) data as well as historical

EO data, including for example satellite data, remote sensing imagery, soil maps, vegetation indices, such as NDVI, EVI, NDRE, and NDMI. It needs to also get EO metadata, for example, through interfaces compliant with the OGC 13-026r8 specification.[9]

DM4. Representation of livestock data (Mandatory): AIM needs to enable common representation of livestock data and traceability of products including: (1) Modeling of dairy and beef farms and data from farm robots: milk and meat production and quality, milk properties and quality (fats, proteins, somatic cells, and bacterial content), economic data; (2) Modeling of data from cows' wearables: animal ID, location, temperature, pedometer data, movement; (3) Modeling of animals' welfare, behavior, and habits: eating habits, respiration monitoring data, rumination, activity, rectal temperature control data, feed and water consumption data, biomarkers related with animal well-being and welfare (e.g., cytokine markers); (4) Food traceability information of dairy products and pastries (tracking of ingredients and supply chain); (5) Modeling of poultry farms: animal welfare, habits, living conditions, stress levels, medical treatment, feeding patterns, feed origin; (6) Traceability information of poultry products (production, transport, retail) to be used in the product passport information; (7) Modeling of apiary and hives: location of hives, apiary weight.

DM5. Representation of meteorological and open spatial data (Mandatory): AIM needs to enable representation of weather data (e.g., temperature, humidity, wind speed/direction, solar radiation, pressure) and open spatial data modeling. Meteorological data will be collected by interfacing with existing sensors, or new sensors that will be provided for this purpose.

DM6. Representation of agricultural machinery data (Mandatory): AIM needs to enable common representation of agricultural machinery data such as: engine data, fuel consumption, emissions, exhaust gas, NOx-conversion, and exhaust temperatures. The data are defined by Controller Area Networks Bus (CAN) protocol specifications; consequently, it will be necessary to consider that the translation of CAN-Bus Model into AIM involves understanding the specific CAN-Bus information (the message set for subsystem data exchange) of the supplier to the vehicle subsystem into new information according to the AIM communication specifications. Finally, AIM needs to represent entities types and formats, relationships among them, possible range between the values (if any).

DM7. Representation of farmer's preferences and DSS recommendations to them (Mandatory): AIM needs to enable common data model able to interpret farmers' needs and preferences including: (1) farmers' needs related to cost optimization (e.g., linking economical aspects of wholesale and retail prices), production issues (better quality of their products, crop variety per field, optimal date for planting and harvesting), cost/benefit analysis of field operations (irrigation/fertilization), optimization on irrigation/fertilization strategies, disease monitoring, yield analysis (e.g., the estimation of crop yield according to climate conditions), animal

[9] http://www.opengis.net/doc/is/opensearch-eo/1.0

welfare tracking; (2) production preferences (e.g., the use of non-chemical pesticides, attention to animal welfare, transparency to the consumers); (3) any other relevant data input collected during farm operations (related to animal welfare, crop production, product's characteristics). AIM should also enable common representation of recommendations and notifications to farmers, as well as the metadata used for providing recommendations to farmers through the DSS system and analytics tools. In this way, farmers' needs and preferences will be adequately analyzed (data integration and analysis) and decision support (visualization) will be provided.

DM8. Flexible and extensible model representation (Mandatory): AIM needs to support flexibility and extensibility in the representation of AIM through the use of a modular approach, the reuse or alignment with thesauri/classifications available as linked data, the use of property graphs and semantics, the use of appropriate data interchange models (e.g., RDF), knowledge representation languages (e.g., SKOS, RDFS, OWL), and rule languages (e.g., SWRL or OWL-RL), which would enable the semantic querying of data.

DM9. Representation of data quality metrics (Mandatory): AIM needs to include quality metrics in its data model. These data will be used for evaluating the accuracy, precision, granularity, completeness, consistency, timeliness, validity, uniqueness (where applicable) of the agri-food data and will be used by the system.

DM10. Provide a basis for data exchange across stakeholders (Mandatory): AIM needs to enable data exchange across authorized stakeholders. To facilitate this, it needs to include data regarding the supply and usage of agri-data and any other type of data that is stored in the AIM unified ontology including any economic transactions regarding the usage of such data.

DM11. Data Models enabling analysis of large volumes of heterogeneous data (Mandatory): AIM needs to enable the analysis and processing of large amount of heterogeneous data, including their storage and transfer. This is paramount as the agri-food domain involves numerous data sources, some of which collect very large data volumes (e.g., satellite data, remote sensing imagery, soil maps).

DM12. General Model for Interoperability (Mandatory): AIM needs to provide a general model for data interoperability, which should be flexible and extensible for all use cases. More specifically: (1) It will be composed of discrete modules addressing specific "competency questions," following best practices in ontology engineering—allowing these to be adopted standards or tightly managed development efforts with clear testability; (2) It needs to handle interoperability for different implementation aspects; (3) Meta-models, domain models, profiles, and vocabularies need to be handled individually using appropriate specialized modeling mechanisms.

DM13. Simplified Profiles of Data Model (Mandatory): AIM needs to provide simple profiles of the general model suitable for individual pilot cases; these profiles will define "schemas"—or views, while the general model will define semantics what objects can be identified and reused in different views.

DM14. Semantic model that supports scalability and support of legacy systems (Mandatory): AIM needs to implement semantic interoperability in a scalable and sustainable way, for example, by maintaining a dependency graph at the module level within each implementation rather than creating a temporary (project scoped) aggregated knowledge graph with no transparency of scope or provenance. It should support semantic interoperability for data originating from existing systems involved in the pilots (legacy systems). It should publish all domain-specific semantic interoperability resources in a canonical standards-based and interoperable fashion appropriate to the type of resource (e.g., vocabulary, schema, object model, profile, data type).

DM15. Governance Arrangements (Mandatory): AIM needs to specify governance arrangements for each component, determining who, when, and how updates to the included components should be handled. This includes pragmatic project scope governance of temporary resources, as well as requirements for governance of project resources that would ensure future interoperability.

DM16. Abstract model for integrating sensors, processing, and decision support systems (Mandatory): AIM needs to provide an abstract model for the general process of linking sensor data through processing chains into decision support systems, including how intermediate data products relate to sources and outputs. This can be based on an existing general model, or, if necessary, to create something new, to be pushed as an OGC (Open Geospatial Consortium) and/or W3C (World Wide Web Consortium) general model specification.

3.2 Semantic Interoperability Requirements

SI1. Service wrappers and translators (Mandatory): AIM needs to facilitate the development of service wrappers and translators, also known as data providers and consumers, which will enable the different tools/platforms in a (regional/national) AKIS to expose and consume data in interoperable forms.

SI2. Mapping AIM with standard models (Mandatory): AIM should implement (semantic) mappings from standard and/or widely used ontologies/vocabularies with the AIM, enabling the semantic integration of data represented using any of these models. As part of the semantic mapping, AIM needs to identify logical connections between classes, properties, and objects across ontologies. The mappings will deal with cases in which, e.g., a class in one ontology is the intersection (or union) of two classes in another, or the complement of another class, or a simple object needs to be mapped to a complex class in another ontology.

SI3. Semantic Interoperability support (Mandatory): Support semantic interoperability, encompassing semantic integration. This can be realized via the implementation of suitable data provider and consumer services, new ontologies, and the mappings with existing ontologies/vocabularies, as well as the other mechanisms developed to facilitate data integration.

SI4. Semantic mapping best practices (Mandatory): Follow best practices and approaches for generating the mapping between AIM and existing ontologies/vocabularies, including: (1) Transformation of existing ontologies into common format, for example, OWL, use of semantic rules or annotations/punning; (2) Reuse of AIM terms and only extend it if necessary. In the latter case, reuse existing terms whenever possible, and only otherwise create new terms/extensions; (3) Use of appropriate mapping constructs/axioms, such as owl:equivalentClass and owl: equivalentProperty with OWL classes/properties; skos:closeMatch, skos: exactMatch, skos:broadMatch, skos:narrowMatch, and skos:relatedMatch with SKOS concepts; owl:sameAs for individuals, etc.; (4) Treating of the mappings as "first class" components of a modular knowledge graph, making them available in line with FAIR principles, and governing them appropriately and transparently. Consider mappings across different levels of specification granularity as well of abstractions using the appropriate mechanisms in a standardized way, for example, mappings from meta-models to models (OWL subclassing); mappings between concepts at the same level of abstraction; mappings between controlled vocabulary terms; mappings between measurements and classifications (e.g., threshold values for "good,"); soft- vs. hard-typing mappings with classes with a subtype property vs. specific subclasses.

SI5. Tools for generating ontology mappings (semi-) automatically (Mandatory): Identify and select, if possible, suitable tools for the (semi-) automatic mapping of ontologies/vocabularies. Some example tools to be analyzed include the Alignment API, PARIS, Map-On, etc.

SI6. Identify tools to validate mappings (Mandatory): In order to facilitate the mapping between the AIM and existing ontologies, it is necessary to identify and select, if possible, suitable tools to validate the generated mappings. This is necessary because some of the mappings may be quite complex. For example, when a specific schema is mapped to a more general schema, then some schema elements may be replaced by use of a qualifying term in corresponding more abstract elements. In such cases, the coverage of the mappings as well as the result of exercising a mapping against the target model need to be validated. It would also be desirable to define a validation process and a simple reference implementation that can define test procedures to be integrated into traditional development tooling.

SI7. Select relevant existing ontologies to align with AIM (Mandatory): Identify and select relevant standards and/or widely used ontologies/vocabularies to align with the AIM and identify the key terms in each of them that would need to be aligned. Examples of dominant, often standardized, ontologies that need to be aligned to AIM to ensure sufficient semantic interoperability include: Saref4agri, FIWARE, AGROVOC, INSPIRE, SOSA/SSN, FOODIE, rmAgro, drmCrop, Soilphysics, AgriFarm, FOODON, etc.

4 AIM Design

In line with best practices and recommendations (e.g., [23–25]), the specification of DEMETER AIM follows a modular approach in a layered architecture, enabling among others:

- to facilitate the interoperability with existing models by reusing available (standard and/or well-scoped) vocabularies and ontologies in the modules, instead of defining new terms, whenever possible
- to facilitate the mapping/alignment with other models, by module instead of the whole model
- to facilitate the extension of the domain/areas covered in AIM with additional modules
- to facilitate its maintenance, by updating and/or modifying only specific modules
- to facilitate the mapping to top-level/cross-domain ontologies
- to foster the reuse of AIM, by enabling reuse at module level

More in detail, AIM has been designed in three layers, namely the core meta-model layer (Sect. 4.1), the cross-domain layer (Sect. 4.2), and the domain layer (Sect. 4.3). This three-layer approach is inspired by and similar to the NGSI-LD information model [3]. Moreover, AIM adopts the same meta-model as NGSI-LD, as discussed in the next section. However, as opposed to NGSI-LD, AIM exploits semantic referencing to foster interoperability and reuse. Accordingly, the cross-domain and domain models are built by reusing, as much as possible, existing standards and/or well-scoped vocabularies and ontologies with well-defined semantics, in addition to include mappings between these models to enable their interoperability and integration of derived data. Briefly, the cross-domain ontology comprises the set of general concepts that aim at providing common definitions for the whole agri-food domain handled by AIM and at avoiding conflicting or redundant definitions of the same classes at the domain-specific layer. The domain ontologies, on the one hand, model information such as crops, animals, agricultural products as well as farms and farmers just to mention some of the most important concepts. In addition to the three AIM-specific layers, a meta-data schema has been specified to describe various concerns of data exchange in AIM-related settings. For the latter, the AIM defines a profile of the IDS Information Model reviewed in Sect. 2, focusing on those aspects of the concerns specified by IDS ("content," "concept," "community of trust," "commodity," "communication," "context") that are relevant to DEMETER. Figure 1 provides an initial overview of the AIM model.

4.1 Meta-model Layer

A meta-model, as its name implies, is a model of a model. Meta-models are typically used for different purposes. For instance, they can be used for the specification of modeling language constructs in a standardized, platform independent manner [26], to specify and restrict a domain in a data model and systems specification [27], or to provide an explicit model of the constructs and rules needed to build specific models within a domain of interest [28]. In fact, as noted in [28], meta-models can be viewed from three different perspectives: (1) as a set of building blocks and rules used to build models; (2) as a model of a domain of interest; (3) as an instance of another

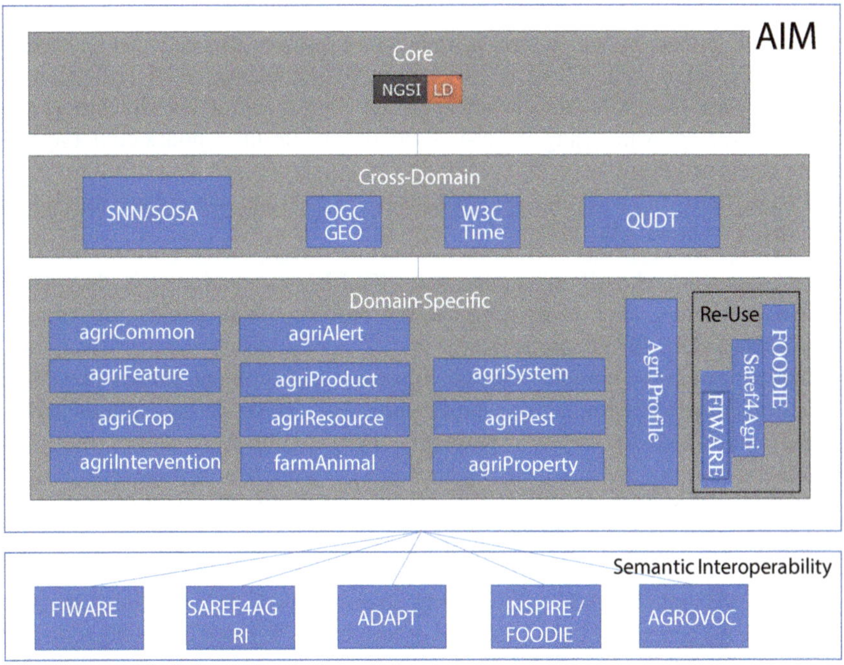

Fig. 1 AIM model overview

model. In the context of the DEMETER meta-model, the first perspective is being considered.

After analyzing different [meta-]models and approaches (e.g., [26, 29, 30–32]), it was decided to follow the meta-modeling approach of NGSI-LD [6], a standard of the European Telecommunications Standards Institute (ETSI) whose mission is to make it easier for end-users, city databases, IoT, and third party applications to exchange information. NGSI-LD is an evolution of the NGSI context interface family, particularly the FIWARE NGSIv2 information model,[10] which was evolved by ETSI ISG CIM initiative to support property graphs, semantics, and linked data by adopting the increasingly popular JSON-LD serialization format (also a priority in DEMETER project).

Property graphs are composed of nodes (vertices), relationships, and properties, where nodes may have properties in the form of key-value pairs, keys are strings, values are arbitrary data types, and relationships are arc (uni-directional, i.e., directed edges) that have an identifier, a start node, and an end node. Like nodes, relationships can have properties attached to them. Linked data, on the other hand, is based on the RDF model, which expresses data as triples of the form <subject, predicate, object>. RDF, however, does not support directly attaching properties to predicates

[10]https://fiware-tutorials.readthedocs.io/en/latest/

(relationships or other properties). Nevertheless, this limitation can be circumvented using RDF reification, which turns a triple into a resource that can then be used as subject of another triple. There are different ways to implement RDF reification. In NGSI-LD, and thus AIM, the chosen approach is via blank nodes, which is particularly convenient when using JSON-LD serialization, as these nodes do not appear in the actual serialization.

Hence, following a similar approach to NGSI-LD, and particularly reusing its meta-model, allows AIM to:

- be compliant and easily integrated with NGSI-LD data and models, thus, facilitating the integration of existing datasets based on these models that may be relevant to DEMETER project.
- represent rich and complex context information of different entities (e.g., systems/platforms/environments) typical within IoT (or WoT) applications, where the context includes the set of properties characterizing these entities, the set of relationships that enmesh them together, and the properties of these relationships and properties. This was the main motivation of NGSI-LD, and it is also a very important aspect of DEMETER project.
- perform back and forth conversion between datasets based on the property graph model and linked datasets, thus being able to exploit the benefits of the two worlds: the benefits of linked data and underlying RDF-based reasoning tools and querying (enabling data integration, knowledge discovery, etc.), and the richer expressivity of property graphs (using predicates as subjects of other predicates).
- exploit semantic referencing, through the use of JSON-LD contexts, where elements in the model can be matched to entities in well-known and/or standard ontologies.

4.2 Cross-Domain Layer

Cross-domain ontologies are defined as a set of common concepts, situations, or constraints which are aimed at avoiding conflicting or redundant definitions of the same concepts in the domain-specific layer [3]. Selecting widely accepted and supported ontologies is the basis for interoperability with other information systems and tools, so in general "canonical" ontologies managed by standardization bodies are preferred although "de facto standards" in widespread use may have advantages. Generally, such ontologies need to be profiled with a set of constraints relevant to the problem domain, such as the selection of mandatory and recommended elements, and constraining general elements to reference domain-specific terms. AIM makes use of standardized or widely adopted ontologies to profile a general meta-model similar to NGSI-LD. This allows the adoption of high-level concepts occurring in domain-specific ontologies presented in Sect. 4.3, without introducing a wide range of alternative structural models. Ontology alignments between AIM components and other ontologies can increase interoperability and should be based on ontology alignments between cross-domain ontologies as a foundation.

Different cross-domain ontologies may represent high-level models found useful for different parts of the modeling problem. For example, in situ observations, timeseries, mobile devices, remote sensing, numerical modeling, statistical summaries, relationship graphs, data catalogues all have different patterns that are most effectively and commonly handled in different meta-models. Specific cross-domain ontologies may be required for each distinct pattern in the DEMETER information scope to support efficient processing, even if a flexible common meta-model may be used for data transfer operations.

The NGSI-LD specification proposes a set of common cross-domain concepts for geographical and geometrical properties, temporal properties and time values, and unit-code properties [4]. Considering the different application domains of AIM, the proposed concepts fit well, however, as defined by NGSI-LD they do not convey any taxonomic information and are neither bound to well-known ontologies or standards. Therefore, a set of supplementary ontologies which serve as an extension of NGSI-LD concepts are integrated. Additionally, concepts for sensor and actuator-related data are introduced by reusing the SSN/SOSA ontology [32]. As SSN/SOSA concepts play a role in various domains served by DEMETER, a cross-domain compatibility between those domains is herewith enabled.

In the following, major concepts and standards chosen to be part of the AIM cross-domain layer are described.

Temporal Properties and Time Values: W3C OWL Time [33] conveys temporal information and time values as instants or intervals, like NGSI-LD time properties and values. However, OWL Time is W3C candidate recommendation and has an active working group. Through this, e.g., more complex time-related models such as timeseries may be proposed for standardization through the OGC Time Domain Working Group which could seamlessly be adopted. Furthermore, W3C OWL Time is compatible with SSN/SOSA.

Geo Properties and Geometry Property: OGC GeoSPARQL [34] is a geographical query language, which offers a comprehensive implementation specification and has been adopted by several tools already. It comes with a definition of supported geographical and geometrical linked data representation to which the DEMETER AIM model should adhere. The GeoSPARQL specification is currently being considered for revision by OGC, so any limitations or desired extensions discovered during DEMETER can be addressed in a locally updated copy and proposed for the next version of the standard.

Units of Measurement: QUDT is one of the most comprehensive units of measurement ontologies [35] and compatible with SSN/SOSA. It is intended as supplement for the NGSI-LD unitCode property.

Sensors and Actuators: The W3C, jointly with the OGC, put forward the SOSA (Sensor, Observation, Sample, and Actuator) and SSN (Semantic Sensor Network Ontology) standard, which is currently in the state of a technical recommendation. *"The Semantic Sensor Network (SSN) and Sensor, Observation, Sample, and Actuator (SOSA) ontologies are set out to provide flexible but coherent perspectives for*

Table 1 GCOS essential climate variables (example)

ID	Area	Variable	Procedure
10106	Atmosphere	Surface wind speed and direction	Wind speed over ocean surface (horizontal)

If the sensor measures wind speed, then:
ObservedProperty: "wind speed"
Procedure: "measured as horizontal component at ocean surface"

representing the entities, relations, and activities involved in sensing, sampling, and actuation. SOSA provides a lightweight core for SSN, whereas SOSA acts as minimal interoperability fall-back level, i.e., it defines those common classes and properties for which data can be safely exchanged across all uses of SSN, its modules, and SOSA." [36] SOSA/SSN, hence, supports different perspectives: observation, actuation, and sampling, which are generally all relevant for DEMETER. However, the integration should particularly focus on the observation perspective as this is the most prevalent use case across DEMETER pilots.

Cross-Domain Integration Process

SOSA provides a basic pattern for metadata for observation processes and can be used to standardize a set of properties in the "property graph" meta-model. Applications of SOSA require the development of approaches for describing ObservedProperty and Procedure subclasses, and instances. These two classes can be used in different levels of abstraction.

For example, consider one of the GCOS "Essential Climate Variables"[11] in Table 1.

Procedures also tend to have spatio-temporal sampling distributions inherent in sample design. Determining how these concerns are modeled and how finely grained descriptions are generated will be a significant activity regardless of which cross-domain model and domain-specific models are chosen—there is always a choice of the boundaries between model, controlled vocabulary, and descriptive text elements that need to be identified.

Integration process of SSN/SOSA with DEMETER AIM for ObservedProperty and Procedure. The following steps are necessary for the integration of SSN/SOSA with the DEMETER AIM:

- Mapping standard SSN/SOSA metadata properties to property names in the AIM meta-model, based on NGSI-LD.
- Defining appropriate subclasses of ObservedProperty and Procedure for agri domain-specific terms and mapping the properties of these subclasses to the property graph model defined by the NGSI-LD meta-model that is extended by AIM.

[11] https://www.ncdc.noaa.gov/gosic/gcos-essential-climate-variable-ecv-data-access-matrix

- Determining controlled vocabularies elements for specific elements of NGSI-LD data models that cannot be covered by the steps above.
- Mapping NGSI-LD data structures onto equivalent common models using controlled vocabularies for constant elements. For example, a schema used by some given source of NGSI-LD data may be specifically for air temperature readings, without any specific element stating that "air temperature" is the observed property, or information on the property observed may be contained only implicitly in metadata about the sampling procedure, etc.

4.3 Domain Layer

In addition to the cross-domain level, AIM also defines the models at the domain level that covers the data needs for the agriculture domain. As the agriculture domain is itself quite broad, in line with the ontology development best practices (see Sect. 4), it was decided to implement multiple modules/profiles, each covering specific aspects of the agriculture domain, instead of defining one large domain ontology. This results in easier maintainability, extensibility, and reusability. Also, as the goal of AIM is to ensure interoperability with existing models and systems, this domain layer is defined based on well-known ontologies and vocabularies related to the agri-food sector, namely Saref4Agri, INSPIRE/FOODIE, and FIWARE Agrifood models. The modules include required extensions to cover DEMETER pilot's needs, as well as alignments between elements in these models.

Following a requirement analysis (see Sect. 4.3.1), a few aspects from the agriculture domain, relevant to DEMETER AIM, were identified. These aspects were then grouped in different areas (or subdomains), each implemented in a separate module. The final domain layer areas (see Sect. 4.3.2) bear similitude to the ones used by the FIWARE Agrifood data models; however, they were extended and/or adjusted to the characteristics of the base models (see Sect. 6).

Domain Layer Requirements

According to the requirement analysis carried out in DEMETER project, AIM needs to enable a common representation of agronomic data including, among others: farm data and data about farm economics; field data, for example, the location and geometry of the fields; data regarding the irrigation and fertilization of fields; soil data such as soil temperature and moisture, soil physical and chemical analysis; data about crops and their treatment, for example, crop type, crop developing stages, crop cultivar or variety, crop health status, pests and pesticides; data related to sensors' measurements such as carbon content. Based on those requirements and on an extensive analysis of the state of the art (see Sect. 2), most of the required concepts and properties were identified in existing ontologies and models, including saref4Agri, FIWARE Agrifood models, INSPIRE/FOODIE, and SSN/SOSA.

More specifically, several classes and concepts are relevant to AIM and, in fact, there are overlaps between various key terms in these ontologies, as described in the following analysis. The following list provides some examples of the relevant classes from the various ontologies to represent the required data:

For crop data:	FOODIE cropSpecies and cropType; FIWARE AgriCrop, Agrifood; Saref4agri s4agri:Crop and s4agri: PlantGrowthStage.
For crop pests:	FIWARE AgriPest, FOODIE Pest.
For farm location and farm parcel/plot data:	FIWARE AgriFarm, AgriParcel; FOODIE Holding, Site, Plot and Management Zone; saref4Agri Farm and Parcel
For watering and fertilization data:	FOODIE Fertilization; Saref4agri WateringSystem.
For soil data:	FIWARE AgriParcelRecord (with many attributes, e.g., soilTemperature, soilMoistureVwc, soilSalinity); Saref4agri SoilMoisture, SoilTemperature; and FOODIE SoilNutrients

In addition, in some cases, these classes are aligned with terms from the SOSA/SSN cross-domain ontology. Moreover, (re-)using the agroVocConcept property from FIWARE models, data represented with AIM can be enriched with information about particular crops, plant products, or pests, by connecting the relevant AGROVOC concepts. As a general observation, in these base models and ontologies, FIWARE tends to have the most attributes associated with each concept (class); however, the models are not available as ontologies, but as JSON-LD contexts and/or JSON schemas. FOODIE has the most classes, but with relatively less documentation. Saref4Agri has a decent number of concepts and also has a good documentation. As a result, the general approach followed was to use Saref4Agri as the main reference given its good documentation, structure, and coverage, and extend with FOODIE and FIWARE entities as well as entities from other ontologies where needed.

AIM Domain-Specific Ontologies

This section provides a summary of the main area groups in AIM domain layer, highlighting how each of the base models was used.

The first area deals with Parcel/Plot data that allows to describe a plot of land that is part of a farm and that is used to plant crops. Now, all three ontologies (FIWARE, Saref4agri, and Foodie) have an equivalent class for this. So, the Saref4Agri concept was aligned with the equivalent ones from FOODIE, while also extending with extra properties that are present in the AgriParcel class of FIWARE Agri. This was done to align the properties and remaining classes defined.

Second, the next specific domain data regards a common representation of livestock data (also useful for the traceability of products). Again, the information regarding data for parcels and farms was used as in the previous area (farm data), but here each parcel is used for raising livestock. In addition, it was necessary to include data regarding the animals and any sensors (e.g., wearable) on them; regarding milk and meat production and quality, milk properties, and quality. Additionally, information such as livestock number, birth date of the animal, sex, weight, or species was also relevant. Similarly, it was necessary to model different types of animals such as poultry, apiary, and hives. A number of classes and concepts from existing ontologies were found relevant to model such information. For instance, for animal data the following classes were relevant from various ontologies: FIWARE Animal; s4agri[12]: Animal, s4agri: MilkingSensor, s4agri: ActivitySensor; INSPIRE: Animals and animals' health. Regarding sensors and wearables on animals, concepts from SSN were also reused in addition to s4agri classes.

Third, the next domain needs to enable a common representation of agricultural machinery data such as engine data, fuel consumption, and emissions as well as information about the machines such as synchronous speed, mechanics scheme, rotational speed, and synchronous pull-out torque. For this domain, relevant ontologies (in addition to general ontologies that deal with sensors such as SSN) are the FOODIE Transport data model and the AFarcloud hierarchy of robotic vehicles (UGVs, UAVs).

Fourth, DEMETER needs to enable the representation of current earth observation (EO) data as well as historical EO data. For this, AIM uses OGC GeoSPARQL and its connection with Saref4Agri as it already is using SAREF as basis for a number of other specific domain ontologies. However, as this spans all the domains it was decided that this requirement to be handled in the cross-domain ontology.

Fifth, DEMETER needs to enable the representation of weather data (e.g., temperature, humidity, wind speed/direction, solar radiation, pressure, sound pressure, sound intensity) and open spatial data modeling. For this, AIM is based on the relevant ontologies from FIWARE together with the classes and attributes present in saref4agri: FIWARE Weather Observed, Weather Forecast, Weather Alert and s4agri:AmbientHumidity, s4agri:AirTemperature.

Sixth, DEMETER needs to include food traceability information of a number of products such as dairy products and pastries or poultry products (production, transport, retail) to be used in the product passport information. For this, AIM takes some concepts and terms from FOODON; however, this introduces a huge ontology, and thus, for the time being, it was decided to include only some of its main terms and include GPS locations (and not state/country info as in FOODON) using the WGS84 Geo Positioning ontology instead. Moreover, the GS1 EPCIS standard is being investigated to be exploited by the DEMETER AIM in support of agri-food product traceability.

Finally, DEMETER needs to enable a common data model able to interpret farmers' needs and preferences including: farmers' needs related to cost optimization (e.g., linking economical aspects of wholesale and retail prices), production issues (better quality of their products, crop variety per field, optimal date for planting and harvesting), cost/benefit analysis of field operations (irrigation/fertilization),

---

[12] Saref4agri namespace prefix

optimization on irrigation/fertilization strategies, disease monitoring, yield analysis (e.g., the estimation of crop yield according to climate conditions), animal welfare tracking; production preferences (e.g., the use of non-chemical pesticides, attention to animal welfare, transparency to the consumers), any other relevant data input collected during farm operations (related to animal welfare, crop production, product's characteristics). So far no dominant ontology that models such data has been identified, and thus, AIM does not includes this domain in the current version; it looks like such an ontology would have to be developed, depending on the needs of the farmers identified in the pilots and included in the second release of AIM.

Agriculture Profile ontology: The demeterAgriProfile ontology imports (so it consists) of the following ontologies: agriCommon, agriFeature, agriCrop, agriIntervention, agriAlert, agriProduct, agriProperty, agriSystem, agriPest, farmAnimal. The initial/core classes are the following: ActivityComplex, Agent, Agri Farm, Agri Parcel, AgriParcelOperation, AgriParcelRecord, AgriPest, AgriProductType, Alert, Animal, AnyFeature, Codelist, Datatype, Deployment, EconomicActivityNACEValue, Feature of interest, FeatureType, ID, Measure, Measurement, Period, Platform, Property, skos:Concept, SpatialObject, System, taxonomic rank, Unit of measure.

Agriculture Commons ontology: In this module, a number of basic classes and concepts are defined: Agent and its subclass Person are gathered from the FOAF ontology and a subclass Farmer is defined as well for use in the Demeter ontology. Also, the general class is subclassed by the FarmHolding class.

Agriculture Features ontology: It consists of the following basic classes: ActivityComplex, AgriFarm, SpatialObject, AnyFeature, AgriParcel, PropertyType, skos:Concept, Codelist, EconomicActivityNACEValue, TractorType, Crop and MachineType.

Agriculture Crops ontology: This entity contains a harmonized description of a generic crop. This entity is associated with the agricultural vertical and related IoT applications. There are three equivalent classes, which are: AnyFeature, Feature, and FeatureType, that encapsulate the necessary subclasses enabling interoperability among existing ontologies.

Agriculture Interventions ontology: This entity has a harmonized description of generic operations performed on a parcel of land. It is primarily associated with the agricultural vertical and related IoT applications.

Agriculture Alerts ontology: This model aims to support the generation of notifications for a user or trigger other actions, when getting alerts. An alert is generated by a specific situation. The main features of an alert are that it is not predictable and that it is not recurrent data. An alert could be, for example, an accident or an extremely high level of measure.

Agriculture Product ontology: This entity contains a harmonized description of a generic agricultural product type. It is primarily connected with the agricultural vertical and related IoT applications. The AgriProductType includes a hierarchical structure that allows product types to be grouped in a flexible way.

Agriculture Pests ontology: AgriPest.ttl ontology contains the agriPest class. It describes the agricultural pest and is basically connected with the agricultural vertical and related IoT applications. It has only one object property, the hasAgriProductType that is a reference to recommended types of products. The latter can be used to treat this pest. This ontology contains the following three data properties: alternateName, description, and name. By using the agroVocConcept property (from the common module), individuals of the agriPest class can be connected to the equivalent pest concept from AGROVOC.

Farm Animals ontology: It describes the proposed animal data model that has been produced from a more general point of view, by adjusting it to the information coming from the devices and sensors used to monitor or record the animals, their status, their relationships, and properties in general.

5 Semantic Interoperability

European Interoperability Framework (EIF) arranges interoperability in six layers [37]: interoperability governance, integrated public service governance, legal interoperability, organizational interoperability, semantic interoperability, and technical interoperability. Of those, DEMETER focuses mostly on the last two. Technical interoperability covers applications and infrastructures used to link computer systems and services. The understanding of the data, relationship between them and the capability to process it by organizations is a focus of the semantic interoperability layer. It includes defining data structures and data elements to describe data exchanges. One of the underlying keys to semantic interoperability is syntactic interoperability, which describes the capability to exchange data between the systems.

The challenges of syntactic interoperability are based on schema (structure) translation. There is limited variability in patterns and techniques that are available that can be easily implemented. Semantic interoperability, on the other hand, represents a much broader challenge, covering issues such as ambiguity of terms, discoverability of semantic grounding, formalism of descriptions, and the context of the intended audience [38].

The FAIR principles (findable, accessible, interoperable, reusable) have particular relevance for semantic descriptions. The AIM model is being designed to maximize its relevance against these principles:

- **Findable:** as an RDF-based model, all concepts are uniquely identified by URIs (stable Web addresses, which resolve to online resources appropriate to the specific concept being identified). These resources can be indexed in a Web context, and the use of Linked Data principles mean that references to related objects are also Web links that can be followed by either a human reader or via machine readable forms.
- **Accessible:** open publication of individual definitions as Linked Data is augmented by the availability of component the AIM model in different forms, such as a package containing complete modules of AIM, and technology-specific

profiles of AIM such as OWL, JSON schema, and SHACL that express those aspects of AIM supported by such technologies.
- **Interoperable:** Interoperability is enabled by both the modular nature of AIM based on existing semantic standards as well as using interoperable publishing mechanisms to maximize flexibility.
- **Reusable:** This is enabled by the Accessibility and Interoperability aspects and facilitated by open publication processes. At a practical level, reusability is supported through the use of formal interoperability profiles, meaning that users can assess reusability at a coarser grained level than analysis of each individual constraint and reference to semantic descriptions. Profiles give identity to interoperability requirements, which make them easier to assess, validate, and reuse.

6 Implementation

AIM modules have been implemented as OWL ontologies[13] and serialized in Turtle format.[14] The ontologies are available in GitHub,[15] and they are also published in AgroPortal.[16] The ontologies were validated using the Pellet reasoner in Protégé,[17] thus, verifying the logical consistency of the ontologies. This was particularly useful to identify issues across different modules. For example, with the logical validation of DEMETER AIM, which imports over ten modules, it was possible to identify and solve some inconsistencies that could not be spotted manually.

Additionally, for each ontology module, the corresponding JSON-LD[18] context was generated. JSON-LD enables the encoding of linked data in JSON, one of the most commonly used formats to exchange data between services, and it helps JSON data to interoperate at Web-scale. The context is used to map terms, i.e., properties with associated values in a JSON document to URIs identifying, for instance, OWL entities in AIM as in our case. The JSON-LD contexts can be easily generated from the ontology modules using existing tools like owl2jsonld[19] or profileWiz[20] (see Sect. 7).

Similarly, for each module, the corresponding SHACL[21] shape has been generated. SHACL shapes allow the validation of data (e.g., JSON payload) against the

[13] https://www.w3.org/TR/owl2-overview/

[14] https://www.w3.org/TR/turtle/

[15] https://github.com/rapw3k/DEMETER/tree/master/models

[16] http://agroportal.lirmm.fr/ontologies/DEMETER-AIM

[17] https://protege.stanford.edu/

[18] https://www.w3.org/TR/2014/REC-json-ld-20140116/

[19] https://github.com/stain/owl2jsonld

[20] https://github.com/RDFLib/profilewiz

[21] https://www.w3.org/TR/shacl/

target model (AIM in this case), making sure that it is compliant with the model and its constraints. The SHACL shapes were generated using Astrea Web Service [39].

All the ontologies, as well as the corresponding JSON-LD contexts, use persistent identifiers that are resolvable. This facilitates both the sharing and usage within different applications, through time. In particular, AIM uses the w3id service for permanent identifiers on the Web,[22] maintained by the W3C permanent identifier community group.[23] Moreover, the ontology modules are available via the OGC Definitions Server,[24] and using content negotiation, or profile parameters,[25] different representations of the module can be retrieved (e.g., turtle, RDF/XML, HTML, JSON). All these distinct representations are generated using the profileWiz tool (see Sect. 7), and they can be listed from the default ontology module Web page by clicking *Alternate Profiles.*[26] In the case of JSON-LD contexts, they can be retrieved also from OGC definition server using parameters "_profile=jsoncontext&_mediatype=application/ld+json", or they can be accessed from the AIM context URL.[27]

Regarding mappings between different vocabularies/ontologies, the AIM defines them in each module by including appropriate ontology axioms, such as equivalent classes (*owl:equivalentClass*), equivalent properties (*owl:equivalentProperty*), subclasses (*rdfs:subClassOf*), and subproperties (*rdfs:subPropertyOf*).

6.1 Meta-model Implementation

As described in Sect. 4.1, AIM adopts and reuses NGSI-LD meta-model, which provides a formal basis for representing "property graphs" using RDF/RDFS/OWL. In particular, the meta-model defines the following entities (adapted from NGSI-LD [3]):

- Entity: An *Entity* is defined as an NGSI-LD Entity, which is the informational representative of something that is supposed to exist in the real world, physicallyor conceptually. Any instance of such an entity must be uniquely identified by a URI and characterized by reference to one or more NGSI-LD Entity type(s).
- Property: A *Property* is defined as an NGSI-LD property, which is a description instance that associates a main characteristic (*Value*), to either an *Entity*, a

[22] https://w3id.org/

[23] The base namespace for the AIM is: https://w3id.org/demeter/

[24] http://defs-dev.opengis.net/demeter/vocab/

[25] https://www.w3.org/TR/dx-prof-conneg/

[26] For example: https://defs-dev.opengis.net/def/w3id.org/demeter/agri/agriFeature?_profile=alt&_mediatype=text/html

[27] https://w3id.org/demeter/agri-context.jsonld

Relationship, or another *Property*. It must include a *hasValue* property to define its target value.
- Value: A *Value* is defined as an NGSI-LD Value that is either a JSON value (i.e., a string, a number, true or false, an object, an array), a JSON-LD typed value (i.e., a string as the lexical form of the value together with a type, defined by an XSD base type or more generally a URI), or a JSON-LD structured value (i.e., a set, a list, a language-tagged string).
- Relationship: A *Relationship* is defined as an NGSI-LD Relationship that describes a directed link between a subject, which must be either an *Entity*, a *Property*, or another *Relationship*, and an object, which must be an *Entity*. It must include the *hasObject* property to define its target object.

The meta-model has been implemented as a JSON-LD context in the same way as the NGSI-LD one. However, instead of defining cross-domain terms in the same context, as NGSI-LD does, the AIM meta-model is limited to the entities enabling the representation of "property graphs" described above. AIM meta-model URI is: https://w3id.org/demeter/core-context.jsonld.

6.2 Cross-Domain Implementation

The DEMETER AIM cross-domain layer was implemented as a suite of ontology modules that reuse in part or as whole the in Sect. 4.2 selected ontologies, along with the corresponding JSON schema and JSON-LD @context documents, which can be incrementally developed and tested individually and combined into high-level context documents for each data profile required. Therefore, required terms and axioms are added and aligned with the cross-domain ontology in analogy with the Integration process of SSN/SOSA with NGSI-LD for ObservedProperty and Procedure explained in Sect. 6.2.

The DEMETER AIM Cross-Domain Ontology acts as a horizontal bridge between different ontologies present in the domain-specific layer. Hence, the integration process started bottom-up with an inspection of the domain-specific layer. Each domain-specific ontology defined in Sect. 4.3 was systematically inspected[28] for occurrences of cross-domain concepts. Classes, object properties and data properties, annotation properties, data types, and individuals belonging to one of the prefixes identified as cross-domain were collected in the cross-domain ontology, represented by a cross-domain.ttl file. The DEMETER AIM Cross-Domain Ontology also provides a vertical integration between the meta- and the domain-specific layer. This is established by:

[28] Inspection of ontologies was supported by the ontology editor Protégé—https://protege.stanford.edu/

- Defining cross-domain classes as rdfs:subClassOf of core#Entity.
- Defining object properties, depending on their nature, as either rdfs: subPropertyOf core#Property or core#Relationship. Relationship, when an objectProperty connects two Entities. Property, when an objectProperty relates an Entity to a value/literal.
- Importing the cross-domain ontology in agriCommons.ttl. Removing definitions of cross-domain class, object properties, data properties, and individuals from all domain-specific ontologies.

An extension of the cross-domain vocabularies with ad hoc terms can happen in future, if proven necessary. The resulting implementation is available online in Turtle format (*.ttl) and the AIM cross-domain layer entry URI is https://w3id.org/demeter/crossDomain.

6.3 Domain-Specific Implementation

As described in Sect. 4.3, the AIM domain modules have been inspired by the FIWARE Agrifood models, but also considered the structure of the main ontologies identified. In particular, the main ontologies and models reused include:

- Saref4Agri (which extends SAREF, and reuses SSN and SOSA ontologies among others).
- FOODIE (which extends INSPIRE agriculture and aquaculture facilities model, and reuses ISO standards).
- FIWARE Agrifood models (which are aligned with the NGSI-LD model). These models, however, are available only as documentation and/or JSON schemes, along with some example instances in JSON and JSON-LD. Also, the terms defined in these (and other) models are available in one large (mostly flat) FIWARE JSON-LD context, which maps terms names to URIs. These URIs, however, are not further defined explicitly in an ontology specifying meaning and semantics in machine-readable format.

The general approach was to use Saref4Agri as main source given its good documentation, structure, and coverage, and extend with FOODIE and FIWARE entities. In order to reuse the entities from these ontologies, best practices in ontology engineering for reusing ontology statements [40][29] were adopted and applied. Reusing ontology statements instead of whole ontologies may be useful in different situations. For instance, reusing large ontologies may become a difficult task because they contain a large amount of knowledge that may not be needed when developing a particular ontology. Also, sometimes, reuse demands to retrieve only pieces of knowledge (e.g., statements) to be integrated in the new ontology being

[29] See summary in http://neon-project.org/web-content/media/book-chapters/Chapter-11.pdf

built rather than to reuse entire ontologies. In the case of AIM, given the modular approach followed, reusing ontology statements was better option to keep modules self-contained and focused, while at the same time facilitating their maintenance and the extensibility. In order to reuse statements, (most of) the statements from the original ontologies were copied into the specific module, and a link to the ontology where the entity is defined was added using the property: r*dfs:isDefinedBy*. This allows getting full definition of the entity, tracing it back to its source, while at the same time allows an easy integration in the module. In the case of FIWARE entities, entities were defined from scratch in the modules, as these were not available in any ontology (as discussed above).

After the reuse (or creation in the case of FIWARE) of the different relevant terms, mappings were added in the module to align the three ontologies/models, as described in Sect. 6. The AIM domain layer entry URI is https://w3id.org/demeter/agri. This entry point is a simple ontology importing all the domain modules, which are:

- agriCommon[30]: module that includes common properties used across all other agri-food modules.
- agriProperty[31]: module focused on the different agri properties measured/observed in agri-food applications (e.g., temperature, humidity) and their connection to the systems used to collect them.
- agriSystem[32]: module including all the entities to represent and describe systems and platforms related to agri-food sector, for example, irrigation system and weather stations., including particular sensors in these systems.
- agriAlert[33]: module enabling the representation of agri-food alerts and their characteristics.
- agriCrop[34]: module focused on the representation of crops and their characteristics.
- agriFeature[35]: module enabling the representation of geo features relevant to agri-food applications, for example, farms and plots.
- agriIntervention[36]: module including entities to represent and describe different agri interventions, for example, fertilization and irrigation.
- agriPest[37]: module enabling the representation of agri pests and their characteristics.

[30] https://w3id.org/demeter/agri/agriCommon
[31] https://w3id.org/demeter/agri-context.jsonld
[32] https://w3id.org/demeter/agri/agriSystem
[33] https://w3id.org/demeter/agri/agriAlert
[34] https://w3id.org/demeter/agri/agriCrop
[35] https://w3id.org/demeter/agri/agriFeature
[36] https://w3id.org/demeter/agri/agriIntervention
[37] https://w3id.org/demeter/agri/agriPest

- agriProduct[38]: module focused on the representation of agri-food products and their characteristics, for example, nutrients and ingredients.
- farmAnimal[39]: module focused on the representation of data about livestock and their characteristics.

7 Methodology for Profiles

AIM is based on a suite of general standards, linked to various domain concepts and controlled vocabularies, and expected to be implemented via various meta-models, such as the NGSI-LD generalized object/property message schema and typical observation aggregations supporting spatial and temporal dimensional viewpoints. The underlying problem for information models is to simultaneously support:

- richness to express all relevant aspects
- implementation in simplified meta-models
- generality of patterns to allow reuse of software
- specificity to capture semantics specific to the application domain

The approach taken by AIM to address the tension between these competing demands for richness and simplicity, generality, and specificity is to formalize the mechanism by which general models are simplified for specific applications—a process generally understood as "profiling." A formal meta-model for profiles is introduced into the AIM meta-model layer, allowing multiple profiles so that:

- parts of AIM can express how they specialized standard model, and by implication it is not necessary for any specific user to handle the complexities of mapping all of AIM to all related standards.
- parts of AIM can be mapped to appropriate implementation meta-models, such as NGSI-LD schema, DCAT dataset descriptions, OGC Timeseries Observations, or any other useful implementation approach.
- resources to support implementation of specific profiles can be discovered via the FAIR semantic interoperability mechanisms:
  - JSON-LD contexts
  - documentation
  - JSON-schema for different meta-models
  - constraints (such as SHACL for RDF serializations)
  - profile relationships between model components

Figure 2 shows how different types of model can be used to describe different aspects of implementation, using a typical system that aggregates sensor data from

[38] https://w3id.org/demeter/agri/agriProduct

[39] https://w3id.org/demeter/agri/farmAnimal

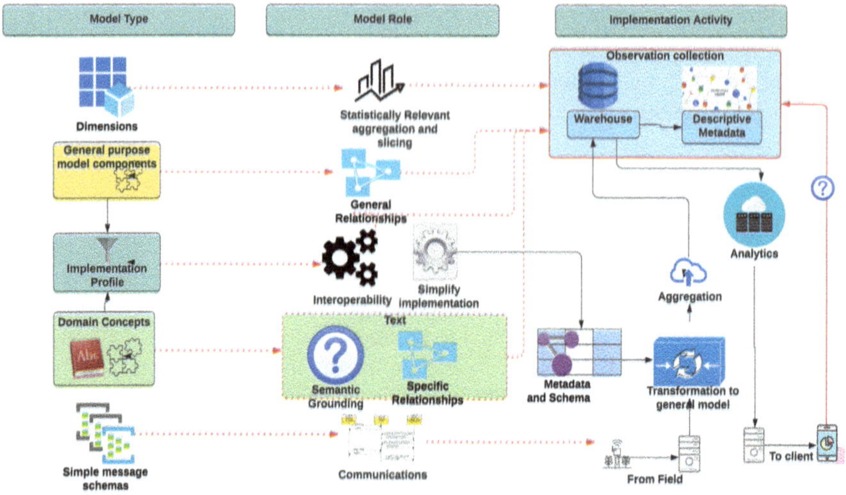

Fig. 2 Model roles in DEMETER

the field and makes it available for analysis. Profiles that formalize the linkages between the different model types mean that the system can provide metadata that links data outputs back to the underlying domain concepts used to specialize the generalized data structures that allow interoperable software to be used to realize the necessary domain-specific semantics.

Profiling provides a pathway from a comprehensive domain model to multiple implementation patterns, but the number of potential resources required, requires tool support to keep implementation and documentation resources in sync with the master model and discoverable in practice. Profilewiz[40] is a toolkit developed to manage the suite of derived model implementation resources according to FAIR principles, and allows AIM to be published as a comprehensive Linked Data resource accessible to multiple platforms.

8 Exemplary Use Cases

AIM is currently under implementation by different components (e.g., services, applications) that are being developed or integrated as part of DEMETER project. AIM provides the reference vocabulary (or lingua franca) enabling different components from different providers to interoperate and exchange data, in order to deliver complete solutions to farmers and other stakeholders that are showcased via a set of pilots. Additionally, AIM is being used as the underlying model of the

[40]https://github.com/RDFLib/profilewiz

data preparation and integration enabler, in order to provide a unified view over heterogeneous datasets using Linked Data as a federated layer.

In order to facilitate the usage of AIM by developers and any other users of the model, a set of guidelines and examples have been prepared for them.

Finding terms and retrieving annotations Developers have different options to find terms from AIM that may be used by their components to produce and/or consume data:

- Load the whole AIM (https://w3id.org/demeter/agri) in an ontology editor like Protege, and then search for terms. This option, however, may be useful only for those having some basic experience with ontologies.
- Open the AIM module URI in a Web browser to open the module entry in OGC definition server. From there it is possible to search terms. However, the search is currently restricted to each module. Alternatively, open one module from the list of modules in the OGC definition server.[41]
- Open AIM in AgroPortal.[42] From there, it is possible to browse all the classes,[43] properties, or mappings (if any). Additionally, it is possible to use recommendations and the annotator functionalities of the portal. These features are also available via API:
  - search for terms in AIM (e.g., Plot[44])
  - search for properties in AIM (e.g., hasAgriParcel[45])
  - get annotations, i.e., potential terms from AIM given an input text (e.g., this parcel crop maize[46])

Validating compliance of data with AIM Developers have different options to validate that the data their components are producing is compliant with AIM, assuming the data is produced in JSON-LD format and it includes the @context declaration that points to AIM JSON-LD context:

- To validate that all elements used in the JSON-LD payload are valid AIM elements, it's enough to use the online JSON-LD playground. This validation, though, is just confirming that all elements used in the payload can be resolved to AIM elements.
- In order to validate the semantics, there are different tools still under evaluation. The goal is to provide a SHACL validation service, reusing existing solutions if

[41] http://defs-dev.opengis.net/demeter/vocab/

[42] http://agroportal.lirmm.fr/ontologies/DEMETER-AIM

[43] http://agroportal.lirmm.fr/ontologies/DEMETER-AIM?p=classes

[44] http://data.agroportal.lirmm.fr/search?ontologies=http://data.agroportal.lirmm.fr/ontologies/-DEMETER-AIM&q=plot

[45] http://data.agroportal.lirmm.fr/property_search?ontologies=http://data.agroportal.lirmm.fr/ontologies/-DEMETER-AIM&q=hasAgriParcel

[46] http://data.agroportal.lirmm.fr/annotator?ontologies=http://data.agroportal.lirmm.fr/ontologies/DEMETER-AIM&text=this+parcel+crop+maize

available, which developers will be able to use from their components.[47] The options under evaluation include:

- use the online SHACL playground.[48] This tool, however, is not checking some restrictions (e.g., datetime format), and it is deprecated as the latest work was moved to a software library.[49]
- use Apache Jena SHACL.[50] This, however, requires downloading the software package and run a command line, which may not be practical for integration inside software components.
- use Astrea Web Service.[51] This is a service that provides a good basis for reusing but is still under testing.

Examples Practically, all DEMETER project pilots are currently implementing AIM in their components. Some examples include the benchmarking component that is under development for Agricolus platform,[52] or the pollination service to facilitate communication between farmers and beekeepers in order to avoid damages on hives due to application of plant treatments. In both examples, a key requirement is the representation of data about a farm.

In the first example, it was necessary to represent basic information about a farm (or holding), e.g., name, geometry, but more detailed information about the parcels (or plots) it manages, as these are the main source of information to the benchmarking component. So, for each of those parcels, in addition to the geometry, it was necessary to capture information about the area, the type of parcel, and the crop planted including the status of the crop, when the crop was last planted. Additionally, the crop description includes a reference to the corresponding AGROVOC concept in order to facilitate integration with other data using AGROVOC for indexing and annotation. A fragment of a JSON-LD of the data payload is listed below, and the complete examples are available in GitHub52.

```
{
 "@context": [
  "https://w3id.org/demeter/agri/agriFeature-context.jsonld",
  "https://w3id.org/demeter/agri/agriCrop-context.jsonld",
  "https://w3id.org/demeter/agri/agriCommon-context.jsonld",
 "https://w3id.org/demeter/agri/agriIntervention-context.jsonld",
  "https://w3id.org/demeter/agri/agriAlert-context.jsonld",
  "https://w3id.org/demeter/agri/agriProduct-context.jsonld",
```

[47] The complete SHACL shapes for AIM is available at https://raw.githubusercontent.com/rapw3k/DEMETER/master/models/SHACL/demeterAgriProfile-SHACL.ttl

[48] https://shacl.org/playground/

[49] https://github.com/zazuko/rdf-validate-shacl

[50] https://jena.apache.org/documentation/shacl/

[51] https://astrea.linkeddata.es/swagger-ui.html

[52 50] https://astrea.linkeddata.es/swagger-ui.html

[51] https://www.agricolus.com/en/

[52] https://github.com/rapw3k/DEMETER/tree/master/models/examples

```
    "https://w3id.org/demeter/agri/agriProperty-context.jsonld",
    "https://w3id.org/demeter/agri/agriSystem-context.jsonld",
    "https://w3id.org/demeter/agri/agriPest-context.jsonld",
    "https://w3id.org/demeter/agri/farmAnimal-context.jsonld"
   ],
  "@id": "urn:ngsi-ld:AgriFarm:72d9fb43-53f8-4ec8-a33c-
fa931360259a",
  "@type": "AgriFarm",
  "name": "Wheat farm",
  "description": "A farm producing wheat",
  "hasGeometry": {
   "@id": "urn:ngsi-ld:AgriFarm:geo:72d9fb43-53f8-4ec8-a33c-
fa931360259x",
   "@type": "Point",
   "asWKT": "POINT(11.3 44.12)"
  },
  "hasAgriParcel":[
   {
    "@id": "urn:ngsi-ld:AgriParcel:72d9fb43-53f8-4ec8-a33c-
fa931360259a",
    "@type": "AgriParcel",
    "hasGeometry": {
     "@id": "urn:ngsi-ld:AgriParcel:geo:72d9fb43-53f8-4ec8-a33c-
fa931360259y",
     "@type": "Polygon",
     "asWKT": "POLYGON (100 0, 101 0, 101 1, 100 1, 100 0)"
    },
   "area": 2012120,
    "description": "Spring wheat parcel",
    "category": "arable",
    "hasAgriCrop": {
     "@id": "urn:ngsi-ld:AgriCrop:df72dc57-1eb9-42a3-88a9-
8647ecc954b4",
     "@type": "AgriCrop",
     "name": "Wheat",
     "alternateName": "Triticum aestivum",
     "agroVocConcept": "http://aims.fao.org/aos/agrovoc/c_7951",
     "description": "Spring wheat"
    },
    "cropStatus": "seeded",
    "lastPlantedAt": "2016-08-23T10:18:16Z"
   } ...
   ]
 }
```


In the second example, the requirements were a little more detailed. So, in addition of the basic information of the farm, it was necessary to capture additional details of the farm structure. In this case, the farm is composed of multiples sites, each with a particular type of activity (e.g., crop growing, animal production) and potentially composed of multiple parcels (each with a particular crop). Moreover, for each parcel, it was necessary to keep the record of crops produced (e.g., the crop species, when the crop was produced, the amount that was produced). The complete example is available also GitHub.

9 Conclusions and Future Work

This chapter presented the Agriculture Information Model (AIM), developed as part of the DEMETER project, which aims at supporting the interoperability of different systems and platforms in the agri-food sector, in order to enable the creation of smart farming solution tailored to the farmers and other stakeholders' needs. AIM provides the common vocabulary that different systems can use to exchange data, and which can be used to access data from different sources in an integrated manner to support the decision-making processes. AIM was designed in three layers, namely meta-model layer, cross-domain layer, and domain-specific layer, and following a modular approach, facilitating the reuse, extension and maintenance of the model. AIM was realized through a suite of ontologies following best practices, and building on the state of the art, thus reusing existing models whenever possible.

This chapter presented an overview of the state of the art, analysis for the development of AIM, the requirements collected to guide its development, and a comprehensive description of the model, including the aspects of semantic interoperability addressed, and the implementation details. Finally, the chapter presented the methodology for profiles that is being applied to facilitate the adoption of AIM, as well as guidelines and examples on how to use the model to support application developers and other users.

Future work includes finalizing some parts of the integration between the different layers as well as the validation of AIM, initially through the set of components currently implementing it as a part of the DEMETER project, but then also with the feedback from the community at large.

Acknowledgments This chapter is based on work carried out under the H2020 DEMETER project (Grant Agreement No 857202) that is funded by the European Commission under H2020-EU.2.1.1.

References

1. Ahmed, N., De, D., & Hussain, I. (2018). Internet of Things (IoT) for smart precision agriculture and farming in rural areas. *IEEE Internet of Things Journal, 5*(6), 4890–4899.
2. Ray, P. P. (2017). Internet of things for smart agriculture: Technologies, practices and future direction. *Journal of Ambient Intelligence and Smart Environments., 9*(4), 395–420.
3. Context Information Management (CIM); Information Model (MODO). Standard, ETSI, France, July 2019.
4. NGSI-LD API: For context information management. Technical report, ETSI, France, January 2019.

5. Context Information Management (CIM). (2019). NGSI-LD API, ETSI GS CIM 009 V1.1.1. Retrieved from https://www.etsi.org/deliver/etsi_gs/CIM/001_099/009/01.01.01_60/gs_CIM009v010101p.pdf.
6. NGSI-LD (Docs) FAQ—Fiware-DataModels. Retrieved from https://fiware-datamodels.readthedocs.io/en/latest/ngsi-ld_faq/index.html.
7. NGSI-LD How to—Fiware-DataModels. Retrieved from https://fiware-datamodels.readthedocs.io/en/latest/ngsi-ld_howto/index.html
8. FIWARE DATA MODELS. Retrieved from https://www.fiware.org/developers/data-models/.
9. SMART AGRIFOOD THE NEW REVOLUTION IN FARMING. Retrieved from https://www.fiware.org/community/smart-agrifood/.
10. SmartM2M; Extension to SAREF. (2020). Part 6: Smart Agriculture and Food Chain Domain, ETSI TS 103 410-6 V1.1.2. Retrieved from https://www.etsi.org/deliver/etsi_ts/103400_103499/10341006/01.01.02_60/ts_10341006v010102p.pdf.
11. ADAPT: AG DATA APPLICATION PROGRAMMING TOOLKIT. Retrieved from https://adaptframework.org/.
12. International data spaces information model, 04 2020.
13. Otto, B., Steinbuß, S., Teuscher, A., Lohmann, S., et al. (2019). Ids reference architecture model. Technical report, International Data Spaces Association.
14. Bader, S., Pullmann, J., Mader, C., Tramp, S., Quix, C., Müller, A. W., Akyürek, H., Böckmann, M., Imbusch, B. T., Lipp, J., Geisler, S., & Lange, C. (2020). The international data spaces information model—An ontology for sovereign exchange of digital content. In *The semantic web, lecture notes in computer science*. Springer.
15. Data on the web best practices: Data quality vocabulary, 12 2016.
16. Data catalog vocabulary (DCAT)—Version 2, 2 2020.
17. ODRL vocabulary expression 2.2, 2 2018.
18. IDS information model SHACL integration, github, 4 2020.
19. GAIA-X: Technical Architecture. (2020, June). Retrieved from https://www.data-infrastructure.eu/GAIAX/Redaktion/EN/Publications/gaia-x-technical-architecture.pdf.
20. The Global Data Model. Retrieved from https://www.gs1.org/sites/default/files/gs1_data_excellence_call_for_action_global_data_model.pdf.
21. Dooley, D. M., Griffiths, E. J., Gosal, G. S., Buttigieg, P. L., Hoehndorf, R., Lange, M. C., Schriml, L. M., Brinkman, F. S. L., & Hsiao, W. W. L. (2018). FoodOn: A harmonized food ontology to increase global food traceability, quality control and data integration. *NPJ Sci Food, 2*, 23. Retrieved from https://www.nature.com/articles/s41538-018-0032-6.
22. Palma, R., Řezník, T., Palomares, M. E., Charvat, K., & Mazurek, C. *An INSPIRE-based vocabulary for the publication of agricultural linked data* (pp. 124–133). Springer. https://doi.org/10.1007/978-3319-33245-1-13
23. Rudnicki, R., Smith, B., Malyuta, T., & Mandrick, C. W. (2016). Best practices of ontology development. October 25, 2016. Technical report.
24. Suárez-Figueroa, M. C. (2012). *NeOn methodology for building ontology networks: Specification, scheduling and reuse*. IOS Press.
25. Bergmanm M. (2016). A reference guide to ontology best practices. Technical report.
26. Haase, P., Palma, R., & d'Aquin, M. (2009, February) Updated version of the networked ontology model.neon project deliverable. Deliverable D1.1.5.
27. Ivanov, P., & Voigt, K. (2011). Schema, ontology and metamodel matching—Different, but indeed the same? In L. Bellatreche & F. M. Pinto (Eds.), *Model and data engineering* (pp. 18–30). Springer.
28. Pidcock, W. (2003, January). *What are the differences between a vocabulary, a taxonomy, a thesaurus, an ontology, and a meta-model?*
29. Henderson-Sellers, B. (2011). Bridging metamodels and ontologies in software engineering. *Journal of Systems and Software, 84*(2), 301–313.
30. Saeki, M., & Kaiya, H. (2007). On relationships among models, meta models, and ontologies. In *Proceedings of the 6th OOPSLA Workshop on Domain-Specific Modeling*.

31. Gasevic, D., Djuric, D., & Devedzic, V. (2006). The Ontology Definition Metamodel (ODM).
32. Pan, W. L., & Liu, D. (2010). ORM-ODM: Ontology definition metamodel for object role modeling. In *2010 3rd international conference on computer science and information technology, Vol. 6*, pp. 511–515.
33. Time ontology in OWL, 2017.
34. OGC GeoSPARQL—A geographic query language for RDF data, 2012.
35. QUDT catalog—Quantities, units, dimensions and data types ontologies, 2020.
36. Semantic sensor network ontology. Technical report, OGC/W3C, USA, January 2017.
37. The new european interoperability framework. Technical report, EIF, Brussels, 2017.
38. Davies, J., Welch, J., Milward, D., & Harris, S. (2020). A formal, scalable approach to semantic interoperability. *Science of Computer Programming, 192*, 102426.
39. Cimmino, A., Fernández-Izquierdo, A., & García-Castro, R. (2020). Astrea: Automatic generation of SHACL shapes from ontologies. In A. Harth, S. Kirrane, A.-C. N. Ngomo, H. Paulheim, A. Rula, A. L. Gentile, P. Haase, & M. Cochez (Eds.), *The semantic web* (pp. 497–513). Springer International Publishing.
40. Suárez-Figueroa, M. C. (2010). *NeOn Methodology for building ontology networks: Specification, scheduling and reuse*. PhD thesis, Facultad de Informática, Universidad Politécnica de Madrid, Spain.

Development of a Framework for Implementing IoT-A on the Beef Cattle Value Chain

Gustavo Marques Mostaço, Roberto Fray Silva, and Carlos Eduardo Cugnasca

1 Introduction

Value chains (VCs) can be defined as a series of interconnected processes or services, going through production, storage, processing, transportation, and marketing of a specific product or commodity, adding value as it moves through the VC [1]. Those commodities can be marketed for domestic or international markets and involve a complex series of stakeholders and products.

Smart farming services refer to modern Information and Communication Technologies (ICT) applied to a specific agricultural VC. They can deliver a more productive and sustainable production by integrating processes and applying innovative techniques to them. In the case of the beef cattle VC, smart services can use Precision Livestock Farming (PLF), Management Information Systems (MIS), and agricultural automation knowledge in order to provide better decision-making and more effective operations [2]. Additionally, it becomes possible to comply with higher level food-quality standards and sustainability labels, increasing their products' value, thus achieving higher value markets.

The paradigm of the Internet of Things (IoT) technologies[1] [5] involves the interconnection of all relevant devices present in the VC. The beef cattle VC aims

[1]From a semantic point of view, IoT can be considered as a worldwide network of interconnected objects, uniquely addressed, based on standard communication protocols, and involving a large number of heterogeneous objects in the process. The increase of devices in this network, in which sensors and actuators coexist, imperceptibly in the environment, and information is distributed on different platforms, is creating the IoT [3, 4].

G. M. Mostaço (✉) · R. F. Silva · C. E. Cugnasca
Department of Computer Engineering and Digital Systems, Escola Politécnica da Universidade de São Paulo (USP), São Paulo, Brazil
e-mail: gmostaco@usp.br; carlos.cugnasca@usp.br

D. D. Bochtis et al. (eds.), *Information and Communication Technologies for Agriculture—Theme III: Decision*, Springer Optimization and Its Applications 184, https://doi.org/10.1007/978-3-030-84152-2_2

to provide full coverage of the processes by collecting, transmitting, analyzing, and storing data from the entire agroecosystem. One example in this VC is a system that registers and controls inputs purchase and consumption throughout the production processes, encompassing breeding, growing, finishing, slaughtering, meat processing, transportation, and logistics, until its products reach the retailers and wholesalers, and finally, the consumer's table.

The IoT reference models provide standard notation and concepts that form the basis for the development of architectural models [4, 6]. The IoT-A project [7] provides a high level of abstraction for developing architectural reference models. It has very detailed documentation, including use case examples, and is the most comprehensive one found in terms of its description of functionalities, domain, and information.[2]

Aspects of the IoT paradigm are vital for the Smart Beef Cattle VC to work properly. With the aid of IoT, one can run the workflow among farmers, service providers, logistics providers, markets, and consumers synchronously all together. That means establishing contact with each participant of the VC, bringing data, and collecting information about their processes, increasing the possibilities for controlling and improving the efficiency of their tasks [5]. Further, without applying IoT concepts, beef cattle farming may continue to work but will not achieve higher performance levels and mitigate quality and environmental problems timely. Therefore, this work aims at setting the ground requirements for the beef cattle VC to operate under the IoT percepts.

Reducing interoperability problems in a VC can help to improve its transparency and coordination. This can positively impact, among other aspects: (1) the agents' results, through decreasing costs and improving revenue; (2) the traceability of the products; (3) the products quality control throughout the VC, especially in the case of product recalls; and (4) the response time when contamination or other kinds of anomalies occur.

The food chain worldwide is highly multi-actor based and distributed, with numerous different actors involved. To study a VC, it is essential to consider the groups of actors that have the same roles in the chain. In agri-food VCs, these groups (also called VC links) are: farmers, industries, shipping companies, wholesalers and retailers, and distributors.

It is important to emphasize that this work has two main contributions: (1) identifying the requirements and services needed to improve the information flow of the beef cattle VC, considering the implementation of IoT technologies; and (2) proposing a Domain Model for implementing IoT on this VC, considering the IoT-A

[2]The IoT-A framework was developed between 2010 and 2013 by the EU. It focuses on multiple domains and provides a high-level abstraction of all components, providing all the necessary information to develop architectural reference models. It has very detailed documentation with several use cases for its different models and components. Some of its components are: domain model, context view, physical-entity view, communication model, information model, functional model, deployment, and operation view. For more detailed information, please refer to: Bauer et al. [7], Preventis et al. (8), Pöhls et al. [9], and Rahimi et al. (10).

framework. This is an essential step for improving VC coordination and reducing the occurrences and the impacts of interoperability problems.

This work is organized in six sections, as follows: Sect. 1 contains the introduction and motivation of the work; Sect. 2 contains the gathering and argumentation of related works; Sect. 3 contains the description of the methodology used; Sect. 4 contains the report of the main results obtained; Sect. 5 contains the discussion concerning the results, the implementation aspects and the difficulties regarding the models; and Sect. 6 contains the conclusions, the observed limitations and plans for future works.

2 Related Work

This section gathers and describes development initiatives on two topics: the IoT frameworks directed to agricultural VCs, in Sect. 2.1; and the IoT-A framework, in Sect. 2.2.

2.1 Frameworks for IoT in Agri-food Value Chains

Wolfert et al. [1] have developed a conceptual framework for applications comprising Big Data in the context of Smart Farming, which is composed of the following factors: agricultural processes, agricultural management, data cycle, organization, and technologies for the business chains. This framework provides the basis for an in-depth analysis of the use of different technologies related to Big Data in agri-food VCs, such as artificial intelligence, machine learning, IoT, among others.

With the analysis of the proposed framework, it was comprehended that the scope of the applications involving Big Data in Smart Farming goes beyond the primary production inside the farm, possibly influencing the whole chain of food supplies. In this sense, the main advantages point in the following directions: gaining predictive insights in agricultural operations, making real-time operational decisions, and redesigning processes and business models. Data privacy and security aspects are also discussed. An important view is set out: potential risks arise as emerging companies gain more power by concentrating all the data generated, possibly leading to a reversal of powers in the current market scenario.

Finally, the main challenges are listed: Data Ownership and related privacy and security issues; Data Quality, which becomes even more critical with real-time data; Intelligent Processing and Analytics require interaction between experts from different domains (agri and data science); Sustainable integration of data from different sources; Sustainable and attractive business models for all stakeholders; and Open Platforms.

Kamilaris et al. [11] propose "Agri-IoT," a semantic framework for IoT-based Smart Farming applications, which supports the reasoning of various streams of heterogeneous, real-time sensor data. For its development, the authors took advantage of concepts and already proposed structures for applications in Smart Cities since it has the closest requirements compared to agribusiness applications, in aspects such as scalability, support of heterogeneous data flows, multiple actors/ users, analytics, and real-time reasoning, decision support, and embedded systems services.

The framework is composed of multiple layers, both lower level layers (device, communication plans), intermediate layers (data, analytics), and higher layers (application, end-user plans). Various software components perform specific operations related to data acquisition, modeling, analysis, or visualization at each layer. Each software component acts as a single entity, with its own open API, which provides a flexible, distributed architecture in which applications can integrate components from different layers, based on their specific needs. In this way, the components become plug and play and can be used selectively according to the specific requirements of agricultural applications.

In agricultural systems, specific ontologies such as AgOnt [12] and AGROVOC[3] are used for the semantic description of the sensors' data flows in real-time, their metadata, the agricultural products to which they are referring, and possible events that occur on the farm. The use of an ontology is critical to deal with systems that are highly complex and involve multiple components and their inter-relations [13].

However, in a more general analysis of the proposed framework, one can see the lack of detail concerning the stages inherent to the different aspects of agricultural production. This may have been due to the authors wishing to undertake a very broad and generalized initial approach regarding the types of agricultural production, both plant and animal, for food or non-food purposes. Another issue is the lack of integration of heterogeneous data flows (sensors, social media, government alerts, among others) in the analysis performed since it only works the simulations of single flows of sensor data.

For Verdouw et al. [14], the study of logistics chains in agribusiness comes from challenges such as dealing with perishable products, sudden inventory changes, and the strict requirements of food safety and sustainability in production. In their work, potential technologies that can meet the requirements of the agri-food chains are presented. Among them, IoT and cloud computing are central issues, increasingly observed in works on this subject. Also, cloud computing alone cannot address the crucial aspects of this type of chain: reliability, efficiency, scalability, and security, requiring an initial hybrid approach between cloud and edge computing.

The authors propose a reference architecture for logistic information systems of agri-food chains. The survey of functional requirements and system components that can meet these requirements resulted in an extensive list, in which four main categories are observed:

[3]FAO AGROVOC Thesaurus—TaxoBank. http://www.taxobank.org/content/agrovoc-thesaurus

1. **Real-time virtualization:** virtual objects must adequately represent the identity, location, and properties (intrinsic and dynamic) of the physical objects of interest in a reliable way, with timely availability. Also, topics such as security and privacy must be unquestionable.
2. **Connectivity:** requires a robust and secure infrastructure to exchange information of objects, as well as standards for identification and exchange of product data and uninterrupted logistics.
3. **Intelligence**: to allow the anticipated knowledge of interruptions or unexpected events, to predict the consequences of these changes after the product reaches the destination and to meet legal and social requirements.
4. **Configuration**: the possibility for the design and implementation of alternative configurations at a low cost, ease of connection with information systems, use of highly customizable software adhering to standardization.

Non-functional requirements were labeled as efficiency, compatibility, usability, reliability, security, ease of maintenance, and portability. These requirements are essential to build systems that can be implemented in real case scenarios, as they consider aspects necessary to maintain the security and reliability of the information. They are also responsible for the aspects that will increase the adoption of the proposed systems and services by the VC agents.

2.2 General Frameworks for IoT and the IoT-A

The IoT reference models provide standard notation and a group of concepts that form the basis for the development of architectural models [4, 6]. In this subsection, the IoT-A framework is presented. Then, a comparative analysis between important IoT frameworks and reference models is made [15], to explain the reasons that motivated the choice of the IoT-A as the most adequate for implementation in this work.

The EU developed the IoT-A project between 2010 and 2013 [7]. It provides a high level of abstraction for developing architectural reference models, and its documentation is very detailed, including use case examples. It includes models for different levels of abstraction: the domain model describes high-level applications; the context view describes the main components of the system in the domain context; the physical-entity view describes the relations between the physical entities (PE) and the virtual entities (VE); the communication model describes the interaction between the devices and objects; the information model describes how data is generated, processed, and transmitted; the functional model describes the principal functionalities of the system, and how they interact with each other; additionally, the deployment and operational view describes the main characteristics and implementation aspects to be considered [7].

The ITU-T reference model was proposed by the International Telecommunication Union (ITU) and is composed of four layers: application, service and application support, network, and device. These layers interact with two groups of capabilities: security and management [5]. However, it is less comprehensive than the IoT-A model and does not specify a clear functional model. Also, there are fewer guidelines for architectural model development.

The Casagras project is another important IoT reference model [16]. It is an initiative between the EU, Japan, China, the USA, and Korea. The following layers compose the model: physical interface; interrogator-gateway; information management, application, and enterprise; wider area communications and Internet. Although there is a description of the layers in detail, this model is less comprehensive than the other two, increasing the difficulty of using it for developing an architectural model.

Unlike the models cited above, the oneM2M is driven by industries, and its objective is to standardize machine-to-machine (M2M) communications [17]. It considers the different stakeholders' multiple views and can be divided into three main layers: application, services, and network. However, it lacks a detailed description of the baselines to develop an architectural model.

The IEEE P2413 standard is still in development and aims at building an architectural framework for IoT, including descriptions and abstractions for different IoT domains [18]. Nevertheless, this model is a generic framework and is still being discussed by workgroups, having no implementation or further documentation.

The IoT-A reference model is the most comprehensive one regarding its description of functionalities, domain, and information. Besides, it fulfills all the requirements for the domain context and presents very detailed documentation. It is also believed that there will be indications for its future adoption, such as the application on different research projects and the continuity of investments by the EU on projects related to the IoT-A [15].

3 Methodology

Figure 1 illustrates the methodology used in this work.

The methodology was divided into three steps. These are:

1. **VC description:** this step aimed at describing the main stages, stakeholders, processes, and the informational flow of the beef cattle VC, based on a thorough literature review. The description was developed considering the state-of-the-art

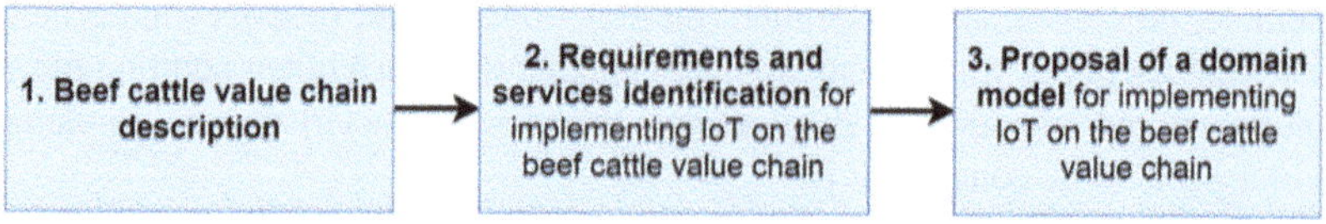

Fig. 1 Steps sequence for the methodology used in this work

concepts of supply chain and agricultural VC design and were focused on addressing the domestic facet of those VCs. The results of this step are presented in Sect. 4.1.

2. **Requirements and services identification:** the main functional and non-functional requirements and services needed to manage the information flow in the beef cattle VC were identified, considering the increasing implementation of IoT technologies in agricultural VCs. This step was based on: (1) periodic meetings with experts from the International Telecommunication Union-Standardization Sector (ITU-T), and (2) an in-depth literature review. The results of this step are presented in Sect. 4.2.
3. **Proposal of a domain model for implementing IoT on the beef cattle VC based on the IoT-A framework:** in this step, a Domain Model to fulfill the requirements and services that were identified in step 2 was designed. This Domain Model also considered the VC description of step 1 with the identified stages, stakeholders, processes, and the informational flow of the beef cattle VC, and was mainly based on the IoT-A framework. The results of this step are presented in Sect. 4.3.

4 Results

This section contains the main results of the research. It is divided into three parts: Sect. 4.1 provides an overview of the beef cattle VC; Sect. 4.2 contains the identified requirements and services to implement the IoT percepts to this VC; Sect. 4.3 presents the IoT frameworks identified as the most adequate to be applied to agri-food VCs.

4.1 Overview of the Beef Cattle Value Chain

The beef cattle VC contains several important components: quality control, efficiency control, certification, sustainability, logistics, business economics, marketing channels, and informational flows [19]. All of these are essential to maintain sustainable flows of products and information throughout the VC.

The commodities inside such chains can be marketed for domestic (national, intra-state, or inter-state) or international markets, involving a complex network of stakeholders and various products. These stakeholders, also referred to as agents on the supply chain literature [20], can respond to ICT changes while are subjected to local, national, and international laws and requirements. One main concern with the present work is to address the domestic facet of those VCs, leaving the international issues for future research. Nevertheless, all the requirements and services identified (Sect. 4.2) and the model proposed (Sect. 4.3), can be adapted for use on the international aspects of those VCs.

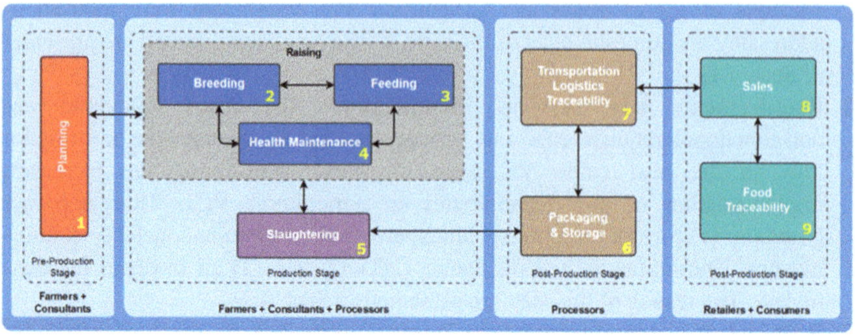

Fig. 2 Overview of the beef cattle value chain

The main stakeholders involved in the beef cattle VC are: Inputs and Services Suppliers; Farmers; Processing and logistics agents (transport, slaughterhouses, industries); Market agents (warehouses, distributors, traders and retailers, groceries); and Consumers. Figure 2 shows an overview of a generic beef cattle VC. It represents the main stages and steps during the planning, production, processing, transportation, and sale of animal products.

The phases and steps characterizing the generic beef cattle VC presented in Fig. 2 are described below [21]. Those are an essential part of the requirements gathering, described in Sect. 4.2.

There are three main stages on this VC. The first is the Pre-Production Stage, in which the farmers and consultants act together to plan the Production Stage, defining the number of animals that will be bred on the farm, which and how much of each input will be used in the Production Stage, and the processes that will be implemented throughout the Production Stage. The second one is the Production Stage itself, in which the animals are bred, fed and, when close to reaching the optimal development stage, go to slaughter. The last one is the Post-Production Stage, in which the meat is packaged, and the cold chain starts, aiming at maintaining product quality until its consumption. The product is stored, transported to the retailers and wholesalers, bought and consumed by the consumers. Each of those stages is divided into phases, which will be described below. It is important to note that the Raising Phase is the only one that is further divided into three steps.

The **Planning** Phase (number 1 in Fig. 2) is the main component of the Pre-Production Stage. At this phase, the resources and raw materials are allocated according to the activities, workforce, and production capacities. It includes such things as land, buildings, equipment, supplies, processes, and laws and regulations (related to both the environment and product quality) that affect the businesses. In this step, the requirements and demands for correct operations are brought up, and the system's capabilities are applied to meet them.

Breeding or Reproduction (number 2 in Fig. 2) is the first step inside the Production Stage. It refers to herd increase through animal reproduction inside the farm or by artificial insemination or embryo transfer techniques. In this step, the main requirements are the animal identification and breeding records, which is a demand for consumers of higher quality products.

Feeding (number 3 in Fig. 2) constitutes one of the most resource-intensive activities in the Production Stage. It considers mainly animal feeding and water consumption and directly affects the final product results and the resource use efficiency. For ruminants, quality grass consumption is a significant concern, but in many climates or rearing methods, hay, silage, and energy and protein-rich foods can be added to the diet [22]. In this step, it is essential to record details about food components, the quality and quantity of ingested food by the animals, and their weight gain. For this step, the most common demands are: Animal identification, Productivity monitoring, Health management, and Animal welfare.

Health maintenance (number 4 in Fig. 2) is directly connected to the raising activities. Together with animal welfare, it implies great concerns among consumers due to their potential impact on product quality and safety. These represent a possible threat to human health and lead to direct production losses and resource use inefficiency. These concerns can commonly include how animals are kept and treated inside the farm, how they are transported, and how they are slaughtered [23, 24].

Slaughtering (number 5 in Fig. 2) considers the reception of the living animals at the slaughterhouse and encompasses the activities related to killing them and processing the carcasses, resulting in the final products that are sold to the market. Due to sanitary concerns, animal slaughtering is only allowed in specific quality-controlled facilities. Some countries may require that these establishments ensure that the animals are handled and slaughtered humanely, and in some cases, under specific cultural processes [25]. In this step, the most important demands are Animal identification, Climate control, and Animal welfare.

Packaging and Storage (number 6 in Fig. 2) occurs after the transformation of the carcass into the main secondary product (meat cuts) and the other by-products (bones, blood, leather, among others). During the packaging phase, each package might be uniquely identified through a slaughterhouse or food industry batch code, containing information such as the production day, the animal and farmer identification, and the list of raw materials used. It is important to note that, for most products, a product package contains parts of several animals. Therefore, tracing it back to its individual animals (or batch of animals) is essential for maintaining traceability on the VC.

Transportation (number 7 in Fig. 2): Once packaged and labeled, the product can be released for the transportation and distribution phase. In the beef cattle VC, the delivery time of fresh products is usually very short due to its high perishability. To deal with this issue, refrigerated storage is generally present right after packaging, increasing the lifetime of the products. There are several processes involved in maintaining the temperature within limits that will preserve product quality as long as possible. This is referred to as maintaining the cold chain.

Sales (number 8 in Fig. 2) is the last part of the distribution activities, where the products are delivered to retailers who perform the sale. The consumer is the end-user of the VC, he/she is the last one to buy the product and demands traceable information on quality standards, animal information, farmer register, country origin, production methods, among others.

Logistics is present throughout the whole VC. It is vital to provide the necessary amount of raw materials and inputs at the correct time during production. It is also responsible for providing a constant flow of animals for the slaughterhouse to maintain its optimal working flow. Incorrect animal transportation and handling can result in product quality reduction and, in severe cases, can lead to death [26]. The most critical logistics requirements are Animal identification, Climate control, Health management, and Animal welfare.

Traceability or product tracking and tracing is the capability to follow the path of a specified production batch throughout the VC as it moves from one organization to the next, allowing the identification of critical quality control points. In the case of food contamination, it helps identify the product batches that may have been contaminated [27]. In general, food businesses engaged in the wholesale supply, production, or food import must have a well-defined system, including production records, clearly described in a written document, to ensure that a fast and efficient recall is possible and timely (the so-called internal traceability).

This information should be readily accessible for both Governmental agencies and the end consumer to identify what batches/product units must be recalled and their location in the VC [14]. It should also be accessible for other companies in the chain to maintain a description of the whole path of the product in the chain (the so-called external traceability). Recording can be done through barcodes, Quick Response (QR) codes, Radio-Frequency Identification (RFID) tags, and other tracking media.

Considering the VC description provided in this section, the next section contains a description of the requirements and services identified for implementing IoT on the beef cattle VC.

4.2 Requirements and Services Identification

The main services and their associated IoT requirements were identified based on the demands of the generic beef cattle VC (Sect. 4.1) and on the research papers by Verdouw et al. [28] and Kamilaris et al. [21]. They are described below:

Animal identification is one of the key information that should be maintained throughout the beef cattle VC since this register accompanies the animal throughout its life inside the farm and should remain attached to all products derived from it. It is crucial for allowing product traceability and has several quality control points. One of these, already referred to in the last section, is the need to trace the finished products back to the individual animals or animal batches. The identified requirement is the need to use any electronic identification (internal, external, or subcutaneous) on the animal's body, allowing its easy and instantaneous identification. Besides the sensor or tag, a proper device or antenna may be needed. For example,

RFID tags can be used on the skin or earrings, but they need an RFID reader to be registered and operated.

Productivity monitoring is mainly applied to the raising set of tasks on the Production Stage, which demands the management of feeding and the main processes inside the farm. This is important because the animals may experience optimal weight gain (based on feed conversion ratio and energy expenditure, two crucial metrics for this VC). The requirements in this category are related to the correct association of important production parameters with each animals' ID, such as weight and food consumption over time (in order to obtain feed conversion ratio), and movement (to estimate energy expenditure). The analysis of the time series generated by those data is also an essential process for improving the planning of future activities and improving the quality of predictions such as weight gain and disease impacts.

Health management is vital to guarantee product quality and safety. As indicated in Sect. 4.1, this category of requirements can be responsible for productivity losses or even threats to human health. Animal health monitoring can be improved using internal and external sensors capable of sensing and communicating essential parameters to the server or the cloud at a time interval appropriate for decision-making. Vital information such as the control of vaccinations, exams, pests, diseases, bruises, and other health-related records should be registered on the management system and at the animal's electronic identification device.

Climate control, such as the ambient temperature and relative humidity, can considerably affect farm animals' metabolism, leading to thermal stress. This results in efficiency loss, increasing the animals' energy use. Other environmental factors may also influence the animals' thermal sensation and heat dissipation, such as air velocity and radiation [29]. These factors can also hasten the final product quality reduction during transport, processing units, or distributors and retailers, as referred to in Sect. 4.1. Regarding this issue, climate control can be improved with environment sensors, sensing and communicating important parameters at an appropriate time interval, and actuators capable of changing essential factors on time. Ideally, these should be integrated on wireless sensor networks and have an alarm system to detect and warn the responsible agent if any anomaly related to environmental variables occurs.

Animal welfare is related to the reduction of animal suffering. When the animal is considered to suffer for any reason, its efficiency in converting feed into muscle decreases leading to direct losses for the farmer. This category of requirements is interrelated to the climate control and health monitoring categories. At this point, systems capable of detecting abnormal behavior should be adopted through GPS monitoring (for extensive farming) or image recognition (for confinements or intensive animal farming) to mitigate the issues related to poor animal welfare.

Information and Communication Security regards the proposal of IoT devices integration on the VC to seamlessly connect all stakeholders, providing the flow of important, strategic, and sometimes crucial information. For this to happen, it is crucial to guarantee the continuity, integrity, and security of communications between the stakeholders and the identified services. Some principles that should be met are: confidentiality, integrity, availability, authentication, lightweight solutions, heterogeneity, policies, and key management systems [30]. Confidentiality and integrity are mandatory when dealing with livestock tracking, due to the sensitive nature of location data, especially in real-time solutions. It is essential to mention that all the communication between users and the devices on animals, machines, or places should be supported in real-time by the IoT infrastructure. This would result in better and faster decision-making, improving product quality throughout the VC, and reducing the probability of widespread product contamination and damage to human health.

The services, sub-services, and activities for the Smart Beef Cattle VC, derived from these groups of requirements, are organized in Table 1. This table also presents their relation to the steps of the generic VC, presented in Fig. 2. Nevertheless, these requirements, services, and sub-services must consider a framework to be implemented. If this is not the case, the interoperability problem will increase, as each agent or VC link will develop and implement its group of solutions, which may not seamlessly interact with the other agents' solutions. For this reason, in the next section, a Domain Model for the implementation of IoT on the cattle beef VC is proposed, considering the IoT-A framework as its basic structure. This is the first step in proposing a useful framework that can be implemented in real-life scenarios. Future works are related to further improving and describing the other parts of the IoT-A model framework applied to this VC.

4.3 *IoT-A for the Beef Cattle Value Chain*

Based on the IoT-A guidelines [7], on the beef cattle VC description (Sect. 4.1), and the requirements and services identified (Sect. 4.2), in this section, the domain model for implementing IoT on the beef cattle VC is proposed.

The objective of the domain model is to describe the system on a higher level, as several other models describe its specific components. This high-level view contains the main elements of the system and how their interactions occur, and a description of the main users, physical and virtual entities, and services. This is an initial step for developing a framework for the beef cattle VC and is illustrated in Fig. 3.

The physical entities are the real objects that generate relevant information for the stakeholders and agents of the VC. After being collected, it will be stored in local and cloud databases. Those agents are referred to in the IoT-A framework as users. Three main physical entities were identified, considering complete internal and external traceability on the VC:

1. **The live animals** themselves, which must be tagged since they are born to allow for complete traceability. In some cases, problems in reading or storing information on the tags can occur, impacting on the individual animal's traceability. This is true, especially for RFID earrings. Nevertheless, machine learning techniques and logical inferences may be used to fill in the missing information on the local

Table 1 Smart Beef Cattle VC main services and sub-services or activities

Services	Sub-services or activities	Steps related
Animal monitoring	Animal identification	ALL
	Animal tracking	2, 3, 4
Animal health maintenance	Prevention and control of diseases	4
	Increasing productive lifetime	2, 3, 4
	Increasing disease resistance	2, 4
	Animal welfare	2, 3, 4, 5
Nutrition and productivity control	Forage quality improvement	3, 4
	Dietary improvements and substitutes	3, 4
	Feed supplements	3, 4
	Digestibility control (rumen microbiome)	3, 4
	Precision feeding	3, 4
Breeding and genetic improvement	Animal selection	2, 3, 4
	Genomic selection	2, 3, 4, 5
	Increasing performance on low-quality feed	3
	Low-methane production	2, 3, 4
Pasture management	Pasture quality and quantity management[a]	3
	Carbon sequestration increase[a]	1, 3
	Integrated and mixed systems[a]	1, 3
Waste handling and treatment	Manure collection[a]	4
	Biogas digesters efficiency management	1
	Gas emission control[a]	1
	Fertilizer production	1, 3
Climate control	Climate sensing and control for living animals	4, 5
	Final product temperature monitoring	6, 7, 8, 9
Logistics management	Distribution recording	7, 8, 9
	Transport environment control	7
	Market analysis alerts	1, 8
Traceability management	Production recording	1, 2, 3, 4
	Processing control	5, 6
	Product tracing	ALL
	Recall management	ALL

[a]Possible sustainability and environmental impacts

and cloud databases, reducing the impact of this problem, as proposed on the work by Silva et al. [31];

2. **The trucks transporting the live animals** to the slaughterhouse, which will have several sensors to gather information related to the quality of transportation in terms of climate and occurrences. This is essential to evaluate the cause of possible injuries to the animals, leading to welfare and quality losses. Besides the quality of the cargo, monitoring the location and integrity of the trucks is also

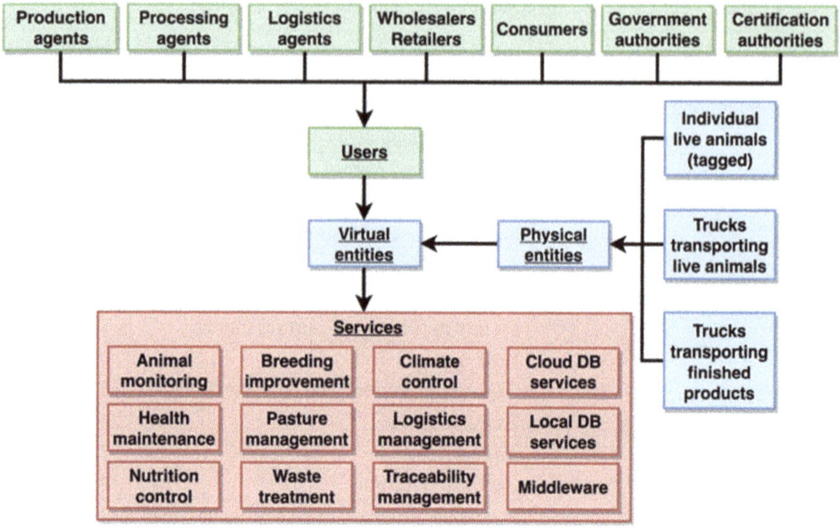

Fig. 3 Domain model for implementing IoT on the beef cattle VC based on the IoT-A framework

essential. This is done mainly through the use of GPS technology and can have the following impacts on the transportation activities: (1) reduction in animal theft; (2) data generation for evaluating possible routes that may reduce transportation costs and injuries to the animals; and (3) better prediction of the time of arrival at the slaughterhouse, which has a direct impact on its production scheduling. Yet, although most of the transportation of live animals is done by trucks, it is essential to notice that there may be different means of transportation along the VC, such as trains, ships, and airplanes, those should also respect the requirements mentioned above;

3. **The trucks transporting finished products**, which can range from simple products such as meat cuts to more complex ones, such as pizzas, sauces, and other products that use the slaughterhouses' products as inputs. In this case, sensors must be placed on the trucks to avoid quality losses and to maintain the cold chain, as well as to reduce product theft. Unlike the other two entities, in this one, actuators have an important role. These are used for regulating temperature and relative humidity inside the cargo area of the trucks. The data collected on these entities must be made available in real-time for the decision-makers to minimize possible quality problems. Although most of the transportation is done by trucks, it is essential to consider that alternative transportation modes, such as trains, ships, and airplanes, may be used. These should also respect the requirements mentioned above.

Each physical entity has a virtual counterpart, which contains all their data and information, denominated virtual entity. It is important to incorporate external data to the virtual entities and use it to make predictions related to: (1) product quality; and (2) logistics relevant performance indicators. This could be achieved by applying the virtualization concepts, known as digital twins, as described in Verdouw et al. [32].

The physical and virtual entities are well described in the IoT-A framework on a model referred to as the entities view. This model, which will be further described in future works, contains, among other aspects: (1) the description of the data lifecycle for the physical and virtual entities; and (2) the resources needed to capture, store, and process the data gathered.

The main users of the system will access information related to the virtual entities and, in the cases in which it is relevant, and the user has the necessary authorization, activate actuators on the real entities. It is important to note that: (1) each user or a group of users has access only to the relevant information to that group, not to all the database in the cloud; (2) the database in the cloud must periodically receive data from the local databases, converting it to the right format using a middleware; and (3) two new stakeholders are introduced as users: government authorities and certification authorities.

Government authorities are related to all the processes involved in regulation, quality control, inspections, and customs-related activities. With a system built on top of the model proposed in this work, it is possible to improve those users' processes, reducing the time and costs related to product exporting, customs, and inspection activities.

Certification authorities are users that provide the list of guidelines needed to obtain specific quality and certification labels and that inspect the facilities and processes in the VC to guarantee that the processes follow those guidelines. They are essential for certain markets with specific demands on production and processing activities, such as religious groups, consumers with health restrictions, and specific high-end consumer segments.

The services in the domain model encompass all the services built on top of the IoT-A framework for this specific use case. In addition to the domain-related services, described in Sect. 4.2, three more services must be incorporated into the domain model so that all the requirements are fulfilled: (1) the infrastructure, resources, and processes to use a cloud database, also considering security, authorization, and privacy issues; (2) the infrastructure, resources, and processes of the local databases, their integration with the cloud, and the reasoning on which information must be sent to the cloud database; and (3) middleware that can identify the type and format of received data, and then convert it to the correct format and insert it in the cloud database. This is essential for the beef cattle VC because, as was already described, different technologies and solutions will continue to coexist after implementing IoT technologies. Lastly, the standard services described in the IoT-A framework should all be implemented on the domain model for this use case.

The next section contains a discussion of the aspects related to the implementation of this model on the beef cattle VC, some of its main advantages and disadvantages, and the next steps on proposing a thorough framework for the beef cattle VC based on the IoT-A framework.

5 Discussion

A domain model was proposed to implement IoT in the beef cattle VC in Sect. 4, describing all its elements and functionalities. Thus, the IoT-A reference model was considered a base framework, as proposed by Silva et al. [15]. This is important to reduce the interoperability problems that may happen if a reference model is not used during the implementation of IoT technologies. A list of requirements and services for the beef cattle VC was also provided, which can be adapted for other agricultural VCs.

On another model that will be described in future works, the IoT-A functional view, these are further described in terms of: (1) devices needed to gather the data; (2) the organization of services; (3) the applications needed for the users to access and process the data; (4) how security must be addressed on the system; (5) how the system must be managed; and (6) how to guarantee communication among all agents and their databases.

An important aspect to consider while building domain models, according to Bauer et al. [7], is that they should consider the evolution of technologies in the context of the use case. This is especially important for this VC since multiple technologies will coexist for the different stages of data collection, processing and storage, used by different agents. This is more evident in the Farm stage due to its heterogeneous nature compared to the other VC links. This aspect was addressed by providing a generic model that can encompass all available and future technologies in communication, data collection (considering product traceability and product and processes monitoring), data processing, and data storage.

With the adoption of the proposed domain model, several traceability problems can be reduced or even solved. Some of its possible impacts are: (1) it will be faster and easier to identify the origin of specific batches of animals separately; (2) it will be possible to identify which processes and products were used during the VC stages; (3) product recalls will be faster, as product batches and their paths on the VC will be easily accessible, reducing the impact of foodborne diseases; (4) internal processes will be more organized, especially inside the Farm; and (5) external traceability will be possible, due to the established connection between animals and meat products.

It can be expected that both farm management and VC processes and logistics will significantly benefit from access to real-time data, real-time forecasting, and tracking of physical items, combined with IoT technologies. This can be further developed to achieve more automation and autonomous operations of the farms and better coordination among VC agents.

Another important topic for further investigation is related to the balance between implementation and operational costs and the financial return of implementing the proposed model. This is important because some agents or links can have significant influence over others. These are called focal links, and, in the case of the beef cattle VC, the slaughterhouses and food industries are considered focal links. Both can enforce the implementation of the abovementioned technologies on their suppliers, generating a problem in terms of division of costs and benefits of the implementation of IoT technologies among the chain's agents [15].

Both the Government (of the producing country) and the VC agents should share the infrastructure costs [15]. The Government can contribute by developing loans with low taxes and more extended payback periods, facilitating the more fragile agents' implementation. However, the operational costs should be the agents' responsibility since they will financially benefit from adopting the proposed model.

Their increased competitiveness will provide the necessary resources to maintain the system and make investments in further automation. Notwithstanding, there should be a third-party organization focused on the regulation of this cost distribution. Otherwise, the focal companies may force other VC links to pay for most of the costs.

The focus of the present work was to promote an overview, identify the main requirements and services, and promote an initial domain model for the beef cattle VC. Therefore, the development of a framework considering those aspects and the implementation aspects, must be further researched and discussed with both Farm administrators and stakeholders from other stages of the VC. In this sense, this work is an initial step towards this discussion, as it provides subsidies for the development of a framework for Smart Beef Cattle Services based on IoT technologies. It also sets the ground rules for the development of similar research for other agricultural and livestock domains. It is expected that farm management and operations research will significantly benefit from the resulting panorama and requirements.

Future works are related to: (1) the development of a framework considering the most important IoT-A models (entity, context, functional, and information) [7]; (2) the validation of the framework with different VC agents; (3) the development of a simulation model to analyze its impacts in comparison to the current situation; and (4) the analysis of aspects such as security and privacy. An extension to other livestock VCs, creating a Smart Livestock Farming infrastructure, is also being researched. This can also be expanded to other agricultural VCs.

6 Conclusions

The growing need for traceability on the beef cattle VC and the increasing gains related to the use of IoT technologies will lead to the increasing adoption of those technologies. This will generate an increasing amount of heterogeneous data, which will not be able to improve decision-making if not processed and used correctly. Good use of this data will also: (1) improve animal welfare; (2) reduce the waste of supplies; (3) reduce quality losses; and (4) better fulfill the customer demands.

In this chapter, an overview of a typical beef production chain is presented, containing the stakeholders involved, its processes, and the informational flow.

After, the requirements and main services for this VC were identified and organized in 9 main services and 31 sub-services or activities through an extensive literature review. A framework for the implementation of IoT technologies in this VC was proposed based on the IoT-A. Lastly, the domain model and its components were presented. It is possible to observe that the proposed model fulfills the identified requirements, while it also attends to essential aspects related to interoperability and security.

The main limitations observed were: (1) the lack of data available on beef cattle traceability; (2) the lack of frameworks designed explicitly for agricultural VCs that are well documented and consider their requirements and services; and (3) the lack of openness of the agents in the VC, which makes it more challenging to evaluate the implementation of new technologies. Future work is related to: (1) further developing the frameworks' models and their components; (2) validating the framework with agents of the VC; (3) simulating the implementation of the proposed model; and (4) analyzing security and privacy aspects.

Acknowledgments This work was supported by the National Council for Scientific and Technological Development (CNPq), and by the Itaú Unibanco S.A. through the Itaú Scholarship Program, at the Centro de Ciência de Dados (C^2D), Universidade de São Paulo, Brazil.

References

1. Wolfert, S., et al. (2017). Big data in smart farming—A review. *Agricultural Systems, 153*, 69–80.
2. Banhazi, T. M., et al. (2012). Precision livestock farming: An international review of scientific and commercial aspects. *International Journal of Agricultural and Biological Engineering, 5* (3), 1–9.
3. Ahmed, N., De, D., & Hussain, I. (2018). Internet of Things (IoT) for smart precision agriculture and farming in rural areas. *IEEE Internet of Things Journal, 5*(6), 4890–4899.
4. Gubbi, J., et al. (2013). Internet of Things (IoT): A vision, architectural elements, and future directions. *Future Generation Computer Systems, 29*(7), 1645–1660.
5. ITU-T. ITU Telecommunication Standardization Sector. (2012). *Y.2060—Overview of the Internet of Things*. Recommendation ITU-T.
6. Atzori, L., Iera, A., & Morabito, G. (2010). The Internet of Things: A survey. *Computer Networks, 54*(15), 2787–2805.
7. Bauer, M., et al. (2013) Internet of Things—Architecture IoT-A Deliverable D1.5—Final architectural reference model for the IoT v3.0.
8. Preventis, A., Stravoskoufos, K., Sotiriadis, S., & Petrakis, E. G. (2016). IoT-A and FIWARE: Bridging the Barriers between the Cloud and IoT Systems Design and Implementation. In *Proceedings of the 6th international conference on cloud computing and services science (CLOSER 2016), v.2*, pp.146–153.
9. Pöhls, H. C., et al. (2014). RERUM: Building a reliable IoT upon privacy-and security-enabled smart objects. In: *2014 IEEE wireless communications and networking conference workshops (WCNCW)*, pp. 122–127.
10. Rahimi, H., Zibaeenejad, A., & Safavi, A. A. (2018). A novel IoT architecture based on 5G-IoT and next generation technologies. In *2018 IEEE 9th annual information technology, electronics and mobile communication conference (IEMCON)*, pp. 81–88.

11. Kamilaris, A., et al. (2017). Agri-IoT: A Semantic Framework for Internet of Things-Enabled Smart Farming Applications. In: *2016 IEEE 3rd world forum on internet of things (WF-IoT)*, pp. 442–447.
12. Hu, S., et al. (2010). AgOnt: Ontology for agriculture internet of things. In *International conference on computer and computing technologies in agriculture* (pp. 131–137). Springer.
13. Abrahão, E., & Hirakawa, A. R. (2018). Complex task ontology conceptual modelling: Towards the development of the agriculture operations task ontology. In *Proceedings of the 10th international joint conference on knowledge discovery, knowledge engineering and knowledge management (IC3K 2018)—Volume 2: KEOD*, pp. 285–292. Science and Technology Publications.
14. Verdouw, C. N., et al. (2018). A reference architecture for IoT-based logistic information systems in agri-food supply chains. *Enterprise Information Systems, 12*(7), 755–779.
15. Silva, R. F., Mostaço, G. M., & Cugnasca, C. E. (2019a) Requirements identification and proposal of a domain model for implementing IoT in the sugar supply chain. In *XII Congresso Brasileiro de Agroinformática (SBIAGRO), 2019*, Indaiatuba, Brasil.
16. Smith, I. (2012). The Internet of Things 2012: New Horizons CASAGRAS2; 2012.. ISBN 9780955370793. http://www.internet-of-thingsresearch.eu/pdf/IERC_Cluster_Book_2012_WEB.pdf.
17. ONEM2M (2019). Functional architecture, Document n. TS-0001-V3.15.1. https://www.onem2m.org/images/files/deliverables/Release3/TS-0001-Functional_Architecture-V3_15_1.pdf.
18. Weyrich, M., & Ebert, C. (2016). Reference architectures for the internet of things. *IEEE Software, 33*(1), 112–116. https://doi.org/10.1109/MS.2016.20
19. CEMA. (2017). CEMA—European Agricultural Machinery—Innovative Livestock Technologies: Making Livestock Farming More Animal-Friendly, Sustainable & Competitive.
20. Chopra, S., & Meindl, P. (2013). *Supply chain management: Strategy, planning and operation* (5th ed.). Pearson Education.
21. Kamilaris, A., Fonts, A., & Prenafeta-Boldú, F. X. (2019). The rise of blockchain technology in agriculture and food supply chains. *Trends in Food Science & Technology, 91*, 640–652.
22. Coleman, S. W., & Moore, J. E. (2003). Feed quality and animal performance. *Field Crops Research, 84*(1–2), 17–29.
23. Humane Farm Animal Care (HFAC). (2018). *Animal care standards: Beef cattle*. Encyclopedia of Reproduction.
24. WSPA. World Animal Protection. (2011). Universal Declaration of Animal Welfare (UDAW)—GlobalAnimalLaw.Org.
25. Grandin, T., & NAMI. (2017). *Recommended animal handling guidelines & audit guide: A systematic approach to animal welfare*.
26. Sheridan, J. J., et al. (1991). *Guidelines for slaughtering, meat cutting and further processing*. FAO.
27. Dabbene, F., Gay, P., & Tortia, C. (2014). Traceability issues in food supply chain management: A review. *Biosystems Engineering, 120*, 65–80.
28. Verdouw, C. N., et al. (2019). Architecture framework of IoT-based food and farm systems: A multiple case study. *Computers and Electronics in Agriculture, 165*, 104939, 26p.
29. Babinszky, L., Halas, V., & Verstegen, M. W. A. (2011). *Impacts of climate change on animal production and quality of animal food products*. InTech.
30. Mahmoud, R., et al. (2015). Internet of things (IoT) security: Current status, challenges and prospective measures. In: *Internet technology and secured transactions (ICITST-2015), 10th international conference for IEEE*, pp. 336–341.
31. Silva, R. F., Mostaço, G. M., Xavier, F., Saraiva, A. M., & Cugnasca, C. E. (2019b). Comparison of the k-means and self-organizing maps techniques to label agricultural supply chain data. In *Conference of the European Federation for Information Technology in Agriculture, Food and the Environment (EFITA), 2019*, Rhodes, Greece.
32. Verdouw, C. N., et al. (2016). Virtualization of food supply chains with the internet of things. *Journal of Food Engineering, 176*, 128–136.

Food Business Information Systems in Western Greece

Vasileios Mitsos, Grigorios Beligiannis, and Achilleas Kontogeorgos

1 Introduction

In our days, companies try to improve their productivity by using information systems (IS) to be competitive in the twenty-first century digital global market. The application of ISs significantly improves the operational efficiency of businesses. The purpose of this chapter is to describe what kind of information systems food businesses of Western Greece use to manage their five basic functions such as human resources, financial and accounting, sales and marketing, operations and production. This chapter also describes the extent of the information systems' application. In a dedicated literature review section, we tend to present relevant studies that have been written at national and international levels in the past. The studies are presented in chronological order, from 1990 until today. After this, in Sect. 2 the methodology applied is presented and discussed, while in Sect. 3 the empirical results are presented. Finally, in Sect. 4 the results and conclusions are discussed.

V. Mitsos (✉) · G. Beligiannis · A. Kontogeorgos
Department of Business Administration of Food and Agricultural Enterprises,
School of Economics and Business, University of Patras, Patras, Greece
e-mail: vasileiosmitsos@upatras.gr; gbeligia@upatras.gr; akontoge@upatras.gr

D. D. Bochtis et al. (eds.), *Information and Communication Technologies for Agriculture—Theme III: Decision*, Springer Optimization and Its Applications 184, https://doi.org/10.1007/978-3-030-84152-2_3

2 Literature Review

2.1 *Studies from 1990 to 2000*

- Cross [1] examined the implementation and the development of a computerized management information system (Dairymis II) of Irish dairy farmers. He claimed that dairy farm income is influenced by many factors including dairy herd management. Dairymis II is a recorder-based system, which helps farmers to compare their own performance with that of other dairy farmers.
- Salin [2] focused his survey on information technology in the Agri-Food supply chain. He claimed that high-tech information systems could offer competitive advantage to Agri-Food enterprises. Lewis [3] promoted the diffusion of computer-based management innovation by gaining a clearer understanding of the evolution of farm management information systems. Characteristics of farm businesses, their financial management information systems (FMIS), and the factors related to the level of sophistication of FMIS were assessed. Results confirmed that the evolution of FMIS is a contiguous process of increasing levels of sophistication gained, by using information from farm record systems (FRS) in decision-making and the adoption of computers in farm management. He also found that the level of sophistication of FMIS is related more to common business factors rather than to factors that are specific in farming activities.
- James et al. [4] examined three questions: (1) who adopted computers, what they and their farms were; (2) what the characteristics of non-adopters were; and (3) what tasks producers wanted from computers to perform. Their results confirmed that most of the variables in earlier studies were identified as influential on computer adoption and still have an impact. These variables include farm size (sales and acres), farm tenure, ownership of livestock, and off-farm employment exposure to computer use.
- Everdingen et al. [5] focused their survey on the adoption of ERP by midsize European enterprises. They supported that big companies had already adopted ERP—systems while medium size companies are an attractive and interesting market for ERP vendors. Lynch et al. [6] investigated the application of intelligent support systems in agriculture. As it is reported a considerable money and effort have been spent on the growth of intelligent support systems (expert systems and decision support systems) for use by farmers. Hence argued that there is a low adoption rate of intelligent support systems.

2.2 *Studies from 2001 to 2005*

- Kuhlmann and Brodersen [7] declared that aspirations are pessimistic for the fast diffusion of demanding and complex information technology (IT) aids and decision support systems (DSSs) among farmers. This view arises from some results of the new institutional economics, recent results of empirical decision research, data from farmers applications of decision models, as well as experiences introducing farm-level DSSs by our own working group at Giessen. For some areas of decision-making, the only hope is that the use of models heightened problem awareness by the user, thus providing additional insights into the usually complex decision space. If, however, farmers wanted to increase the economic efficiency of their marketing processes and production by decreasing friction and waste, then knowledge-based DSSs must continue to be developed and refined.

- Cox [8] reviewed developments in technology which were contributing to global improvements in livestock production and crop, in terms of product quality, environmental considerations, and the welfare of people and livestock.
- Xia and Sun [9] argued that computational fluid dynamics (CFD) had been applied in the food processing industry. CFD is a simulation tool which uses strong computer and applied mathematics for the prediction of heat, mass, optimal design, and momentum transfer in industrial processes. This chapter presented the application of CFD in food industries including refrigeration, drying, refrigeration, mixing, and sterilization.
- Hill and Scudder [10] focused on electronic data interchange (EDI), an important class of IT used for interorganizational information transfer in the supply chain. Data from a survey of the food industry is used to examine if the use of EDI is with respect to interfirm coordination activities involving customers and suppliers. The influence of demographic characteristics on EDI use is also investigated. The results suggested that firms view EDI as a tool for improving efficiencies rather than as a tool for facilitating supply chain integration. There is also a surprising difference among firms in use of EDI with customers vis-à-vis suppliers. Firms tended to be much more accommodating of their customers' desires than of their suppliers'.
- Otles and Onal [11] engaged in computer-aided engineering food industry software. They described numerous of software engineering tasks which applied in the food industry.
- Verbeke [12], firstly focused on individual characteristics that shape information needs, and then discussed information provision through labelling and mass media. It emerged that the consumer needs for information cannot be taken for granted. The provision of even more and too detailed information entailed a risk of information overload, resulting in consumer indifference or loss of confidence. Instead, targeted information provision and segmentation are proposed as potential solutions to market failure from information asymmetry.

2.3 *Studies from 2006 to 2010*

- Wang et al. [13] presented an overview on recent development of wireless sensor technologies and standards for wireless communications as applied to wireless sensors. Examples of sensor and wireless networks were applied in food production and agriculture for precision agriculture and environmental monitoring. A machine to machine (M2M)-based machine and radio frequency identification (RFID)-based traceability systems was also discussed. In this work, it was also discussed that advantages of wireless sensors and obstacles prevent their fast adoption. Finally, based on an analysis of market growth, the chapter discussed future trend of wireless sensor technology development in food industry and agriculture. Finally, not only the advantages of using CFD were discussed but also the future of CFD applications is outlined.

- Wen [14] presented a knowledge-based intelligent e-commerce (KIES) system for selling agricultural products. The KIES system not only provided agricultural products sales, sales forecasting, and financial analysis, but also it provided feasible actions or solutions. The intelligent system integrated a database, a model, and a rule based to create a tool which could be used by the managers for the process of decision-making problems via the Internet.
- Sorensen et al. [15] defined and analyzed the system boundaries and relevant decision processes for such a novel farm management information system (FMIS) as a prerequisite for a dedicated information modelling. The scope and boundaries of the system were described in terms of functionalities and actors, where actors are entities were interfacing with the system (e.g., managers, databases, software). Their research showed the benefit of using dedicated system analysis methodologies as a preliminary step to the actual design of a novel farm management information system compared with other more rigid and activity-oriented system analysis methods.
- Nikkila et al. [16] examined the farm management information systems (FMIS) for precision agriculture. They found that FMIS became more complex as they include new technologies in Internet connectivity. Moreover, FMIS for precision agriculture had certain additional requirements to traditional FMIS. Research aimed to identify the requirements posed by precision agriculture on FMIS.
- Bojnec and Ferto [17] provided an adapted gravity model to measure the impact of the number of the Internet users on food industry trade between developed organizations for economic cooperation and development countries using both panel and cross-sectional data. It was found that there is a significant, positive, and over time increasing effect of the Internet on food industry exports, confirming that market-specific entry costs for food industry exports were reduced by the Internet. The significant positive effect pertained to the Internet was found in the importing countries. The significant positive effects on food industry exports depended on the country's economic size and bilateral common proximities and features. The Internet mitigated the countries' proximities but increased the distance between the countries.
- Corso et al. [18] dealt with interorganizational knowledge management (KM) in the Italian food industry. Based on case studies, the need for knowledge was defined by sharing among supply chain members in different collaborative activities. Then a framework is presented, in order to investigate Information Technology (IT)-based solutions for supply chain management (SCM) and fulfilling the knowledge management needs of the companies.

2.4 Studies from 2011 to 2015

- Sorensen et al. [19] employed the core-task analysis (CTA) method involving a combination of practice-based modelling, science-based modelling, and integrated information modelling. They aimed to identify the requirements which were posed by precision agriculture on FMIS and then evaluated a modern Web-based approach to the implementation of an FMIS that fulfilled these additional requirements.
- Kumar [20] used a case study of the Indian tobacco company's (ITC) *e-Choupal* initiative. In the chapter, the role of information delivery through information and communication technology (ICT) enhancing decision-making capabilities of Indian farmers is empirically analyzed. Users of *e-Choupal* show significantly better decision-making aptitudes, as compared to non-users, on various agricultural practices across the agricultural supply chain. Further, the users' socio-demographic background such as income levels, the social category they belong to, education levels, and landholding size, also played a significant role in impacting decision-making aptitudes. The impact is particularly prominent in production planning and post-harvest as also in marketing-related decisions. Policy implications of these findings are discussed. The study emphasized the importance of designing ICT-enabled information systems to suit the socio-demographic profile of the user groups.
- Hedman and Henningsson [21] in their research presented a model explaining industry-wide information systems integration. They found that the IS integration was inhibited by incompatible value integration and that product sensitivity, continuous production process, and the presence of "value chain captains"—powerful actors dominating the industry—led to higher levels of integration.
- Caudill [22] in his study examined the concepts of information systems which are introduced for use in innovative practices. Specifically, he explored the role of technology in facilitating innovation as well as specific technologies. Some open source tools are introduced to support open innovation.
- Berger and Hovav [23] examined the impact of dairy management information systems such as AfiMilk on farm performance. They claimed that dairy management information systems improved operational efficiency, reduced labor costs and product defects, improved the balance between supply and demand and optimize product mix.
- Fountas et al. [24] presented current advancements in the functionality of commercial and academic FMIS. The study focused on open-field crop production and centers on farm managers as the primary users and decision-makers. The data analyzed were core system application domains and architectures, adoption and profitability, and FMIS solutions for precision agriculture, as also the most information-intensive application area.
- Berera et al. [25] attempted to highlight the importance of ICT in improving marketing activities of retail business in agricultural areas in Indian economy. This chapter also discussed vast potential of implementing ICTs in Indian

agricultural business activities with some success stories and models for justification of the importance of ICT in Agriculture Retail Marketing.

2.5 *Studies from 2016 Until Today*

- Rose et al. [26] examined the factors that affect the use and the uptake of decision support tools by advisers and farmers in the UK. Decision support tools are usually software-based and play an important role in farm decision making. They found that there are numerous agricultural decision support tools in operation in the UK. Furthermore, they claimed that 15 factors influenced advisers and farmers to use decision support tools, which included performance, cost-effectiveness, and usability.
- Bilali and Allahyari [27] explored the contribution of information and communication technologies (ICTs) to transition towards sustainability along the food chain (production, consumption, processing, distribution). A particular attention is devoted to precision agriculture as a food production model integrates many ICTs. ICTs can contribute to Agrofood sustainability transition by increasing resource productivity, decreasing management costs, reducing inefficiencies, and improving food chain coordination. The chapter also explored some drawbacks of ICTs, as well as the factors limiting their uptake in agriculture.

3 Methodology

The aim of the current research is to examine the usability of food business information systems in Western Greece. This questionnaire-based study took place between January 25th and February 15th of 2017 from a PhD student at University of Patras, Greece. Primary data was collected through a questionnaire interview. The respondents were businessmen, ordinary employees, and chief executive officers. It was qualitative research with a questionnaire consisting of close answers. The sample size was 31 food businesses out of 73 that where in total, namely 42%, representative of the use of information systems. The aim of the study is to draw conclusions for a rather small specific area (prefecture of Aitolia and Akarnania in Greece) so the size of the sample is too small. Due to the rather small sample, we cannot generalize these conclusions for the entire Greece. The questionnaires were completed 80% by telephone interviews, 10% with physical interviews, and 10% by email. Almost all firms replied to the questionnaire. Data statistical analysis was done using SPSS statistical program.

4 Results

4.1 Adoption of Human Resources Information Systems

The human resource information system (HRIS) is "the composite of databases, computer applications, software and hardware necessary to collect, store, record, manage, manipulate data, present and deliver for human resources" [28]. In this section, the implementation of human resources information systems in food businesses of Western Greece is analyzed. Figure 1 presents that 38.5% of enterprises apply human resources information systems while the vast majority does not.

From Table 1 it is evident that most food processing businesses do not use any software program to manage human resources functions while a fewer business use six different well-known software packages. Moreover, 40% use software packages

Fig. 1 Application of human resources information systems

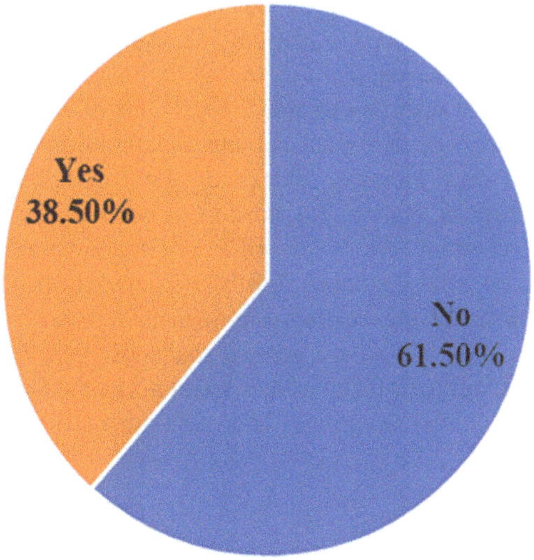

Table 1 Application of software packages in human resources functions

Software	Rewards		Staff selection		Human resources planning		Staff evaluation	
	F	P (%)	F	P (%)	F	P (%)	F	P (%)
None	19	60	27	87	26	83.5	26	83.5
ERP	7	22.5	3	9.5	3	9.5	3	9.5
HRM	1	3.5	1	3.5	1	3.5	1	3.5
Business Plus	1	3.5	–	–	–	–	–	–
Paycheck Ultra	1	3.5	–	–	–	–	–	–
E-Banking	1	3.5	–	–	–	–	–	–
Tailor Made	1	3.5	–	–	1	3.5	1	3.5
Total	**31**	**100.0**	**31**	**100.0**	**31**	**100.0**	**31**	**100.0**

F frequency, *P* percentage

for rewards, 13% for staff selection, 16.5% for human resources planning, and 16.5% for staff evaluation. Lastly, ERP programs is the most widespread tool for managing human resources functions, while tailor made software packages are used by only one company and only for human resources planning, staff evaluation, and rewards.

4.2 Adoption of Accounting and Financial Information System

An accounting information system (AIS) can be defined as a computer-based system that processes supports decision tasks and financial information in the context of coordination and control of organizational activities [29]. In this section, the implementation of financial and accounting information systems in food businesses of Western Greece is analyzed. To start with, Fig. 2 depicts that 90.5% of companies apply financial and accounting information systems.

The analysis of Table 2 shows that most food companies use software programs to manage their financial and accounting functions. For this purpose, they use five software packages. Moreover, 91% of survived firms use software packages for financial reports, for costing as well as for budgeting and 87.5% for general accounting. Moreover, ERP systems are used more for managing financial and accounting functions. Lastly, it is evident that MS Excel is used by only one firm for budgeting, costing, and financial reports.

Data of Table 3 present that most food businesses use software packages to manage their financial and accounting functions such as payable accounts, accounts receivable, securities, and cash management. Totally, they use five software

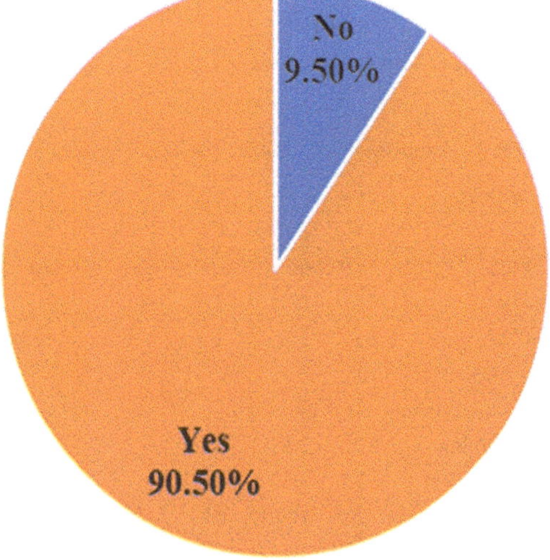

Fig. 2 Application of accounting and financial information systems

Table 2 Application of software packages in financial and accounting functions

Software	Financial reports		General accounting		Budget		Costing	
	F	P (%)	F	P (%)	F	P (%)	F	P (%)
None	3	9	4	12.5	3	9	3	9
ERP	24	77	24	77	24	77	24	77
Business Plus	1	3.5	1	3.5	1	3.5	1	3.5
Pylon Business Accounting	1	3.5	1	3.5	1	3.5	1	3.5
Calculus Ultra Suite	1	3.5	1	3.5	1	3.5	1	3.5
MS Excel	1	3.5	–	–	1	3.5	1	3.5
Total	**31**	**100.0**	**31**	**100.0**	**31**	**100.0**	**31**	**100.0**

F frequency, *P* percentage

Table 3 Application of software packages in accounting and financial functions

Software	Payable accounts		Accounts receivable		Securities		Cash management	
	F	P (%)	F	P (%)	F	P (%)	F	P (%)
None	4	12.5	3	9.	4	12.5	4	12.5
ERP	24	77	24	77	24	77	24	77
Business Plus	1	3.5	1	3.5	1	3.5	1	3.5
Pylon Business Accounting	1	3.5	1	3.5	1	3.5	1	3.5
Calculus Ultra Suite	1	3.5	1	3.5	1	3.5	1	3.5
MS Excel	–	–	1	3.5	–	–	–	–
Total	**31**	**100.0**	**31**	**100.0**	**31**	**100.0**	**31**	**100.0**

F frequency, *P* percentage

packages (ERP, Business Plus, Pylon Business Accounting, Calculus Ultra Suite, MS Excel). Moreover, 87.5% of food businesses use software packages for payable accounts and 91% of firms use software for accounts receivable. Also, 87.5% of examined enterprises use software packages for securities and cash management at the same rate. Additionally, ERP systems are the most common for managing financial and accounting functions. Lastly, one company uses MS Excel only for accounts receivable function.

4.3 *Adoption of Sales and Marketing Information Systems*

A marketing information system is as a structure that consists of equipment, procedures and people for collecting, analysis, storing, and evaluation. It can determine needs accurate and timely for marketing decision-makers. [30]. This section analyses the implementation of marketing and sales information systems in food businesses of Western Greece. At first, Fig. 3 shows that 71% of companies apply marketing and sales information systems.

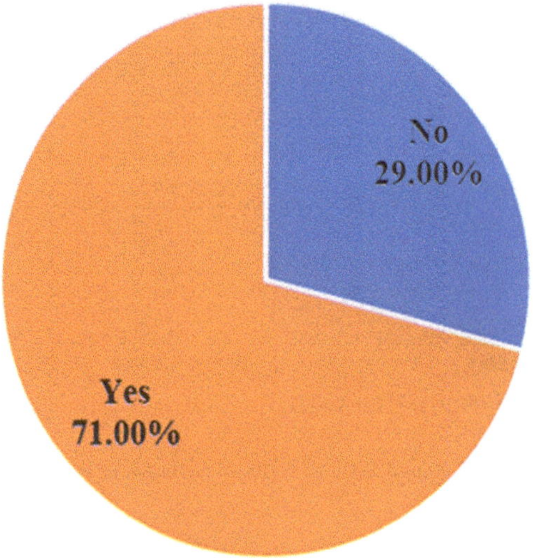

Fig. 3 Application of marketing and sales information systems

Table 4 Application of software packages in marketing and sales functions

Software	Sales forecast		Monitoring sales prospects		Product management		Customer management	
	F	P (%)	F	P (%)	F	P (%)	F	P (%)
None	15	48.5	15	48.5	13	42	9	29
ERP	13	41	13	41	16	51	18	58
Business Plus	1	3.5	1	3.5	1	3.5	1	3.5
MS Excel	1	3.5	1	3.5	1	3.5	–	–
SRS	1	3.5	1	3.5	–	–	–	–
PBS	–	–	–	–	–	–	1	3.5
CRM	–	–	–	–	–	–	2	6
Total	**31**	**100.0**	**31**	**100.0**	**31**	**100.0**	**31**	**100.0**

F frequency, *P* percentage

As it can be seen in Table 4, about half of the food businesses examined use software to manage marketing and sales functions while most use them for customer management. Totally, six ISs (ERP, Business Plus, MS Excel, SRS, PBS, and CRM) are used by food businesses to manage their marketing and sales functions such as sales forecast, monitoring sales prospects, product management, and customer management. At start, 51.5% of enterprises use software packages for sales forecasts and for monitoring sales prospects. Moreover, 58% of food businesses use software packages for product management and 71% use them for customer management. Additionally, ERP systems used mainly by food businesses for their marketing functions. Lastly, only one business uses MS Excel and two businesses use the CRM system to manage their customer.

4.4 Adoption of Operational Information Systems

This section presents the implementation of operational information systems in food processing companies in Western Greece. At first, Fig. 4 reveals that almost half of food businesses apply operational information systems.

Table 5 reveals that the high percentage of food enterprises don't apply any software package to manage the operational functions. Additionally, food businesses for this purpose apply six different software packages. Firstly, 52% of food businesses use software packages for orders management, 45.5% for ordering, 49% for customer service, and 46% for management of finished products. In this case, ERP systems are the most commonly used. Furthermore, one company uses the MS Excel for management of finished products and orders management and two businesses

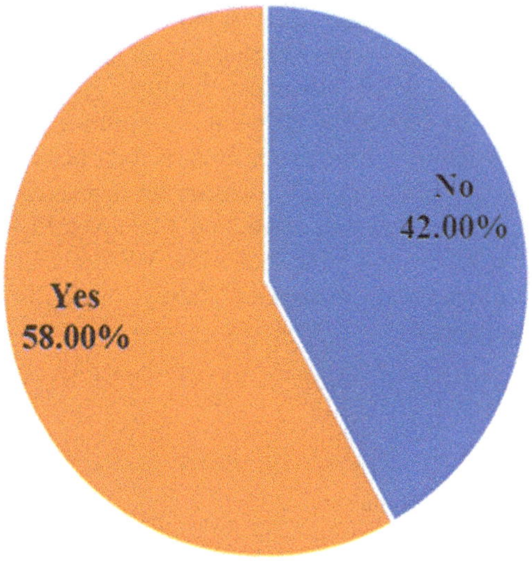

Fig. 4 Application of operational information systems

Table 5 Application of software packages on operational functions

Software	Orders management		Ordering		Customer service		Management of finished products	
	F	P (%)	F	P (%)	F	P (%)	F	P (%)
None	15	48	17	54.5	16	51	17	54
ERP	11	35	11	35	10	32	9	29
XVan	1	3.5	1	3.5	1	3.5	1	3.5
CRM	–	–	–	–	1	3.5	–	–
Business Plus	1	3.5	1	3.5	1	3.5	1	3.5
MS Excel	1	3.5	–	–	–	–	1	3.5
Tailor made	2	6.5	1	3.5	2	6.5	2	6.5
Total	**31**	**100.0**	**31**	**100.0**	**31**	**100.0**	**31**	**100.0**

F frequency, *P* percentage

apply tailor-made programs. Lastly, one of the surveyed companies use the CRM program for customer service.

4.5 Adoption of Production Information Systems

Production information system can be defined as a flexible process whereby a finished product is produced as required from assembled components at the lowest cost possible without sacrificing desired quality [31]. This section presents the implementation of production information systems in food companies in Western Greece. Firstly, Fig. 5 reveals that 42% of firms apply production information systems.

Table 6 indicates that the majority of food businesses don't apply software packages to manage production functions. Additionally, ERP systems are the most used for production functions. What is more, they use six different ISs (ERP, Aqua Manager, MRP, MS Excel, Business Plus, Tailor Made) for production functions such as production planning, inventories, production operation, and production scheduling. Furthermore, 36% of food businesses use software packages for production planning, 42% use them for inventories, 33% for production operation, and 33% for production scheduling. Moreover, ERP in this case are used again at a high rate. Lastly, this table shows that MS Excel applied by one business for production planning and inventories. Lastly, three businesses use tailor-made software packages.

Fig. 5 Application of production information systems

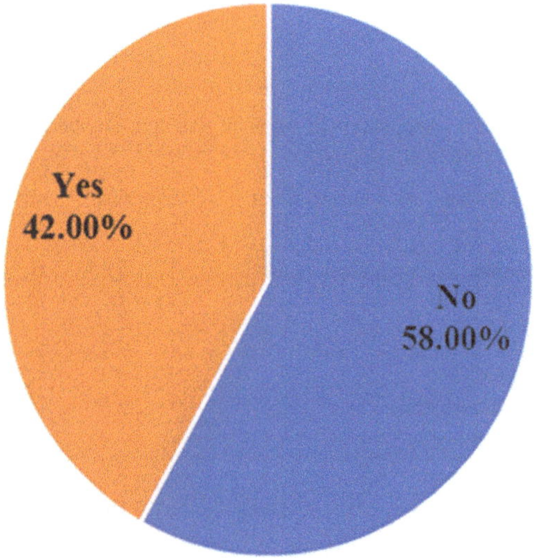

Table 6 Application of software packages on production functions

Software	Production planning		Inventories		Production operation		Production scheduling	
	F	P (%)	F	P (%)	F	P (%)	F	P (%)
None	20	64	18	58	21	67	22	70
ERP	6	19	6	19	5	16	5	16.5
Aqua Manager	–	–	1	3.5	1	3.5	–	–
MRP	1	3.5	1	3.5	1	3.5	1	3.5
MS Excel (MS Office)	1	3.5	1	3.5	–	–	–	–
Business Plus	1	3.5	1	3.5	1	3.5	1	3.5
Tailor made	2	6.5	3	9	2	6.5	2	6.5
Total	**31**	**100.0**	**31**	**100.0**	**31**	**100.0**	**31**	**100.0**

F frequency, *P* percentage

Table 7 Number of software packages that food businesses use

Software packages	Frequency	Percentage (%)
One package (ERP)	19	61
Two packages (ERP, EXCEL)	7	22.5
Three packages (ERP, CRM, SRS)	4	13
Four packages (ERP, AQUAMANAGER, XVAN, MRP)	1	3.5
Total	**31**	**100.0**

Table 8 Number of categories in which software packages are used by food businesses

Categories	Frequency	Percentage (%)
1 (one)	4	13
2 (two)	8	26
3 (three)	7	22.5
4 (four)	7	22.5
5 (five)	5	16
Total	**31**	**100.0**

4.6 Analysis of Software Packages Applications in Food Businesses of Western Greece

Table 7 reveals that most food processing companies use only one software package while less enterprises use more of them.

Table 8 shows that most food processing enterprises use software packages in two categories, while fewer companies use them in more categories. Moreover, four businesses use software packages in one category.

Table 9 reveals that most food companies use only one software for each category while a small number of food businesses use two software packages in one category.

Table 10 shows that the most food companies use the same software package to manage two of their functions and fewer use the same software package to manage only one function. Moreover, even fewer food businesses use the same software

Table 9 Number of software packages used in each category

Software package	Frequency	Percentage (%)
One software package in one category	28	90
Two software packages in one category	3	10
Total	**31**	**100.0**

Table 10 Categories in which the same software packages were used

Categories that use the same software package	Frequency	Percentage (%)
The same software package in one category	7	23
The same software package in two categories	11	35.5
The same software package in three categories	6	19
The same software package in four categories	6	19
The same software package in five categories	1	3.5
Total	**31**	**100.0**

Table 11 Software packages used in businesses with less than two million euros turnover

Turnover	Software package	Frequency	Percentage (%)
<2 mill €	ERP	11	57
	Galaxy CRM of Singular	1	5.5
	Pylon Business Accounting	1	5.5
	Epsilonnet Business Plus	1	5.5
	XVAN	1	5.5
	e-banking	1	5.5
	Excel	2	10
	Tailor Made	1	5.5
Total	**8**	**19**	**100.0**

package to manage three functions. Finally, one enterprise only uses the same software package for all functions.

Table 11 reveals that food companies with turnover less than two million euros use eight software packages. Particularly, the majority of very small food businesses use ERP packages while four firms use the Galaxy CRM of Singular Logic, the Epsilonnet Business Plus (https://www.epsilon-pliroforiki.gr/el/ep-soft/epsilonnet/epsilonnet-business-plus, 2018), the Pylon Business Accounting (https://pcexpertsgreece.com/pylon-business-accounting, 2018) and the XVan software package. Lastly, two of the companies use the MS Excel (https://www.excelsoftware.com, 2018), one uses the e-banking and the other one uses tailor-made software packages.[1]

According to Table 12, food enterprises with a turnover of more than two million euros and less than ten million euros use seven software packages. Specifically, most

[1] https://www.answers.com/Q/What_is_tailor_made_software, 2018.

Table 12 Software packages used in businesses with more than two million euros and less than ten million euros turnover

Turnover	Software package	Frequency	Percentage (%)
<10 mill €	ERP	7	55
	Calculus Suite of Prosvasis	1	7.5
	Paycheck Ultra of Prosvasis	1	7.5
	Epsilonnet Business Plus	1	7.5
	Prosvasis Business Suite	1	7.5
	Aqua Manager	1	7.5
	Tailor Made	1	7.5
Total	**7**	**13**	**100.0**

Table 13 Software packages used in businesses with more than ten million euros turnover

Turnover	Software package	Frequency	Percentage (%)
>10 mill €	ERP	7	54.5
	Material Requirements Planning (MRP)	1	7.5
	Business Payment HRM of Epsilonnet	1	7.5
	Retail System of Singular Logic (SRS)	1	7.5
	Excel (MS Office)	2	15.5
	Tailor Made	1	7.5
Total	**6**	**13**	**100.0**

Table 14 Export businesses

Exports	Business	Percentage (%)
Yes	18	58
No	13	42
Total	**31**	**100.0**

small food businesses use the ERP software. In addition, five businesses use the Calculus Suite of Prosvasis (https://www.prosvasis.com/calculus, 2018), the Paycheck Ultra of Prosvasis (https://www.prosvasis.com/paycheck-14, 2018), the Epsilonnet Business Plus, the Prosvasis Business Suite (https://www.prosvasis.com/pbsone, 2018), and the Aqua Manager (https://www.aqua-manager.com, 2018) software packages. Lastly, only one small business uses tailor-made software package.

As we can see in Table 13, food enterprises with a turnover of more than ten million euros use six different software packages. More specifically, the majority of medium-sized firms use ERP software, three of them use the Material Requirement Planning (MRP), the Business Payment HRM of Epsilonnet (https://www.epsilonnet.gr/product, 2018), and the Retail System of Singular Logic (SRS) software packages. Finally, two businesses use MS Excel, and one uses tailor-made software packages.

As shown in Table 14, the use of software packages based on the export character of food businesses. More specifically, most of the food enterprises that use software packages are exported companies.

Table 15 Software packages used in exporting businesses

Exports	Software packages	Frequency	Percentage (%)
Exporting	ERP	15	46
Companies	Pylon Business Accounting of Epsilonnet	1	3.5
	Paycheck Ultra of Prosvasis	1	3.5
	CRM of Insightly	1	3.5
	Prosvasis Business Suite (PBS)	1	3.5
	Calculus Ultra Suite of Prosvasis	1	3.5
	Business Payment HRM of Epsilonnet	1	3.5
	Material Requirement Planning	1	3.5
	Aqua Manager	1	3.5
	XVan	1	3.5
	Excel	3	9.5
	Tailor Made	4	13
Total	**12**	**31**	**100.0**

Table 16 Software packages used in non-exporting businesses

Exports	Software packages	Frequency	Percentage (%)
Non-exporting companies	ERP	10	72
	Galaxy CRM of Singular	1	7
	Retail System of Singular Logic (SRS)	1	7
	Epsilon net Business Plus	1	7
	Excel	1	7
Total	**5**	**14**	**100.0**

Table 15 shows that food businesses that export use 12 different software packages. More specifically, most exporting enterprises use ERP systems while nine companies use other kind of software packages. Lastly, three companies that export their products use MS Excel while the rest use other tailor-made packages.

Table 16 presents that non-exporting food companies use five software packages. Particularly, the highest number of non-exporting firms use ERP software packages, three businesses use the Galaxy CRM, the Retail System of Singular Logic, and the Epsilonnet Business Plus software packages. Finally, there is one non-exporting company that uses the MS Excel.

5 Conclusions

To conclude, food processing companies in Western Greece apply information systems in different segments to manage their functions. Specifically, 90% apply information systems in finance and accounting, 71% in marketing and sales, 58% in operations, 42% in production, and 39% for human resources management. Moreover, some tailor-made packages are used by exporting companies, and MS Excel is used only by small non-exporting firms. Furthermore, big companies that export their products use information systems to a greater extent. The overwhelming majority uses the ERP software, and the rest use different software packages like HRM, CRM, MRP, SRS, XVan, Business Plus, Aqua Manager, and MS Excel. Apart from this, few businesses apply tailor-made packages mainly for handling operations, production, and human resources. In addition, ERP systems are manufactured by different firms engaged in software like Singular Logic, Entersoft, Ergologic, Prisma Win, Altec, Pegasus, Megasoft, TESAE, Data Communication, and BMS. We cannot generalize the results for the whole Greece because the sample of survey is small. Taking into consideration all the prementioned, it is prevalent that there is lack of information systems in human resources, in production, and in operational functions. To conclude, the gathered information comprises a serious indication that food industry in Western Greece is attractive for information systems programmers [5] and vendors. Consequently, there is great potential for information systems to be used more often by companies in Western Greece in the future.

References

1. Cross, S. (1991). Development and implementation of a computerized management information system (DAIRYMIS II) of Irish dairy farmers. *Journal of Computer and Electronics in Agriculture, 6*(2), 157–173.
2. Salin, V. (1998). Information technology in agri-food supply chains. *Journal of International Food and Agribusiness Management Review, 1*(3), 329–334.
3. Lewis, T. (1998). Evolution of farm management information systems. *Journal of Computers and Electronics in Agriculture, 19*(3), 233–248.
4. James, C., Ascough, I. I., Hoag, D. L., Marshall Frasier, W., & McMaster, G. S. (1999). Computer use in agriculture: An analysis of Great Plains producers. *Journal of Computers and Electronics in Agriculture, 23*(3), 189–204.
5. Everdingen, Y., Hillegersberg, J., & Waarts, E. (2000). ERP adaption by European midsize companies. *Journal of Communication of the ACM, 43*(4), 27–31.
6. Lynch, T., Gregor, S., & Midmore, D. (2000). Intelligent support systems in agriculture: How can we do better? *Australian Journal of Experimental Agriculture, 40*(4), 609–620.
7. Kuhlmann, F., & Brodersen, C. (2001). Information technology and farm management: Developments and perspectives. *Journal of Computers and Electronics in Agriculture, 30*(1–3), 71–83.
8. Cox, S. (2002). Information technology: The global key to precision agriculture and sustainability. *Journal of Computers and Electronics in Agriculture, 36*(2-3), 93–111.
9. Xia, B., & Sun, D. (2002). Applications of computational fluid dynamics (CFD) in the food industry: A review. *Computer and Electronics in Agriculture, Ireland, 34*(1-3), 5–24.

10. Hill, C. A., & Scudder, G. D. (2002). The use of electronic data interchange for supply chain coordination in the food industry. *Journal of Operations Management, 20*(4), 375–387.
11. Otles, S., & Onal, A. (2004). Computer-aided engineering software in the food industry. *Journal of Food Engineering, 65*(2), 311–315.
12. Verbeke, W. (2005). Agriculture and the food industry in the information age. *Journal of European Review of Agricultural Economics, Belgium, 32*(3), 347–368.
13. Wang, N., Zhang, N., & Wang, M. (2006). Wireless sensors in agriculture and food industry—Recent development and future perspective. *Journal of Computers and Electronics in Agriculture, 50*(1), 1–14.
14. Wen, W. (2007). A knowledge-based intelligent electronic commerce system for selling agricultural products. *Journal of Computers and Electronics in Agriculture, 57*(1), 33–46.
15. Sorensen, C. G., Fountas, S., Nash, E., Pesonen, L., Bochtis, D., Pedersen, S. M., Basso, B., & Blackmore, S. B. (2010). Conceptual model of a future farm management information system. *Journal of Computers and Electronics in Agriculture, 72*(1), 37–47.
16. Nikkila, R., Seilonen, I., & Koskinen, K. (2010). Software architecture for farm management information systems in precision agriculture. *Journal of Computers and Electronics in Agriculture, 70*(2), 328–336.
17. Bojnec, S., & Ferto, I. (2010). Internet and international food industry trade. *Journal of Industrial Management & Data Systems, 110*(5), 744–761.
18. Corso, M., Dogan, S. F., Mogre, R., & Perego, A. (2010). The role of knowledge management in supply chains: Evidence from the Italian food industry. *International Journal of Networking and Virtual Organisations, Italy, 7*(2/3), 163–183.
19. Sorensen, C. G., Pesonen, L., Bochtis, D. D., Vougioukas, S. G., & Suomi, P. (2011). Functional requirements for a future farm management information system. *Journal of Computers and Electronics in Agriculture, 76*(2), 266–276.
20. Kumar, J. A. S. (2011). Information and communication technologies (ICTs) and farmers' decision-making across the agricultural supply chain. *International Journal of Information Management, 31*(2), 149–159.
21. Hedman, J., & Henningsson, S. (2012). *Information systems integration in the food industry* (pp. 145–160). Department of IT Management, Copenhagen Business School.
22. Caudill, J. G. (2013). *The role of information systems in innovative food and beverage organizations* (pp. 343–355). Woodhead Publishing Series in Food Science, Journal of Technology and Nutrition.
23. Berger, R., & Hovav, A. (2013). Using a dairy management information system to facilitate precision agriculture: The case of the Afimilk® system. *Journal of Information Systems Management, 30*(1), 21–34.
24. Fountas, S., Carli, G., Sorensen, C. G., Tsiropoulos, Z., Cavalaris, C., Vatsanidou, A., Liakos, B., Canavari, M., & Wiebensohn, J. (2015). Farm management information systems: Current situation and future perspectives. *Journal of Computers and Electronics in Agriculture, 115*, 40–50.
25. Berera, B. S., Panda, B., & R.A., Nayak N., Behera A. C., Jena S. (2015). Information communication technology promoting retail marketing in agriculture sector in India as a study. *Journal of Procedia Computer Science, 48*, 652–659.
26. Rose, D., Sutherland, W., Parker, C., Lobley, M., Winter, M., Morris, C., Twinning, S., Ffoulkes, C., Amano, T., & Dicks, L. (2016). Decision support tools for agriculture: Towards effective design and delivery. *International Journal of Agricultural Systems, 149*(1), 165–174.
27. Bilali, H. E., & Allahyari, M. S. (2018). Transition towards sustainability in agriculture and food systems: Role of information and communication technologies. *Journal of Information Processing in Agriculture, 5*(4), 456–464.

28. Ngai, E. W. T., & Wat, F. K. T. (2006). Human resource information systems: A review and empirical analysis. *Personnel Review, 35*(3), 297–314.
29. Nicolaou, A. (2000). A contingency model of perceived effectiveness in accounting information systems: Organizational coordination and control effects. *International Journal of Accounting Information System, 1*(2), 91–105.
30. Hosami, F., Tahvildari, K., Moradi, M., Taheri, A., Nasab, A. H., & Jan, N. K. L. (2013). Marketing information system in organizational processes. *Interdisciplinary Journal of contemporary research in business, 4*(11), 720–724.
31. Boggs, R. (1987). Implications of formal production information systems for production management. *International Journal of Operations & Production Management, 8*(4), 22–33.

Part II
Primary Production

From Precision Agriculture to Agriculture 4.0: Integrating ICT in Farming

Lefteris Benos, Nikolaos Makaritis, and Vasileios Kolorizos

1 Introduction

Agriculture 4.0 can be regarded as the advancement of precision agriculture benefiting from Information and Communication Technology (ICT), which constitutes, nowadays, an integral element of it. ICT is an umbrella term for describing any tool, device, or application which allows for data gathering, exchange, and communication via transmission and interaction [1]. It contains everything including radio and satellite systems, computers, network software and hardware, mobile phones, electronic money transfers along with several applications and services related to them. The evolution of ICT is facilitating the transformation of conventional farming practices. This trend applies not only in economically developed countries, but in less developing ones as well. In the latter, the adoption of ICT (access to internet, use of smartphones, etc.) takes place at an accelerating rate and could turn out to be beneficial in the near future for seasonal drought forecasts, for example, contributing to the climate-smart agriculture [2]. ICT improves the farmers' access to markets, knowledge, information, and technology skills, while it increases the resilience and sustainability of the farming systems in order to come up against new challenges. In general, agricultural stakeholders must be aware of information regarding the crops, production techniques, equipment and agricultural inputs, market and other aspects including expert device, weather forecast, and credit availability [3]. ICT helps farmers, producers, and researchers to efficiently communicate and interact with the interchange of knowledge that is essential, in particular, when problems emerge.

L. Benos (✉) · N. Makaritis · V. Kolorizos
Institute for Bio-Economy and Agri-Technology (iBO), Centre for Research and Technology Hellas (CERTH), Thessaloniki, Greece
e-mail: e.benos@certh.gr

D. D. Bochtis et al. (eds.), *Information and Communication Technologies for Agriculture—Theme III: Decision*, Springer Optimization and Its Applications 184, https://doi.org/10.1007/978-3-030-84152-2_4

According to Sørensen et al. [4], ICT direct or indirect implementation in agriculture extents (but not limited) to:

- Farm Management Information Systems (FMIS)
- Information on production techniques
- Precise geo-spatial data and 3D maps with elevation information
- Visualization and integrated display
- Telematics
- Photometry
- Ambient environment sensing
- Extended sensors network computing
- Satellite data sensing
- Wearable sensing and computing
- Unmanned aerial systems
- Automation and robotics, equipment process monitoring
- Big data at different scales and difference sources
- Traceability systems
- Access to financial services

ICT can be used either directly or indirectly [5]. The former refers to the case of ICT being used directly to precision agricultural practices. To this end, remote sensors are used which are supported by satellite technology GIS and agronomics. ICT also enables farmers to track and respond to the changes of weather. Furthermore, meteorological stations on the farmland may be connected with the computers or mobile phones of farmers with the intention of sending useful information regarding a variety of parameters such as air humidity, wind speed, rainfall, air temperature, solar radiation, and soil moisture [6, 7]. As far as the indirect services of ICT in agriculture are concerned, supporting during the decision-making stage is the most recognized indirect benefit. Towards meeting the requirements of precision agriculture, the following essentials are necessary: (a) observations originated from sensors for the georeferenced data acquisition, (b) computerized systems that can visualize and process the data, such as GIS and software for generating 3D models or maps, (c) decision support systems, and (d) best agricultural practices having the potential to be implemented by agricultural machinery.

An illustrative example of how precision agriculture is succeeded is given in Fig. 1. The process starts through the acquisition of crop, environmental and soil data (Phase 1). To collect data, visual observations, sensors, and conventional samples are required, which are georeferenced via global navigation satellite systems. After the acquisition of the data, an essential process is to extract valuable information (Phase 2). This information is used in the phase of decision-making (Phase 3), which corresponds to the decisions for management operations; for example, if a variability exists that recommends non-uniform handling. This implies, for instance, how to tackle the different parts of the farmland concerning the doses of fertilizers, pesticides, and irrigation. This phase is regarded as the most challenging one. Finally, the necessary operations are carried out in the field including crop protection, irrigation,

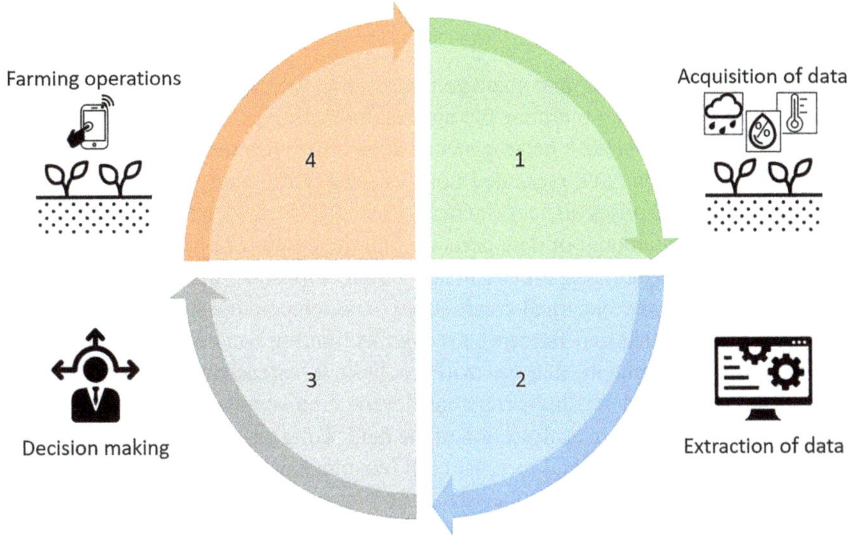

Fig. 1 The four phases in farming operations control

and fertilization (Phase 4) in accordance with the decision-making phase and the entire cycle can start again [8].

Subsequently, some of the most important parts of ICT are described in short by mentioning some indicative applications in agriculture.

2 Agriculture 4.0 Constituents

2.1 Internet of Things (IoT)

After the emergence of the World Wide Web and mobile internet, currently, the so-called IoT has immerged, which is an integral part of ICT and encompasses a family of technologies that can offer several solutions in the direction of the modernization of agricultural sector and contribute to ecological footprint reduce (precise pesticides usage, efficient irrigation, etc.) and food safety. As the first term (Internet) shows, the capability of networking constitutes a key feature of IoT. Although internet is, generally, a human end-user's system, IoT corresponds to things, namely to non-human entities. In effect, machine-to-machine interaction occurs. As mentioned in the relative literature [9, 10], the infrastructure of IoT has three main layers:

- The perception layer, which is associated with sensing.
- The network layer, which is associated with data transfer.

- The application layer, which is associated with the storage and manipulation of data.

The IoT data analysis and management can be utilized to predict situations, automate processes, and optimize the agricultural activities. Additionally, the need for interoperability between heterogeneous apparatuses has resulted in the development of tools, which have generated new services offering an added value to data flows [10]. IoT comprises devices (including sensors, cameras, and weather stations) that gather a great amount of data pertaining to the environmental and crop performance. Mapping of soil properties can also be accomplished, which involves useful data for soil moisture, electrical conductivity, and temperature that can offer information to agronomists and farmers to recognize limiting factors of the crops yield. Smart farming equipment, that has wireless links to extensive data gathered from ground sensors or/and satellites, can consider the crop conditions, for instance, and regulate the application in each section of the field. Time series are created which can subsequently be analyzed and forecast crop recommendations.

The increasing demand for the use of IoT systems in agriculture implies the necessity for robust, accurate, and interoperable logistics traceability systems. Indeed, the technologies existing under the umbrella term of IoT offer the proper tools in order to develop and keep such kinds of services for supporting supply chains [11, 12]. IoT is being implemented in several parts of the storage and transportation of food; ranging from the producer to the reseller up to the consumer. Also, from the optimization of post-harvest treatment, management, and monitoring of the storage facilities, shipping containers of controlled environment to delivery drones [13]. As recent studies have suggested, IoT is a key factor towards establishing a sustainable agriculture [2, 14]. Today, the IoT domain has been evolved by providing low-cost systems like commercially available accurate sensors (e.g., ground, climate, and radiation sensors). Potential benefits from IoT in agriculture includes also improved performance, accuracy, scalability, visibility, cost-effective services, transparency, comprehensive status information, and environmental protection.

2.2 *Artificial Intelligence (AI)*

AI stands for the simulation of the intelligence of human in machines, which are programmed to mimic human actions and think like them. The term also extends to any device that displays traits related to a human mind including features such as problem-solving and learning. The ideal AI can rationalize and subsequently take actions which have the higher possibility of reaching a specific objective. AI includes branches such as Artificial Neural Networks, Fuzzy Logic, Probabilistic Methods, and Machine Learning. Agriculture is turning to AI solutions to organize data, control pests, detect diseases, monitor soil, yield healthier crops, and other tasks within the whole food supply chain [15–18]. Investigations have indicated that approximately 40% of the global food production can be lost by virtue of diseases, insects, and weeds. Therefore, chemical insecticides, fungicides, pesticides, and herbicides are widely used which, in high concentrations, can also be dangerous for both human health and the environment [19]. AI sensors, for example, can identify and target the undesired weeds and afterwards select which herbicides should be applied [20]. This can contribute to prevent excessive use of the above-mentioned chemicals that may be transferred to the production.

2.3 *Machine Learning (ML)*

ML, which is a subset of AI, can extract useful information from field data captured by sensors and carry out specific tasks that were typically executed by humans in the past [21]. It combines advanced algorithms to understand complex systems, optimize specific scenarios, identify, and predict rare events. One major class of Machine Learning, called Deep Learning [22], is based on multiple Artificial Neural Networks architectures that can deal and analyze large amounts of data. More specifically, Convolutional Neural Networks (CNN) [23], a greatly adopted subclass of Deep Learning, are used in image recognition problems since they can learn various patterns that appear in images. ML techniques have been implemented to a variety of agricultural applications, mostly to crop management [24]. Indicative examples involve plant disease detection [25], weed detection for spraying processes [26], and yield prediction in fields [27]. By implementing such techniques, farmers can be informed about cropping conditions and take crucial decisions, resulting in the improvement of their products and the increase of the crop yields. Moreover, farmers can decide about essential financial issues of their crops, examine the performance of the developed hybrids, identify specific areas of fields with crop diseases, as well as apply precise treatment to their fields. Other applications where ML is also exploited include, plant species identification [28], soil moisture monitoring [29], and water stress status estimation [30].

2.4 *Big Data Analytics*

Modern digital technologies which are implemented in agriculture produce large amounts of data under an unprecedented rate [31]. However, this phenomenal situation indicates the need for collection and storage at large-scale levels, as well as editing, analyzing, and interpreting of these "big data" connected with several heterogeneous sources. Analysis of big data is the widely accepted term to portray the new practices that have the potential to make value for decision-makers [32–35]. Even though the analysis of big data was popular in several sectors, it has been adopted in the agricultural domain only recently. Recent studies [35–37] review the relevant literature. According to the analysis of Kamilaris et al. [36], for example, the big data can be originated from several dissimilar sources such as:

- The farmlands (ground sensors including weather stations, chemical detection sensors, biosensors, etc.)
- Previous data collected by third-party and governmental organizations (reports, statistical yearbooks, alerts, regulations from public bodies, etc.)
- Online web services and repositories
- Airborne sensors (e.g., UAVs and satellites)
- Real-time data originated from private companies, crowdsourcing-based practices by mobile phones (evidence about crops, plants, yields, transportation weather conditions, etc.), social media (diseases recognized in many farms, occurrence of natural hazards)

As Wolfert et al. [37] highlighted, big data is only a difficulty of data volume rather than the capability of searching, aggregating, visualizing and cross-referencing the huge datasets over a reasonable timescale. It is about the proficiency to extract insights and information where it once was technically or economically not feasible [32, 33, 35].

2.5 Wireless Sensor Networks (WSN)

The term WSN refers to a set of spatially distributed sensors intending to monitor and record the environmental conditions and organize the data that is collected at a central site. The environmental conditions include magnitudes like humidity, wind, temperature [38], among others. Sensor nodes entail sensing, data processing as well as communication components. Especially in agricultural production, applications, algorithms, and protocols of sensors network should provide self-organizing capacity [39]. Another distinctive characteristic of WSNs is their cooperative capabilities [40]. Each sensor node is equipped with an on-board processor. Consequently, in place of transmitting raw data to the nodes which oversee the fusion, they exploit their own processing capacity to perform simple calculations and send only the needed and, to some extent, processed data. Furthermore, the collected data can be utilized for examining the temporal changeability of environmental impacts that are particularly vital in the light of conventional crop production conversion towards a sustainable one [41]. As a result, the application of WSNs gives the opportunity to reduce the total energy consumption, taking into consideration that the existing cropping systems are strongly responsible for high fossil energy consumption [42, 43]. The tiny-sized, low-power, multi-functional and self-organized nodes, which constitute a cost-effective and scalable technology, render WSNs an impending steppingstone in the framework of precision agriculture automation in farming.

2.6 Blockchain

Blockchain is one of the distributed ledger technologies (DLT) applied in various sectors. Essentially, blockchain is a digital system that maintains a sequence (or chain) of data records (or blocks) along with values that correspond to the current state of the block as well as the previous one [44]. By applying this technology in agriculture, products can be tracked in real-time, rendering supply chain more secure and smarter. In this way, farmers, manufacturers, and distributors around the globe are connected through a blockchain network, leading to the sharing of traceability information concerning the origin of the products or the allergens they might contain [45]. In order to guarantee the security and quality of food products, blockchain solutions ensure that all requirements are met through every stage of production, transportation, and preservation [46]. A blockchain framework can also be implemented in supply chain related to animal products, where product traceability can prevent unsustainable practices [47]. Supply chain can be carried out by two separate chains; a) The chain that contains information about the business users who take part in every process, and b) The chain that records data from all transactions [48].

2.7 Cloud Computing

Cloud computing is often mentioned as "the fifth utility service" since it provides continuous network access to a wide variety of shared computing resources, such as servers, storage, networks, and services [49]. The combination of cloud computing technologies along with IoT systems can optimize the storage, management, and access of information produced from various farming activities [50]. Acquisition of a large amount of data and high storage capacity are some of the cloud computing tools that provide continuous sharing of information among all stakeholders [51]. Moreover, instead of owning expensive ICT equipment, farmers can access on-demand numerous ICT services from the cloud, by being charged each time they use these services [52]. For example, farmers can be timely informed about severe weather events which may occur and affect their cultivations [53]. Public or private organizations can also take advantage of the automatic store and easy access to large-scale land records, as they are able to retrieve beneficial information about the owners of the farmlands, the soil status, and the production history, among others.

2.8 Automated Guided Vehicles

Unmanned Ground and Aerial Vehicles, usually abbreviated as UGVs and UAVs, respectively, are utilized in agriculture for automating a variety of resource-demanding, time-consuming, and labor-intensive operations [54]. In contrast with heavy robotic systems, automated guided vehicles are a more flexible technology for performing complicated operations without manual operation [55, 56]. Advanced UGVs and UAVs use several kinds of cameras and sensors for self-navigation within the working ecosystems. In summary, they are equipped with Inertial Measurement Units (IMUs), Real-Time Kinematic Global Positioning System (RTK-GPS) receivers, Light Detection and Ranging (LiDAR), thermal, Red Green Blue (RGB), and Red Green Blue Depth (RGB-D) cameras, to mention but a few. With regard to UGVs, they can drive long distances by carrying loads or/and measuring soil or crop data. They can also navigate within agricultural fields aiming at soil and imagery aggregation as well as weeding, harvesting, irrigation, and spraying [57–59].

The use of UAVs overcomes some major limitations pertaining to previous methods used in the past. More specifically, gathering data from farmlands was tackled directly by specialized experts, or manned aerial vehicles, or satellites. Nevertheless, the long times for data collection, and the necessity for special permissions proved to be some disadvantages [60]. UAVs, on the other hand, are a cost-effective alternative via carrying various kinds of sensors, while they do not need restrictive permissions. Automated guided vehicles can serve also in the prevention of diseases, as they can apply site-specific pesticides and fertilizers [6, 7, 61–66]. Considering the advancement on new sensors along with the establishing of new legal contexts worldwide, the expansion of UGVs and UAVs is expected to continue over the coming decades [67, 68].

2.9 5G Technology

As described above, ICT aims at providing to the end-user remotely access control, easy operation, and configurability. In conjunction with smart farming, it can transform farmers to more productive and informed professionals. However, the synergy of ICT with smart farming necessitates a communication network. In essence, several sensors, drones, and ground vehicles should be connected to perform either automatically or semi-automatically (human intervention) multiple tasks to collect data and analyze them via ML algorithms, for example [24]. Currently, the 4G mobile network provides the path to connect all these devices with the intention of assuring data transactions without delay or loss, leading to substantially reduces in time, labor, and costs. Although 4G connectivity paradigm has displayed encouraging promise, there still exist several limitations constraining the use of ICT to its maximum potential. In brief, the existing wireless networks are

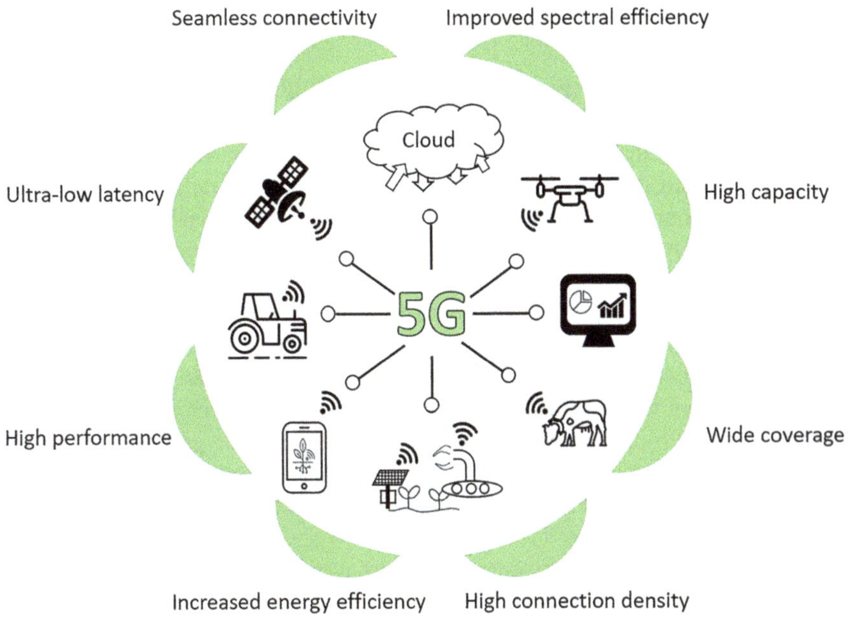

Fig. 2 Indicative applications of 5G in agriculture along with potential benefits

not able to cover some remote areas. Consequently, various networks need to be interconnected. This poses an additional challenge pertaining to changing data rates, resource allocation, handoff problems between heterogeneous networks, as well as poor battery life of the numerous transmitters and antennas.

Considering the ever-increasing usage of ICT, there is an urgent need for more speed, secure communication, and scalability combined, of course, with low costs. Towards this direction, high connectivity coupled with reliable ultralow latency is a prerequisite [69]. This can be accomplished by exploiting 5G (5th Generation), which is an investment in all sections, including smart farming. Some indicative applications along with possible benefits coming from the use of 5G in agriculture are summarized in Fig. 2 according to the recent relative literature [70–72]. Finally, despite 5G advantages, its restricted coverage demands additional infrastructure. In fact, the 5G base station demonstrates a shorter range comparing to 4G; this calls for the installation of additional base stations. Generally, beaming with less power can be succeeded while processing big data, with the installation costs regarding base stations being minimal since they can be positioned anywhere. As with all ICT adoption, over time, 5G technologies are going to be more affordable to small-scale farms [73].

3 Discussion

In summary, considering the plethora of challenges that modern agriculture must face, there is an imperative need to increase the effectiveness of the agricultural practices. To that end, a crucial prerequisite for the agriculture 4.0 realization is ICT adoption. As analyzed above, Agriculture 4.0 includes a wide range of technologies contributing to transformation and transition from the conventional to the digital era of agricultural production. The chapter demonstrated the potential of 4.0 group of technologies penetration in agricultural activities; however, the challenge remains since adoption of such technologies is still in its infancy. A main reason on this, is that for most of the technologies mentioned it is required a high infrastructure investment cost. Overall, the adoption of digital technologies by farmers is a versatile problem with various dimensions. In general, new technologies adoption has been studied by rural sociology, which mainly focuses on profitability coming from these technologies. Also considering the global awareness on environmental issues, environmentally friendly technologies have been examined, especially in developed economies. Nevertheless, ICT adoption should rely on meeting human needs and social responsibility, a fact that is not well established in the agricultural sector [74]. Focusing on the factors affecting the decision of a farmer to adopt ICT, larger farms can adopt easier ICT systems due to their capacity to absorb risks [79]. Moreover, their closeness to significant market locations can provide valuable access to important information, rendering ICT adoption easier. As far as the farmer's personal characteristics are concerned, education and age seem to play a key role. In particular, a high level of education can make the farmer to be more aware of the possible benefits and reduce some adoption cost, while as farmer's age increases, the likelihood of adoption declines. However, older farmers may adopt easier ICT due to their extensive experience on taking management decisions [75].

The technological transformation of the agricultural sector has triggered the uncertainty of farmers with respect to their position in this new digitized era. In fact, a remarkable labor outflow has been observed from agricultural occupations to other sectors, particularly from those occupations related to standardized tasks. The ongoing progress on AI seems to place additional pressure on farmers since also non-standardized tasks previously performed solely by humans are carried out automatically. Throughout this "automatization course", the so-called job polarization appears that is closely related to substitution and, thus, unemployment [76, 77], More specifically, this phenomenon describes the parallel growth of both high- and low-skill jobs at the cost of middle-skill ones. The above consideration, however, shows only one side of the coin. The distinctive characteristics of human regarding decision-making, perception, and acting will still be very important in farming. Complementarity is going to be a key element resulting in the generation of new high-skilled jobs, which can also attract young workers in agriculture. Thus, the matter of "complementarity vs. substitution" between human and technology is going to be consequential in the next decades.

Another aspect of this issue is that investments in technology, science, and engineering are constantly increasing as an opportunity to boost economic growth and innovation. The high-tech jobs that are being created are normally more productive and demanding and, consequently, highly paid. In turn, the high-paid workers are going to give a large amount of the income on local services, which are non-routine in nature. As stated in [78], a single high-tech job creation can generate 2.5–4.4 additional non-routine jobs. Finally, as Lawrence Clark Powell, a noted writer, said:

> "We are the children of a technological age. We have found streamlined ways of doing much of our routine work. Printing is no longer the only way of reproducing books. Reading them, however, has not changed".

Doubtlessly, there is a long way to go to render ICT systems affordable and implemented on a large scale, even in developed countries. Policy makers should concentrate on ICT potential on contribution to enhance agricultural strategies for the sake of both increasing the outcome of each farmer but also for providing a stable environment for achieving the desired sustainable food security.

References

1. Bank, W. (2011). *ICT IN AGRICULTURE Connecting Smallholders to Knowledge, Networks, and Institutions*. Washington.
2. Walter, A., Finger, R., Huber, R., & Buchmann, N. (2017). Opinion: Smart farming is key to developing sustainable agriculture. *Proceedings of the National Academy of Sciences, 114*(24), 6148–6150. https://doi.org/10.1073/pnas.1707462114
3. Sorensen, C. G., Pesonen, L., Fountas, S., Suomi, P., Bochtis, D., Bildsøe, P., & Pedersen, S. M. (2010). A user-centric approach for information modelling in arable farming. *Computers and Electronics in Agriculture, 73*(1), 44–55. https://doi.org/10.1016/j.compag.2010.04.003
4. Sørensen, C. A. G., Kateris, D., & Bochtis, D. (2019). ICT innovations and smart farming. In *Communications in computer and information science* (Vol. 953). https://doi.org/10.1007/978-3-030-12998-9_1
5. Sørensen, C. G., Pesonen, L., Bochtis, D. D., Vougioukas, S. G., & Suomi, P. (2011). Functional requirements for a future farm management information system. *Computers and Electronics in Agriculture, 76*(2), 266–276. https://doi.org/10.1016/j.compag.2011.02.005
6. Bochtis, D. D., Sørensen, C. G., Green, O., Bartzanas, T., & Fountas, S. (2010a). Feasibility of a modelling suite for the optimised biomass harvest scheduling. *Biosystems Engineering, 107*(4), 283–293. https://doi.org/10.1016/j.biosystemseng.2010.05.005
7. Bochtis, D. D., Sørensen, C. G., & Vougioukas, S. G. (2010b). Path planning for in-field navigation-aiding of service units. *Computers and Electronics in Agriculture, 74*(1), 80–90. https://doi.org/10.1016/j.compag.2010.06.008
8. Bochtis, D. D., Sørensen, C. G. C., & Busato, P. (2014). Advances in agricultural machinery management: A review. *Biosystems Engineering*. https://doi.org/10.1016/j.biosystemseng.2014.07.012
9. Sethi, P., & Sarangi, S. R. (2017). Internet of things: Architectures, protocols, and applications. *Journal of Electrical and Computer Engineering, 2017*. https://doi.org/10.1155/2017/9324035
10. Tzounis, A., Katsoulas, N., Bartzanas, T., & Kittas, C. (2017). Internet of things in agriculture, recent advances and future challenges. *Biosystems Engineering, 164*, 31–48. https://doi.org/10.1016/j.biosystemseng.2017.09.007

11. Iakovou, E., Bochtis, D., Vlachos, D., & Aidonis, D. (2015). Supply chain management for sustainable food networks. In E. Iakovou, D. Bochtis, D. Vlachos, & D. Aidonis (Eds.), *Supply chain management for sustainable food networks*. https://doi.org/10.1002/9781118937495
12. Verdouw, C. N., Beulens, A. J. M., & van der Vorst, J. G. A. J. (2013). Virtualisation of floricultural supply chains: A review from an internet of things perspective. *Computers and Electronics in Agriculture, 99*, 160–175. https://doi.org/10.1016/j.compag.2013.09.006
13. Yu, J., Subramanian, N., Ning, K., & Edwards, D. (2015). Product delivery service provider selection and customer satisfaction in the era of internet of things: A Chinese e-retailers' perspective. *International Journal of Production Economics, 159*, 104–116. https://doi.org/10.1016/j.ijpe.2014.09.031
14. Popović, T., Latinović, N., Pešić, A., Zečević, Ž., Krstajić, B., & Djukanović, S. (2017). Architecting an IoT-enabled platform for precision agriculture and ecological monitoring: A case study. *Computers and Electronics in Agriculture, 140*, 255–265. https://doi.org/10.1016/j.compag.2017.06.008
15. Anagnostis, A., Asiminari, G., Papageorgiou, E., & Bochtis, D. (2020). A convolutional neural networks based method for anthracnose infected walnut tree leaves identification. *Applied Sciences, 10*(2), 469. https://doi.org/10.3390/app10020469
16. Anagnostis, A., Tagarakis, A. C., Asiminari, G., Papageorgiou, E., Kateris, D., Moshou, D., & Bochtis, D. (2021). A deep learning approach for anthracnose infected trees classification in walnut orchards. *Computers and Electronics in Agriculture, 182*, 105998. https://doi.org/10.1016/j.compag.2021.105998
17. Jha, K., Doshi, A., Patel, P., & Shah, M. (2019). A comprehensive review on automation in agriculture using artificial intelligence. *Artificial Intelligence in Agriculture, 2*, 1–12. https://doi.org/10.1016/j.aiia.2019.05.004
18. Patrício, D. I., & Rieder, R. (2018). Computer vision and artificial intelligence in precision agriculture for grain crops: A systematic review. *Computers and Electronics in Agriculture, 153*, 69–81. https://doi.org/10.1016/j.compag.2018.08.001
19. Dasgupta, S., Meisner, C., Wheeler, D., Xuyen, K., & Thi Lam, N. (2007). Pesticide poisoning of farm workers-implications of blood test results from Vietnam. *International Journal of Hygiene and Environmental Health, 210*(2), 121–132. https://doi.org/10.1016/j.ijheh.2006.08.006
20. Talaviya, T., Shah, D., Patel, N., Yagnik, H., & Shah, M. (2020). Implementation of artificial intelligence in agriculture for optimisation of irrigation and application of pesticides and herbicides. *Artificial Intelligence in Agriculture, 4*, 58–73. https://doi.org/10.1016/j.aiia.2020.04.002
21. Ferentinos, K. P. (2018). Deep learning models for plant disease detection and diagnosis. *Computers and Electronics in Agriculture*. https://doi.org/10.1016/j.compag.2018.01.009
22. LeCun, Y., Bengio, Y., & Hinton, G. (2015). Deep learning. *Nature, 521*(7553), 436–444.
23. LeCun, Y., Bottou, L., Bengio, Y., & Haffner, P. (1998). Gradient-based learning applied to document recognition. *Proceedings of the IEEE, 86*(11), 2278–2324.
24. Liakos, K., Busato, P., Moshou, D., Pearson, S., Bochtis, D., Liakos, K. G., … Bochtis, D. (2018). Machine learning in agriculture: A review. *Sensors, 18*(8), 2674. https://doi.org/10.3390/s18082674
25. Yang, X., & Guo, T. (2017). Machine learning in plant disease research. European. *Journal of Biomedical Research, 3*(1), 6. https://doi.org/10.18088/ejbmr.3.1.2017.pp6-9
26. Liu, B., & Bruch, R. (2020). Weed detection for selective spraying: A review. *Current Robotics Reports, 1*(1), 19–26.
27. Elavarasan, D., Vincent, D. R., Sharma, V., Zomaya, A. Y., & Srinivasan, K. (2018). Forecasting yield by integrating agrarian factors and machine learning models: A survey. *Computers and Electronics in Agriculture, 155*, 257–282. https://doi.org/10.1016/j.compag.2018.10.024
28. Wäldchen, J., Rzanny, M., Seeland, M., & Mäder, P. (2018). Automated plant species identification—Trends and future directions. *PLoS Computational Biology, 14*(4), e1005993.

29. Ge, X., Wang, J., Ding, J., Cao, X., Zhang, Z., Liu, J., & Li, X. (2019). Combining UAV-based hyperspectral imagery and machine learning algorithms for soil moisture content monitoring. *Peer Journal, 7*, e6926. https://doi.org/10.7717/peerj.6926
30. Gutiérrez, S., Diago, M. P., Fernández-Novales, J., & Tardaguila, J. (2018). Vineyard water status assessment using on-the-go thermal imaging and machine learning. *PLoS One, 13*(2), e0192037. https://doi.org/10.1371/journal.pone.0192037
31. Chi, M., Plaza, A., Benediktsson, J. A., Sun, Z., Shen, J., & Zhu, Y. (2016). Big data for remote sensing: Challenges and opportunities. *Proceedings of the IEEE, 104*(11), 2207–2219. https://doi.org/10.1109/JPROC.2016.2598228
32. Lampridi, M. G., Sørensen, C. G., & Bochtis, D. (2019a). Agricultural sustainability: A review of concepts and methods. *Sustainability (Switzerland), 11*(18). https://doi.org/10.3390/su11185120
33. Lampridi, M. G., Kateris, D., Vasileiadis, G., Marinoudi, V., Pearson, S., Sørensen, C. G., ... Bochtis, D. (2019b). A case-based economic assessment of robotics employment in precision arable farming. *Agronomy, 9*(4). https://doi.org/10.3390/agronomy9040175
34. Lampridi, M., Kateris, D., Sørensen, C. G., & Bochtis, D. (2020). Energy footprint of mechanized agricultural operations. *Energies, 13*(3), 1–15. https://doi.org/10.3390/en13030769
35. Sonka, S. (2016). Big data: Fueling the next evolution of agricultural innovation. *Journal of Innovation Management, 4*(1), 114–136. https://doi.org/10.24840/2183-0606_004.001_0008
36. Kamilaris, A., Kartakoullis, A., & Prenafeta-Boldú, F. X. (2017). A review on the practice of big data analysis in agriculture. *Computers and Electronics in Agriculture, 143*, 23–37. https://doi.org/10.1016/j.compag.2017.09.037
37. Wolfert, S., Ge, L., Verdouw, C., & Bogaardt, M. J. (2017). Big data in smart farming—A review. *Agricultural Systems, 153*, 69–80. https://doi.org/10.1016/j.agsy.2017.01.023
38. Ullo, S. L., & Sinha, G. R. (2020). Advances in smart environment monitoring systems using iot and sensors. *Sensors (Switzerland), 20*. https://doi.org/10.3390/s20113113
39. Garg, V. K. (2007). Wireless personal area networks: Low rate and high rate. In *Wireless communications & networking* (pp. 675–712). https://doi.org/10.1016/b978-012373580-5/50054-5
40. Bai, X., Liu, L., Cao, M., Panneerselvam, J., Sun, Q., & Wang, H. (2017). Collaborative actuation of wireless sensor and actuator networks for the agriculture industry. *IEEE Access, 5*, 13286–13296. https://doi.org/10.1109/ACCESS.2017.2725342
41. Hokazono, S., & Hayashi, K. (2012). Variability in environmental impacts during conversion from conventional to organic farming: A comparison among three rice production systems in Japan. *Journal of Cleaner Production, 28*, 101–112. https://doi.org/10.1016/j.jclepro.2011.12.005
42. Nassi O Di Nasso, N., Bosco, S., Di Bene, C., Coli, A., Mazzoncini, M., & Bonari, E. (2011). Energy efficiency in long-term Mediterranean cropping systems with different management intensities. *Energy*. https://doi.org/10.1016/j.energy.2010.06.026
43. Srbinovska, M., Gavrovski, C., Dimcev, V., Krkoleva, A., & Borozan, V. (2015). Environmental parameters monitoring in precision agriculture using wireless sensor networks. *Journal of Cleaner Production, 88*, 297–307. https://doi.org/10.1016/j.jclepro.2014.04.036
44. Raikwar, M., Gligoroski, D., & Kralevska, K. (2019). SoK of used cryptography in blockchain. *IEEE Access, 7*, 148550–148575. https://doi.org/10.1109/ACCESS.2019.2946983
45. Tian, F. (2016). An agri-food supply chain traceability system for China based on RFID & blockchain technology. In *2016 13th International Conference on Service Systems and Service Management, ICSSSM 2016*. https://doi.org/10.1109/ICSSSM.2016.7538424.
46. Tian, F. (2017). A supply chain traceability system for food safety based on HACCP, blockchain & Internet of things. In *14th International Conference on Services Systems and Services Management, ICSSSM 2017- Proceedings*. https://doi.org/10.1109/ICSSSM.2017.7996119.
47. Marinello, F., Atzori, M., Lisi, L., Boscaro, D., & Pezzuolo, A. (2017). *Development of a traceability system for the animal product supply chain based on Blockchain Technology*.

48. Leng, K., Bi, Y., Jing, L., Fu, H. C., & Van Nieuwenhuyse, I. (2018). Research on agricultural supply chain system with double chain architecture based on blockchain technology. *Future Generation Computer Systems, 86*, 641–649. https://doi.org/10.1016/j.future.2018.04.061
49. Mell, P. M., & Grance, T. (2011). *The NIST definition of cloud computing*. https://doi.org/10.6028/NIST.SP.800-145.
50. Symeonaki, E. G., Arvanitis, K. G., & Piromalis, D. D. (2019). Cloud computing for iot applications in climate-smart agriculture: A review on the trends and challenges toward sustainability. In A. Theodoridis, A. Ragkos, & M. Salampasis (Eds.), *Innovative approaches and applications for sustainable rural development. HAICTA 2017. Springer earth system sciences*. Springer. https://doi.org/10.1007/978-3-030-02312-6_9
51. Du, Y., Wang, Y., & Shang, Y. (2020). Innovative research on the direction of network information based on cloud computing technology. *Journal de Physique, 1648*, 42007. https://doi.org/10.1088/1742-6596/1648/4/042007
52. Prasad, S., Peddoju, S. K., & Ghosh, D. (2013). AgroMobile: A cloud-based framework for agriculturists on mobile platform. *International Journal of Advanced Science and Technology, 59*, 41–52. https://doi.org/10.14257/ijast.2013.59.04
53. Namani, S., & Gonen, B. (2020). Smart agriculture based on IoT and cloud computing. *Proceedings - 3rd International Conference on Information and Computer Technologies, ICICT 2020*, pp. 553–556. https://doi.org/10.1109/ICICT50521.2020.00094.
54. Benos, L., Bechar, A., & Bochtis, D. (2020). Safety and ergonomics in human-robot interactive agricultural operations. *Biosystems Engineering, 200*, 55–72. https://doi.org/10.1016/j.biosystemseng.2020.09.009
55. Bechtsis, D., Moisiadis, V., Tsolakis, N., Vlachos, D., & Bochtis, D. (2019). Unmanned ground vehicles in precision farming services: An integrated emulation modelling approach. *Communications in Computer and Information Science, 953*, 177–190. https://doi.org/10.1007/978-3-030-12998-9_13
56. Tsouros, D. C., Bibi, S., & Sarigiannidis, P. G. (2019). A review on UAV-based applications for precision agriculture. *Information (Switzerland)*. https://doi.org/10.3390/info10110349
57. Bechar, A., & Vigneault, C. (2016). Agricultural robots for field operations: Concepts and components. *Biosystems Engineering, 149*, 94–111. https://doi.org/10.1016/j.biosystemseng.2016.06.014
58. Bechar, A., & Vigneault, C. (2017). Agricultural robots for field operations. Part 2: Operations and systems. *Biosystems Engineering, 153*, 110–128. https://doi.org/10.1016/j.biosystemseng.2016.11.004
59. Bochtis, D., Griepentrog, H. W., Vougioukas, S., Busato, P., Berruto, R., & Zhou, K. (2015). Route planning for orchard operations. *Computers and Electronics in Agriculture, 113*, 51–60. https://doi.org/10.1016/j.compag.2014.12.024
60. Mulla, D. J. (2013). Twenty five years of remote sensing in precision agriculture: Key advances and remaining knowledge gaps. *Biosystems Engineering, 114*(4), 358–371. https://doi.org/10.1016/j.biosystemseng.2012.08.009
61. Bochtis, D. D., & Vougioukas, S. G. (2008). Minimising the non-working distance travelled by machines operating in a headland field pattern. *Biosystems Engineering, 101*(1), 1–12. https://doi.org/10.1016/j.biosystemseng.2008.06.008
62. Faiçal, B. S., Pessin, G., Filho, G. P. R., Furquim, G., de Carvalho, A. C. P. L. F., & Ueyama, J. (2014). Exploiting evolution on UAV control rules for spraying pesticides on crop fields. *Communications in Computer and Information Science, 459*, 49–58. https://doi.org/10.1007/978-3-319-11071-4_5
63. Giles, D. K. (2016). Use of remotely piloted aircraft for pesticide applications: Issues and outlook. *Outlooks on Pest Management, 27*(5), 213–216. https://doi.org/10.1564/V27_OCT_05
64. Jensen, M. A. F., Bochtis, D., Sorensen, C. G., Blas, M. R., & Lykkegaard, K. L. (2012). In-field and inter-field path planning for agricultural transport units. *Computers and Industrial Engineering, 63*(4), 1054–1061.

65. Jensen, M. F., Bochtis, D., & Sørensen, C. G. (2015). Coverage planning for capacitated field operations, part II: Optimisation. *Biosystems Engineering*. https://doi.org/10.1016/j.biosystemseng.2015.07.002
66. Zhou, K., Leck Jensen, A., Sørensen, C. G., Busato, P., & Bochtis, D. D. (2015). Corrigendum to "Agricultural operations planning in fields with multiple obstacle areas" [Comput. Electron. Agric. 109 (2014) 12-22] DOI: 10.1016/j.compag.2014.08.013]. *Computers and Electronics in Agriculture, 116*. https://doi.org/10.1016/j.compag.2015.07.013.
67. Moysiadis, V., Tsolakis, N., Katikaridis, D., Sørensen, C. G., Pearson, S., & Bochtis, D. (2020). Mobile robotics in agricultural operations: A narrative review on planning aspects. *Applied Sciences*.
68. Roldán, J. J., del Cerro, J., Garzón-Ramos, D., Garcia-Aunon, P., Garzón, M., de León, J., & Barrientos, A. (2018). Robots in agriculture: State of art and practical experiences. In *Service robots*. https://doi.org/10.5772/intechopen.69874
69. Tarver, C., Tonnemacher, M., Chen, H., Zhang, J., & Cavallaro, J. R. (2021). GPU-based, LDPC decoding for 5G and beyond. *IEEE Open Journal of Circuits and Systems, 2*, 278–290. https://doi.org/10.1109/OJCAS.2020.3042448
70. Hu, X. (2021). Coupling of agricultural economy and environment based on 5G network and internet of things system. *Microprocessors and Microsystems, 80*, 103569. https://doi.org/10.1016/j.micpro.2020.103569
71. Li, M., & Abula, B. (2020). Evaluation of economic utility of smart agriculture based on 5G network and wireless sensors. *Microprocessors and Microsystems*, 103485. https://doi.org/10.1016/j.micpro.2020.103485
72. Tang, Y., Dananjayan, S., Hou, C., Guo, Q., Luo, S., & He, Y. (2021). A survey on the 5G network and its impact on agriculture: Challenges and opportunities. *Computers and Electronics in Agriculture, 180*, 105895. https://doi.org/10.1016/j.compag.2020.105895
73. Bouras, C., Kollia, A., & Papazois, A. (2017). Dense deployments and DAS in 5G: A techno-economic comparison. *Wireless Personal Communications, 94*(3), 1777–1797. https://doi.org/10.1007/s11277-016-3711-0
74. Rose, D. C., & Chilvers, J. (2018). Agriculture 4.0: Broadening responsible innovation in an era of smart farming. *Frontiers in Sustainable Food Systems, 2*, 87. https://doi.org/10.3389/fsufs.2018.00087
75. Wollni, M., & Andersson, C. (2014). Spatial patterns of organic agriculture adoption: Evidence from Honduras. *Ecological Economics, 97*, 120–128. https://doi.org/10.1016/j.ecolecon.2013.11.010
76. Heyman, F. (2016). Job polarization, job tasks and the role of firms. *Economics Letters, 145*, 246–251. https://doi.org/10.1016/J.ECONLET.2016.06.032
77. Marinoudi, V., Sørensen, C. G., Pearson, S., & Bochtis, D. (2019). Robotics and labour in agriculture. A context consideration. *Biosystems Engineering, 184*, 111–121. https://doi.org/10.1016/j.biosystemseng.2019.06.013
78. OECD. (2016). *Business brief: Jobs in the digital era work differently.*
79. Sansel Tandogan, N., & Gedikoglu, H. (2020). Socio-economic dimensions of adoption of conservation practices: What is needed to be done? Organic Agriculture. https://doi.org/10.5772/intechopen.93198

On the Routing of Unmanned Aerial Vehicles (UAVs) in Precision Farming Sampling Missions

Georgios Dolias, Lefteris Benos, and Dionysis Bochtis

1 Introduction

Certainly, there is a substantial need of agricultural products required to be produced to compensate with the ever-increasing global population. In particular, the Food and Agriculture Organization (FAO) of the United Nations and the International Telecommunication Union (ITU) estimate that food production will be increased up to 50% by 2050, since the population will reach approximately 9 billion [1]. In addition, climate change poses significant challenges affecting agricultural sector. Severe weather events may cause rapid increase in food prices and shortage of water needed for crops [2, 3]. However, part of these challenges can be tackled by applying Precision Agriculture (PA) practices and smart farming technologies.

By adopting methods of PA, farmers can use fertilizers, pesticides, and irrigation water based on the exact needs of each type of crop and crop conditions. This approach can improve products quality, increase crop yields as well as reduce energy consumption and, thus, environmental degradation. One of the main advantages of PA is that not only provides knowledge about cropping variability, but also identifies features that affect crop yields over time.

One of the tools that PA exploits is the Information and Communications Technology (ICT) services [4, 5]. ICT is a prerequisite for PA practice implementation able to provide useful information, e.g., about the soil and crop status, for a more efficient and targeted use of pesticides and fertilizers in agricultural

G. Dolias · L. Benos · D. Bochtis (✉)
Institute for Bio-Economy and Agri-Technology (iBO), Centre for Research and Technology Hellas (CERTH), Thessaloniki, Greece
e-mail: d.bochtis@certh.gr

D. D. Bochtis et al. (eds.), *Information and Communication Technologies for Agriculture—Theme III: Decision*, Springer Optimization and Its Applications 184, https://doi.org/10.1007/978-3-030-84152-2_5

production [6, 7]. Remote Sensing (RS) and image analytics are applied in order to collect and process data, enabling farmers to know which part of their fields exhibits variations [8, 9]. In this way, farmers can adjust their practices, thus, resulting in the increase of the agricultural production, while the ecological footprint can be minimized drastically [10]. Aerial monitoring is one of the most common RS techniques. This technique utilizes satellites, manned aircrafts, and Unmanned Aerial Vehicles (UAVs) for the purpose of capturing crop images.

So far, images taken from satellites have some limitations related to their inadequate resolution and their subjectability to unstable weather conditions [11]. Furthermore, although manned aircrafts can provide images with better quality, the cost of such operations is considerably high. Consequently, UAVs are proved to be an appropriate solution for aerial monitoring, since they provide full coverage of fields on demand and their captured images can be directly delivered for further process.

For the purpose of circumventing some difficulties emerging during such types of operations, UAVs routing is a very important process which guarantees a successful and smooth navigation. UAVs routing includes a network mission which needs to be optimized, by considering that drones must travel from an initial site to a destination one to carry out a specific task. To this end, the sequence of the points to be passed through is determined by variables including distance, time, cost, and energy [12], as it has been applied in various studies for ground agricultural machinery during the last decades [13, 14].

This chapter focuses on UAVs and how they are exploited in terms of PA. In particular, the basic classification of UAVs is presented along with the main usage of them in various PA applications. Subsequently, the methods applied pertaining to safe and optimal drones routing are elaborated as well as the most commonly used algorithms for its realization. Finally, representative demonstrations of the application of these algorithms regarding drone routing in agricultural sector are presented and conclusions are drawn.

2 Types of UAVs and Their Use

In general, UAVs are considered to be aircrafts, which are not controlled by pilots on board as do the ordinary ones. They are autonomous or remotely-controlled vehicles through a ground station for ensuring their smooth navigation over the field. Usually, as far as the former consideration is concerned, UAVs flight mission is created before their operation. UAVs used for civilian operations are divided into two main categories, namely rotary-wing and fixed-wing. One example of each type of drones can be seen in Fig. 1a and b.

Initially, UAVs were mainly utilized for military purposes. However, their use has been extended to a wide variety of sectors. Security surveillance, agriculture, waste management, energy, and telecommunications are some of the domains that

Fig. 1 (**a**) Rotary-wing drone Spreading Wing 1000 by DJI and (**b**) fixed-wing drone eBee X by senseFly

have been greatly impacted by the emerging use of drones. According to the National Purchase Diary Panel group, drones sales in the United States have doubled in 2017 [1]. In addition, based on an economic report by the Association for Unmanned Vehicle Systems International (AUVSI) [15], PA and public safety will take a large bite out of the market in the near future. It is estimated that these two sectors will make up to 90% of UAVs potential markets.

3 UAVs Applications in Precision Agriculture

Until now, farmers have used UAVs in several agricultural applications so as to address issues concerning variations in crops and soil properties and plant diseases. As an indicative example, Chebrolu et al. [16] proposed a technique for tackling visual changes of field and the growing crops in crop images captured by UAVs. This method attempts to reconstruct 3D models of the field over the crop season based on aligned in-field point clouds. The proposed system was evaluated against manually labelled ground truth, resulting in up to 90% matching of images correspondences, on average. In [17], authors also carried out UAV missions aiming at the detection of drainage pipes, since a common problem in agriculture is that farmers have to remove frequently water from soil. Initially, rainfall rate, soil surface water content, and temperature were measured by utilizing a rain gauge and two soil/temperature sensors for in-field testing, respectively. Afterwards, detection of drainage pipes was examined with UAV equipped with visible (VIS), near-infrared (NIR) and thermal infrared (TIR) sensors. Among these three UAV surveys, UAV-TIR imagery was found to detect more effectively drainage pipes.

In [18], an approach was introduced for more effective weed detection in rice fields. More specifically, RGB (Red-Green-Blue) images of high resolution captured by UAV were fused along with low resolution multispectral images in order to detect Gramineae weed in rice fields. Three different detection systems of Neural Networks (NN) were applied, while their performance was tested under two indices, namely M/M_{GT} and MP. The former expresses how much weed area is being detected, whereas the latter indicates the precision of detected weed. It was reported that the best performance in weed detection occurred in the NN where indices ranged as follows: $80\% < M/MGT < 108\%$ and: $70\% < MP < 85\%$.

Christiansen et al. [19] employed a Light Detection and Ranging (LiDAR) sensor with the intention of mapping canopy heights, especially winter wheat fields. Based on the point clouds produced by LiDAR, combined with data from other sensors, the height of the crop was estimated to be between 0.35 and 0.58 m. Moreover, UAVs have also been utilized as a tool for various vegetation indices extraction. Furthermore, in [20] an algorithm was developed in order to estimate Green Area Index (GAI) from images being taken from cameras of four different bands. By using Root Mean Square Error (RMSE) as a measure of uncertainty, the estimated GAI over rapeseed and wheat crops was close to 0.2.

UAVs can be used in conjunction with on-field vehicles so as to monitor and manage crops in a more efficient way. For instance, Vasudevan et al. [21] combined an Unmanned Ground Vehicle (UGV) along with a UAV. Multiple images were collected from the UAV, resulting in the extraction of various indices, such as Normalized Difference Vegetation Index (NDVI), Soil Adjusted Vegetation Index (SAVI), and Optimized Soil Adjusted Vegetation Index (OSAVI). In contrast, UGV mainly provided information about soil condition, such as acidity and pH.

However, in most cases a single UAV alone cannot complete monitoring processes, due to its limited power resources [6]. In fact, the mission time is significantly decreased when UAV is also equipped with heavy payloads, such as specialized cameras or spraying systems. Therefore, utilization of a group of UAVs (multi-UAVs) in PA was tested in monitoring tasks of large-area crops in [22]. Also, multiple UAVs were deployed to create full map of crops by capturing geo-referenced pictures. Based on the state and the capabilities of each aerial vehicle, tasks were assigned to each one of them. Then, a path planning algorithm was executed producing the path that each vehicle has to follow. The authors conducted experiments to test the effectiveness of this system by taking into account also different weather conditions.

4 UAVs Route Planning

As it has already been stated, limited battery life can affect UAVs operations performance. This means that, in many cases, they cannot complete the scheduled tasks within a single mission. Thus, drones have to operate for longer time periods and, thus, the cost of such operations increases remarkably. In addition to this, agricultural drones are used in applications where their payload increases drastically their overall weight. An indicative example is UAVs implemented for spraying operations [23] where, in some cases, pesticide tanks can overcome the volume of 10 L [6]. In particular, spraying process must be carried out carefully, so that agrochemical products are sprayed in specific crop areas. In addition, it is worth mentioning that drones operate in areas where different types of obstacles might block their way. These obstacles must be quickly detected in order to prevent any collision during the flight.

For all the above-mentioned reasons, it is obvious that drones must be routed carefully during their navigation. They should cover the targeted area in a time and cost-efficient manner, by consuming reasonably the available energy and complete their mission without any physical damage. This is one of the reasons why drones routing has been studied greatly in recent years. The main approach of studying drones routing is the Travelling Salesman Problem (TSP). This problem aims to calculate optimal path by allowing drone to visit each point of interest exactly once and return to its initial position. Accordingly, drone is considered to be the salesman and the visited cities are the areas explored by the drone [24]. In case of drones having limited capacity and having to visit specific regions, then TSP problem is described as Vehicle Routing Problem (VRP), or vehicle scheduling, as it has been applied for ground agricultural vehicles; both conventional [25–28] and autonomous ones [29–31], as well as part of their high level control [32].

TSP is a known combinatorial optimization problem which is classified as a Nondeterministic Polynomial time (NP) problem. This means that by increasing the number of cities, the running time of any applied algorithm increases as a polynomial of a high order. In these cases, calculations of optimal solutions require significant computational effort, even for 10 visited cities [33]. Figure 2 illustrates a typical example graph of a TSP solution concerning routing between 24 nodes.

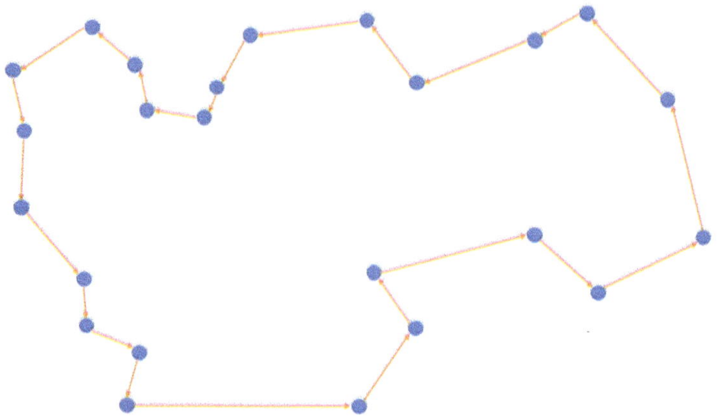

Fig. 2 Solution to a TSP instance of 24 cities

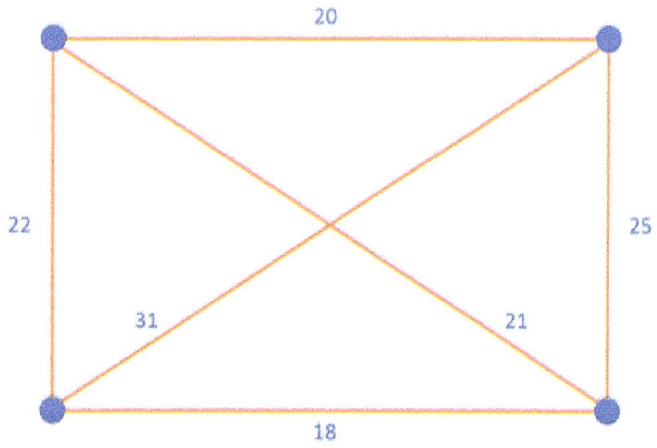

Fig. 3 A typical representation of graph with four nodes. The numbers denote the weight of each edge

TSP is usually modelled as a weighted graph [34]. In graph theory, a graph is defined as a diagram which comprises points (nodes) and lines that connect pairs of points [35]. A typical graph composed of four nodes is depicted in Fig. 3. In particular, cities are expressed as the graph nodes (or vertices), graph edges denote paths between cities and paths distance is represented by edges weight. This graph is described as undirected, since the distance between all pairs of nodes is the same in both directions. In this case, this type of problem is referred to as symmetric TSP. The cost of this graph is the travelling distance covered through the nodes of the graph. This problem is solved by finding minimum path, after starting and ending at the same vertex and visiting each vertex exactly once.

In the next section, a discussion about the various methods utilized for finding solutions to this type of problems is presented.

5 Algorithms for Solving TSP

In order to solve routing problems characterized as TSP, the following two groups of algorithms are mainly utilized:

- *Algorithms for exact solutions*: The approaches of this type guarantee optimal solutions; however, they are doomed to work fast only for graphs with a relatively small number of nodes.
- *Algorithms for sub-optimal solutions*: These approaches produce optimal or near-optimal solutions in reasonable computational times, even for large number of nodes. In some TSP instances of several million nodes, algorithms can lead to very satisfactory near-optimal solutions [36].

Figure 4 presents the approaches for solving TSP analyzed in this section.

5.1 *Exact Algorithms*

In order to find an exact solution, one first approach would be to calculate all possible permutations and check which combination results in the shortest path. Thus, by starting from a random vertex among n nodes in total, there are: $n - 1$ choices as second vertex to visit, then: $n - 2$ as the next one, and so forth. By multiplying all these choices, the required number for all combinations is: $(n - 1)!$. Since this is an undirected graph, travel cost from one vertex to another is the same in both directions, allowing the division of the total number of tours by 2: $(n - 1)!/2$. This value increases extremely fast, rendering TSP quite difficult to be solved, even for graphs with 20 nodes only. In the next paragraphs, various approaches which guarantee optimal solutions to TSP instances will be presented.

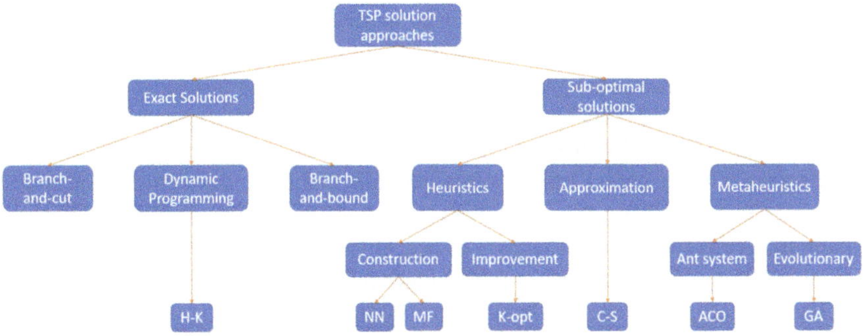

Fig. 4 Different approaches for solving TSP instances; Abbreviations H-K, NN, MF, C-S, ACO, and GA correspond to Held-Karp, Nearest Neighbor, Multiple Fragment, Christofides-Serdyukov, Ant Colony Optimization, and Genetics Algorithms, respectively

Dynamic Programming

Dynamic programming (DP) is considered to be both a computer programming and a mathematical method for optimization. It was introduced by Richard Bellman [37]. This method attempts to solve a problem in a recursive way, by dividing a complex problem into sub-problems which are much easier to be solved. In [38], it was proved that dynamic programming can produce exact solutions for TSP of up to 17 nodes within reasonable computational times. The Held-Karp algorithm [39], an application of dynamic programming, generates exact solutions to the TSP with time complexity of: $O(n^2 2^n)$.

Held-Karp algorithm has been utilized in [40] in order to create routes related to parcels delivery to various locations. This type of dynamic programming generated solutions to the TSP by taking into account the velocity and the range of various types of drones. The results of this work included drones' routes with minimum cost concerning both aerial and land-based vehicle.

Branch-and-Bound Algorithms

One of the most common approaches for finding solutions of optimization problems, like TSP, is the one of branch-and-bound [41]. This method finds optimal solutions by searching all possible solutions in an efficient way [42]. More specifically, branching refers to the way all possible solutions create the solution space [43]. By this method, solution space forms a tree-like structure which helps to pick next visited graph vertex. As expected, branching ensures that all possible solutions will be examined. However, this can increase the computational time, in case of a graph with a considerable number of nodes. At this point, the term bound comes in handy. Bounding sets a bound on the route length, leading to pruning off solutions estimated to perform poorly. Thus, this technique reduces effectively the solution space as well as the computational time.

An illustrative example of how this method is applied to TSP can be seen in Fig. 5. In particular, this figure depicts the consequent steps of branch-and-bound algorithm in order to generate the shortest route in a graph which consists of five nodes. Under every node, a number representing the length of each corresponding route is shown. When branches are incomplete, this number is expressed as a sum of two numbers: $(l + b)$. The first one (l) refers to the distance covered so far, whereas (b) denotes a lower bound based on the length of the route through the nodes that have not been visited yet. This lower bound is generated from the minimum spanning tree method, which takes into account the unvisited nodes as well as the initial and final vertex of every corresponding route.

The first complete route will be the upper bound of the solution tree. Hence, based on the graph of Fig. 5, the first upper bound is 42. When a vertex is added to an uncompleted route, a lower bound is calculated based on the remaining nodes. For instance, concerning the uncompleted route: $A \rightarrow B \rightarrow E$, a lower bound is

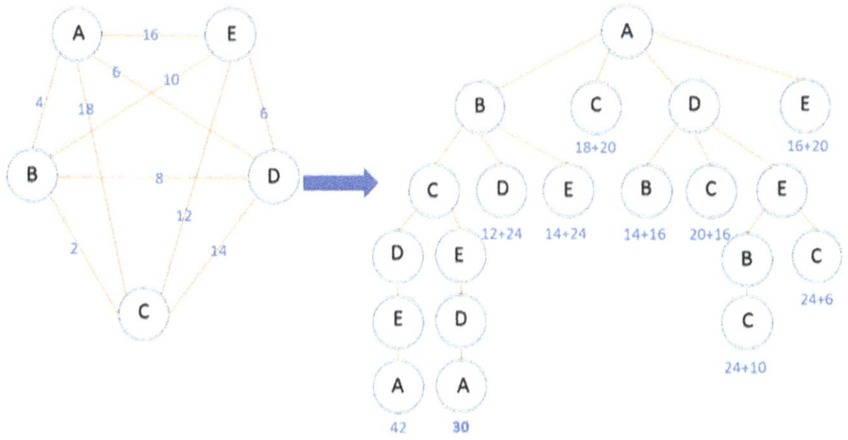

Fig. 5 Branch-and-bound algorithm applied in a TSP instance of 5 nodes

Fig. 6 Minimum spanning tree of uncompleted route $A \rightarrow B \rightarrow E$

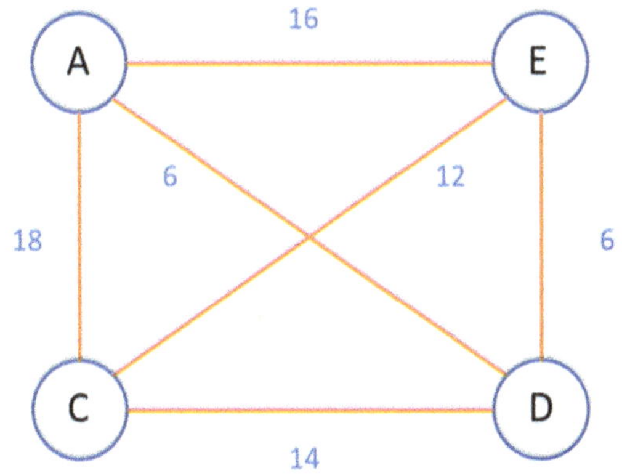

measured, by taking into account initial and final nodes A, E as well as the rest of the nodes C and D. Concretely, a minimum spanning tree containing nodes A, C, D, and E estimates the lower bound of the uncompleted route. Figure 6 illustrates the way minimum spanning tree leads to the calculation of the lower bound. Since lower bound is found to be 24, it is added to the length of the uncompleted route. If sum is greater or equal to the upper bound, this means that this solution is not better than the best solution found so far. In this case, uncompleted route (5) leads to a sum: $14 + 24 = 38$, which is greater than the best solution so far, namely 30. As a result, this branch is pruned, avoiding any future redundant calculations. Therefore, branch-and-bound reduces drastically the solution space, resulting in the acceleration of the search process.

Generally, the time complexity of the branch-and-bound algorithms is calculated by taking into account two factors: branching factor, f, expressing the maximum number of branches per tree node and search depth, d, denoting longest path from the initial point to one leaf [44]. Consequently, the running time of these algorithms is described as: $O(Mf^d)$ as the worst-case scenarios, where M represents a bound of time required to examine a sub-problem. Usage of this method for exact TSP solutions is convenient for a range of nodes from 40 to 60 [45].

In [46], a solution was suggested to a TSP case related to environmental monitoring based on these algorithms. In particular, they examined the optimization of aircraft routes operating simultaneously. A specific branch-and-bound algorithm was modified in order to calculate optimal routes of two-dimensional problems for various control tasks.

Branch-and-Cut Algorithms

The cutting-plane method [47] is combined with branch-and-bound algorithms to improve their performance. This combination comprises the branch-and-cut algorithm [48]. Cutting-plane method is applied in linear programming where, under some assumptions, there is always an extreme or corner point which can be optimal. Then, this extreme point is checked as a possible solution. If this test fails, there is a linear inequality which separates this point from a set of points that satisfy the problem constraints. This linear inequality is described as a cut and it is included in the problem relaxation process whereas non-integer solution is rejected. This method continues repetitively until an optimal solution is found.

Branch-and-cut algorithms are utilized for solving TSP with larger nodes than the previous exact methods. In particular, a solution to the problem of delivering goods via both a truck and a drone in tandem is examined in [49]. By using a specific type of branch-and-cut algorithms, authors found optimal solutions for serving 20 customers in less than an hour.

Nevertheless, exact solutions of problems with up to 2000 cities can be calculated by utilizing these algorithms. As expected, these calculations require intense computational effort. The time complexity of this type of algorithms depends on several factors, such as the branching strategies, the calculated bounds and the applied cutting planes [50]. The current record for finding optimal solution belongs to a TSP instance with 85,900 cities [51] following a branch-and-cut structure.

5.2 *Algorithms for Sub-optimal Solutions*

This kind of approaches aim at finding near-optimal solutions to problems where exacts methods lead to extremely slow solutions. Some of these approaches manage to find solutions efficiently in large TSP instances with millions of cities [45]. In these cases, calculated route can be only 1–2% larger than the optimal solution [52].

Approximation Algorithms

The goal of this type of algorithms is to solve optimization problems, such as TSP, by finding approximate solutions [53]. It is known that in many optimization problems, there are not solutions within polynomial time. Approximation algorithms attempt to calculate how close sub-optimal solutions can be solved in polynomial time. More specifically, they guarantee that the sub-optimal solutions are larger than the optimal ones by an approximation factor [45]. During their analysis, they also prove the worst case of the returned solution as a measure of its quality. In general, this is the main characteristic that differentiates approximation from heuristic algorithms. Even though heuristics generate fairly good solutions, it is not assured whether they can always find solutions to problems, like TSP, or not [45].

Christofides-Serdyukov Algorithm

This algorithm is named after the two scientists who introduced it independently during the 1970s [54, 55]. Christofides-Serdyukov (C-S) algorithm is responsible for rendering approximation algorithms practical candidates for solving intractable problems. It is applied to symmetric TSP instances where triangle inequality is satisfied for any triangle formed from graph nodes. It is proved that this algorithm solutions are at most 3/2 times the optimal solution length. So far, this is the only algorithm that achieves an approximation ratio of 1.5 for this specific type of TSP instances. However, there are approximation algorithms which achieve better solutions for other TSP instances. Figure 7 presents a typical route generated by Christofides-Serdyukov algorithm to a TSP instance of 9 nodes.

A solution to a drone routing problem is exploited in [56] based on C-S algorithm. More specifically, the aim of this study was to examine the minimization of the drone's energy consumption during the collection of data from wireless sensors. Thanks to this approach, drone was able to navigate through the paths in order to gather sensors data by consuming less energy than traditional approaches.

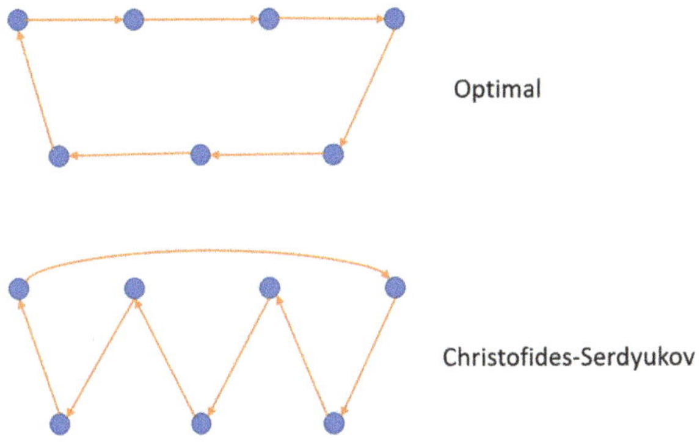

Fig. 7 Solution generated by Christofides-Serdyukov (C-S) algorithm in comparison to the optimal one of TSP for nine nodes. This approximation algorithm leads to a solution that connects the two farthest cities

Heuristics

Heuristics techniques generate solutions faster than exact methods. Nevertheless, these solutions are not always optimal [57]. In order to opt for an heuristic algorithm, there is a number of parameters that need to be taken into account: completeness, accuracy and precision, optimality, and execution time [58]. In general, heuristics are divided into two main classes: construction (or decomposition) heuristics and improvement heuristics.

Construction heuristics refer to processes which start initially from a small instance and add a part to the solution each time a specific number of operations is executed. This repetitive process terminates when a near-optimal complete solution is produced. Due to the fact that, in these techniques, a large problem is divided into smaller sub-problems, they are also referred to as decomposition heuristics. Nearest Neighbor [59] and Multiple Fragment [60] algorithms are two examples of construction heuristics.

On the other hand, a possible complete solution is defined as an initial point of the improvement heuristics. After that, these algorithms iteratively improve the initial solution and continue until a near-optimal solution is reached. A typical example of this kind of algorithms is k-opt or Lin-Kernighan heuristics [61].

Nearest Neighbor Algorithm

Nearest Neighbor (NN) is one of the simplest algorithms applied for solving TSP. In fact, it is among the first heuristics introduced for finding solutions of shortest route. This algorithm is executed as follows:

1. All graph nodes are considered as unvisited.
2. An arbitrary vertex is randomly picked as the initial one.
3. The shortest edge is found that connects an unvisited vertex and the one that was last added. This path is added to the existing one.
4. This vertex is removed from the group of unvisited nodes.
5. If all nodes are visited, last visited vertex is connected with the initial point, this path is added to the total path and the process is then terminated. Else, the procedure returns to step 3.

The implementation of this algorithm can be observed in Fig. 8. Having chosen A as the initial vertex of a TSP instance of five cities, the algorithm calculates distances of A with any other node. The red dashed lines denote all executed calculations, whereas green lines express the shortest edge to each vertex. Since node B is closer to A, the edge AB is added to the drone route. This procedure is executed repetitively until all nodes are visited and the last added vertex, vertex E in Fig. 8, is connected with the initial one, A.

This algorithm checks the nearest neighbor of every vertex and results in a time complexity of: $O(n^2)$; thus, it constitutes an effective and quick approach for solving

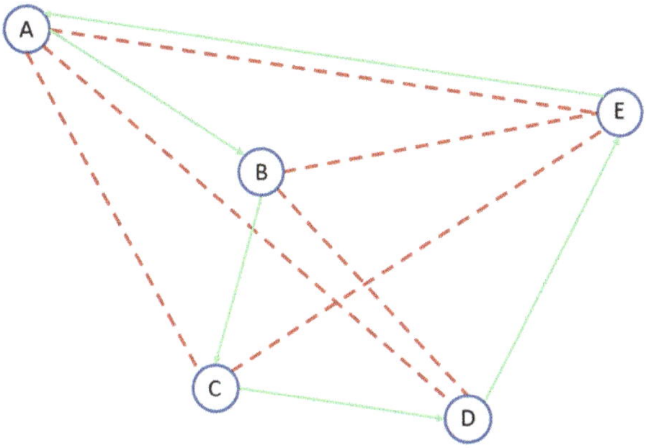

Fig. 8 Nearest neighbor algorithm applied in a TSP instance of five cities

Table 1 Implementation of Multiple Fragment algorithm in a TSP instance of six cities; Generated segments after each iteration

Shortest edge	Edges action	Created segments
–	–	A; B; C; D; E; F;
BC	Join B with C	A; BC; D; E; F;
AD	Join A with D	AD; BC; E; F;
DE	Join E with AD	ADE; BC;F;
FD	Discard	ADE; BC;F;
CF	Join F with BC	ADE; BCF;
EF	Join ADE with BCF	ADEFCB

TSP. However, it does not always consider the shortest paths [62]. In addition, it sometimes fails to reach even a possible solution, even though one may exist [63].

NN algorithm was utilized in [64] for optimal drone routing regarding the distribution of parcels. By applying this heuristic, simulations were conducted with up to 20 nodes and up to 3 drones. Drones flight speed and covered distance were taken into account in these simulations. The results showed a reduction in the delivery cost as well as the energy consumption.

Multiple Fragment Algorithm

This heuristic is named after the technique that structures the graph nodes into sets of fragments. It is also mentioned as the "greedy algorithm" concerning TSP [65]. The first step of the Multiple Fragment (MF) algorithm is to consider each graph vertex as a fragment. The route is initialized by the nodes of the shortest path. At every stage, shortest edges are added to existing fragments. Then, iterative steps of MF connect fragments located closed to each other. Therefore, this heuristic manages to keep multiple fragments of routes as a part of the complete graph. The algorithm stops when the remaining fragment has length equal to the number of the nodes. This approach can be comprehended through the example of Table 1 pertaining a TSP instance of six cities.

The execution time of this algorithm is described as: $O(n^2 logn)$. This means that it requires more time than NN to find solutions for TSP. Nevertheless, computational time in worst cases of Multiple Fragment are better, since NN, on average, leads to routes which are up to 25% greater than the optimal one.

The authors of [66] proposed a solution to an extension of TSP by applying MF algorithm. In this scenario, graph nodes can be visited more than once. The results indicated that this algorithm could lead to near-optimal solutions which are, on average, 8% greater than the optimal ones.

k-Opt or Lin-Kernighan Heuristics

Lin-Kernighan (LK) algorithm constitutes one of the most common iterative heuristics for solving symmetric TSP [67]. It finds optimal or sub-optimal solutions by implementing successive improvements to one initial solution [68]. It is considered to be a local search algorithm [69]. In local search algorithms, iterative improvements are executed locally in small regions of the solution space. These changes are called moves. In k-opt algorithm, k edges are exchanged between nodes of their surrounding area. This process is terminated when the new route is not better than the current one. More specifically, 2-opt moves, a special case of k-opt heuristics, are illustrated in Fig. 9. They change the way four graph nodes are connected and result in an improved shorter route.

In general, LK algorithm initializes an arbitrary feasible route R as a solution to the TSP. Then, it constructs two groups regarding the edges exchanged in the solution space. These two groups are initially empty and constructed one element at a time. The first one: $X = \{x_1, ..., x_k\}$ contains all edges removed from current

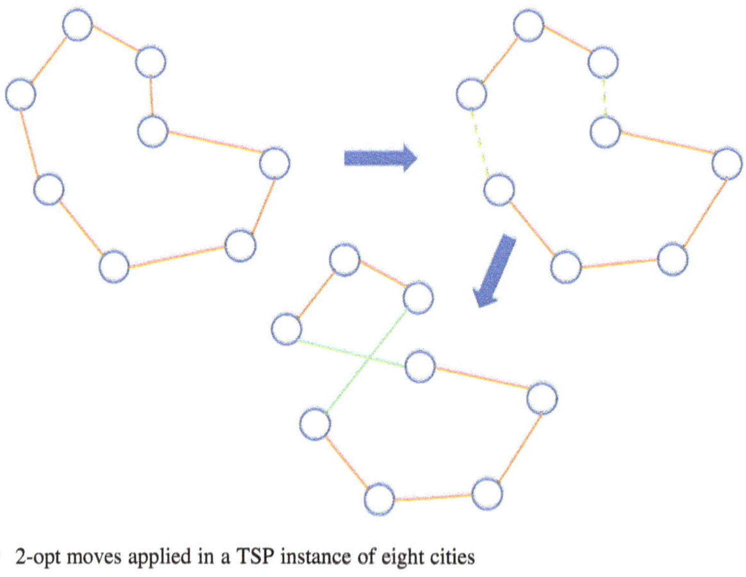

Fig. 9 2-opt moves applied in a TSP instance of eight cities

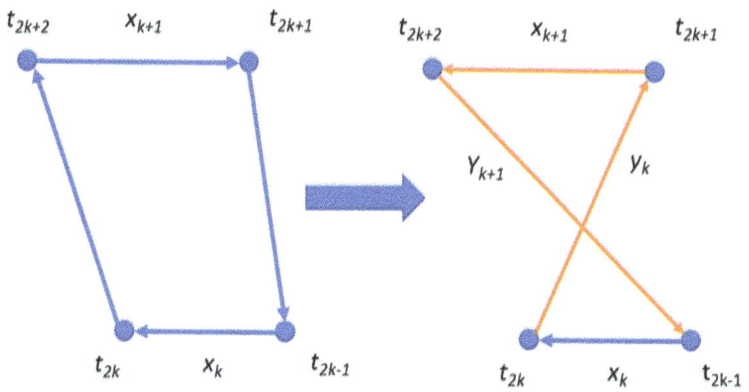

Fig. 10 Restrictions regarding x_k, y_k, x_{k+1}, and y_{k+1} chosen in k-opt algorithm

route. Edges replacing the deleted edges are saved in the second group: $Y = \{y_1, \ldots, y_k\}$. Elements: y_k and y_k denote the pair of edges placed in Y and X in step k, respectively.

Edges chosen for exchanges in LK algorithm must meet the following criteria [67]:

1. The first criterion restricts the pairs: (x_k, y_k) and (x_k, y_{k+1}) to share one endpoint. More specifically, assuming t_1 expresses one of x_1 endpoints, edges added in X and Y have to meet the rules: $x_k = (t_{2k-1}, t_{2k})$, $y_k = (t_{2k}, t_{2k-1})$, and $x_{k+1} = (t_{2k+1}, t_{2k+2})$, $y_{k+1} = (t_{2k}, t_{2k+1})$, where: $k \geq 1$. These restrictions can be seen in Fig. 10. By applying these rules in each step of LK algorithm, a new improved route is generated. This new route is basically closed, since $y_k = (t_{2k}, t_1)$. An exchange similar to this one is described as sequential. Two special cases of k-opt method, namely 2-opt and 3-opt, are considered to be sequential.
2. Edges chosen for exchanges should guarantee that the improved route is a closed tour. This requires that moves based on edges: $x_k = (t_{2k-1}, t_{2k})$, where t_{2k} joins t_1, generate a complete tour. This criterion is referred to as the feasibility criterion and results in the reduction of execution time.
3. The third criterion restricts that y_k are chosen on condition that they always lead to an improved route. This restriction renders LK algorithm highly efficient as it avoids redundant calculations.
4. Groups X and Y must contain completely different edges. Hence, running time is significantly reduced.
5. Every edge must be substituted by the five nearest neighbors. Edge: $y_k = (t_{2k}, t_{2k+1})$ can replace edges which include endpoint t_{2k}.

The problem of calculating optimal drone routes was addressed in [70] regarding remote sensing operations. More specifically, LK algorithm was exploited in order to find solutions to simulations of TSP instances with multiple depots. The results confirmed that this algorithm can be applied effectively in similar activities.

In [71], authors proved that, in some cases, k-opt moves may lead to exponential running time by generating optimal routes. However, this happens rarely in most of the practical applications [72]. Typically, LK algorithm leads to effective and high-quality optimal, or near-optimal solutions, in polynomial time.

Metaheuristics

Metaheuristics constitute processes which modify the implementation of subordinate heuristics in order to produce solutions of higher quality [73]. Instead of just producing random starting solutions, they generate initial solutions to local search procedures in a more sophisticated manner [74]. In this way, high-quality solutions are generated quicker, thanks to their ability to make efficient decisions. These probabilistic decisions are based on previous decisions or prior experience. Substantially, metaheuristics "scan" the search space in order to explore the near-optimal solutions by taking advantage of accumulated search experience [75].

One of the most common metaheuristics applied for solving combinatorial problems, such as TSP, are genetic algorithms (GA) [76] and Ant Colony Optimization (ACO) [77]. These algorithms are described as nature-inspired since their implementation mimics procedures and behaviors of natural world [75]. Furthermore, sub-optimal solutions are explored in the solution space by performing recombination on sets (populations) of points. Thus, they are also defined as population-based algorithms. However, their key characteristic is that they make use of recently visited nodes for finding solutions in the search space. Therefore, this "memory" feature classifies also these algorithms as memory-based algorithms.

Genetic Algorithms

Genetic algorithms (GA) belong to a group of algorithms described as Evolutionary Computation algorithms [78]. Their foundations are based on the biological evolution of living organisms in order to adapt to their habitat changes. Generally, human beings struggle to survive in their environment and produce offspring. The ones who are described by genetic traits of higher quality are likely to beat their competitors. Since their offspring possess their parent's features, they survive and take part in the reproduction process of new offspring. This repetitive process can lead to descendants who perform better than their ancestors and adapt well to their environment [45]. An overall representation of how GA are implemented is depicted in Fig. 11.

As a first step of genetic algorithms, an initial solution is presented as a group of individuals, called genes. This group consists a population named chromosomes [68]. A TSP instance of seven cities represented as a fully connected graph is illustrated in Fig. 12a. In addition, Fig. 12b delineates how GA generate the population containing possible solutions to this TSP instance.

In each iterative process of the GA, the next-generation derives from a set of operations applied to the genes of the current chromosomes. These operations are

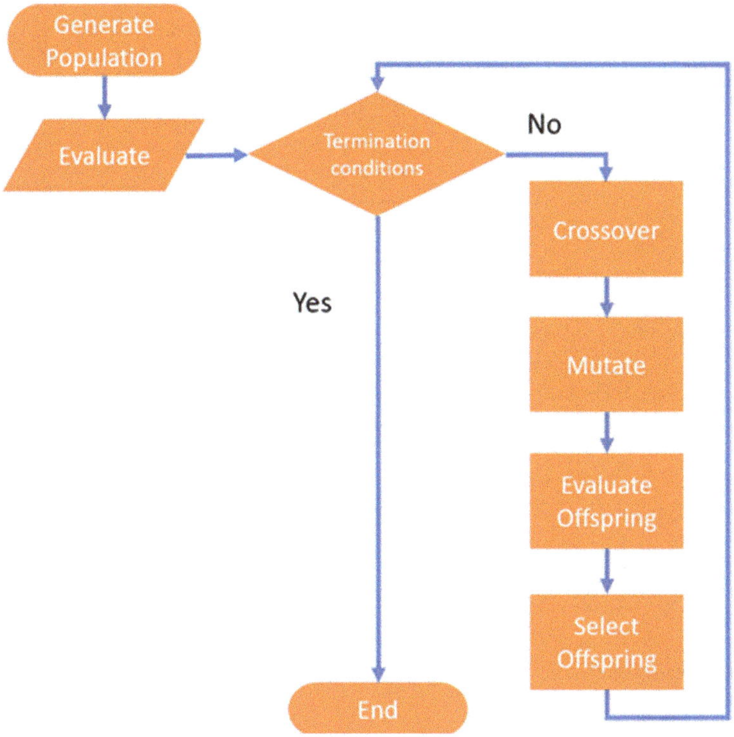

Fig. 11 Flowchart presenting the main processes of genetic algorithms

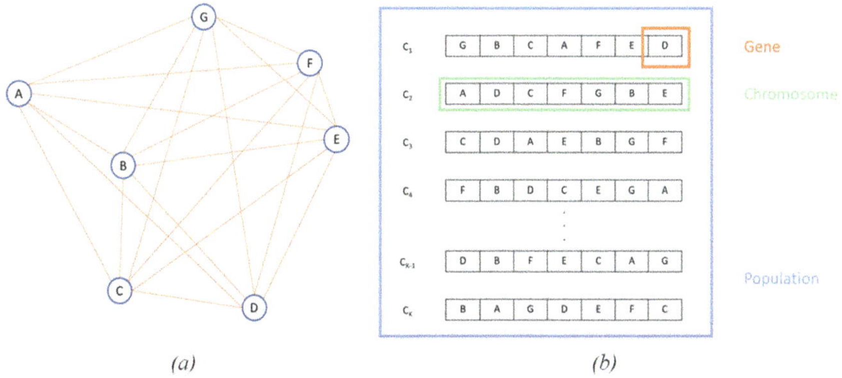

Fig. 12 (**a**) Fully connected graph describing TSP instance of seven cities and (**b**) the respective population created by the genetic algorithms

basically referred to as recombination or crossover, resulting in the creation of the chromosomes offspring. Crossover operations can be applied to one or more points of each chromosome. Figure 13a shows a single-point crossover point resulting in

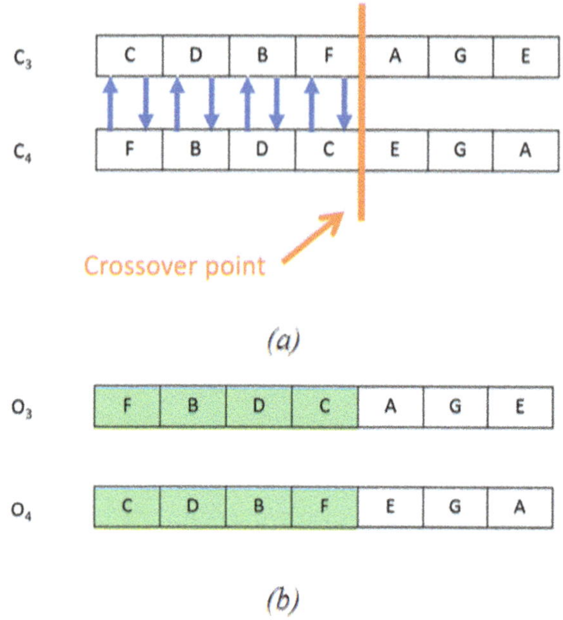

Fig. 13 (**a**) Crossover implementation in two chromosomes and (**b**) the two generated offspring

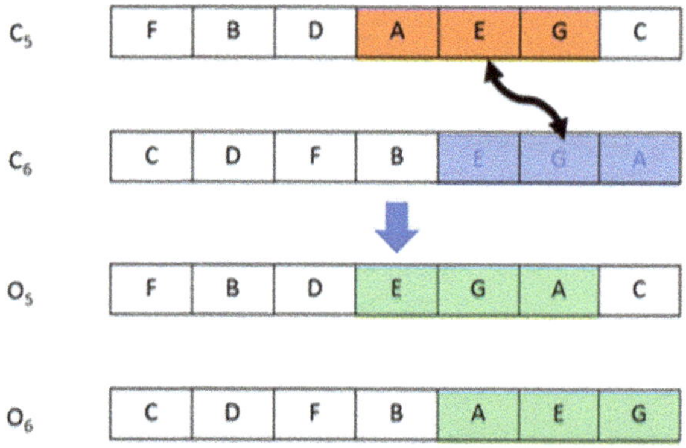

Fig. 14 Mutation applied in two chromosomes and the created offspring

the exchange of the first four chromosomes genes. The two new solutions after the crossover implementation can be seen in Fig. 13b.

Mutation is another operation which is executed in each step and interpreted as a kind of adaption to the problem requirements. A typical example of this operation in two chromosomes and the generated offspring are presented in Fig. 14. The criterion that determines which genes will be chosen as part of the generated population is their fitness to the problem. This criterion resembles the natural selection of offspring who fit the most the evolving environment. Thus, the genes which present higher fitness are more likely to be chosen for the next generation since they result in a shorter route [79].

GA have been widely utilized in finding near-optimal routes through TSP. In [80], for instance, authors investigated how improved drone's navigation over large agricultural lands can result in the increase of fertility. Aerial images obtained by drones play an important role on this. Hence, optimization of drones routing was examined by adopting a specific type of GA. The calculated near-optimal route minimized the flight time, allowing drone to capture more field images, despite its limited energy resources.

Ant Colony Optimization

Ant Colony Optimization (ACO) constitutes a memory-based metaheuristic strategy based on stigmergy [81]. This concept describes the way insects communicate indirectly to organize their activities. In particular, ACO algorithm mimics the way ants cover shortest paths between their nest and food sources [82]. Ants tend to move along paths where most of the colony members have already left trails of a substance called pheromone. In a nutshell, a group of artificial ants is exploited in the solution space, leading to solutions based on the pheromone trails generated through iterative processes.

ACO algorithm is an improved and extended metaheuristic framework of Ant System algorithm [75]. Its main features are presented in Fig. 15. As a first step, pheromone values of all artificial ants are initialized to equal positive values. After that, each ant moves between the graph nodes by depositing pheromone in every edge that they go across. Decisions on how they will move on the graph are taken randomly, based on probabilistic models. These models take into account heuristic values of the graph as well as pheromone values. Therefore, ACO gradually constructs a solution by adding edges of the graph. During the construction of the solution, each ant is able to "remember" the path that lead to its partial solution.

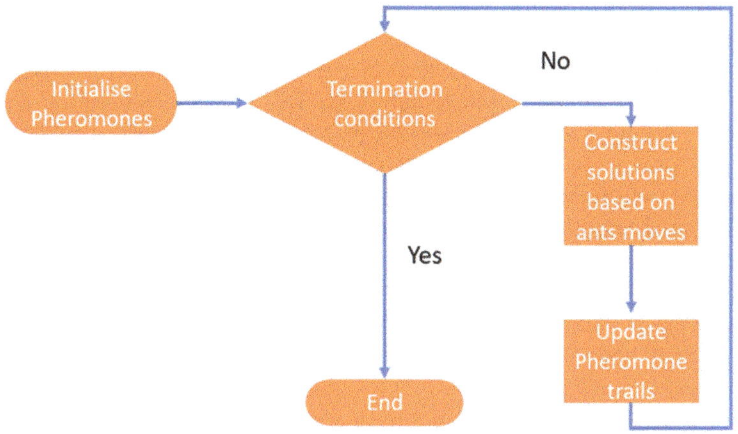

Fig. 15 Overall process of Ant Colony Optimization algorithm

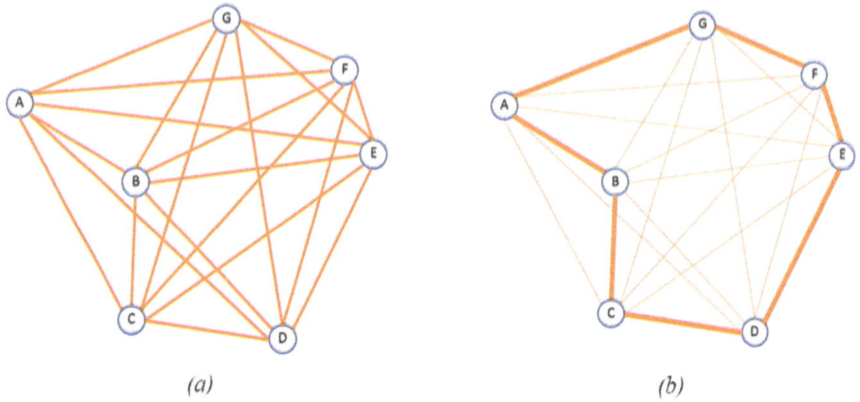

Fig. 16 (**a**) Pheromone initialization for a TSP instance of seven cities and (**b**) updated pheromone trails after iterative processes of Ant Colony Optimization (ACO) algorithm

Every time one ant adds a component into its partial solution, it updates its pheromone trails. However, pheromones are not only updated in every step of their walks. When one ant has covered a complete tour, the length of the constructed solution is tested. If the route leads to a solution with shorter path, then it deposits more pheromone. Since each ant has kept the paths of the solution to its memory, it can also update pheromone trails of specific parts of the solution based on the solutions length. Apart from that, pheromone trails are diminished over time. This operation, called evaporation, is crucial, due to the fact that it prevents the algorithm from converging to a sub-optimal solution too early, without exploring extensively the solution space [75].

There are also overall actions executed in the graph globally. For instance, solutions generated by ants are examined by local search processes. In addition, information regarding the state of the pheromone trails are collected after several iterations. Then, this information can determine if the algorithm will deposit more pheromone, leading the search process into shorter routes.

Concretely, the functionality of pheromones is presented in Fig. 16a which shows how pheromones are initialized equally for a TSP instance of seven cities. After successive iterations of ACO, most of the pheromone trails have been diminished as it can be shown in Fig. 16b. On the contrary, pheromone trails of edges which lead to the shortest route have greater values and are represented with thicker lines.

ACO has been utilized in [83] pertaining to UAVs path planning in a multi-objective TSP model. A combination of ACO with Particle-Swarm Optimization was developed for solving this specific TSP model. This hybrid metaheuristic approach yields effective solutions in reasonable computational times for models of 10 up to 50 nodes.

6 Demonstration of UAVs Routing in Agriculture

After presenting the outlines of several algorithms which provide solutions to TSP, a demonstration of how these approaches are implemented in selected agricultural applications is presented in this section. The performance of these approaches is tested under different conditions. In particular, two arable crop fields and one orchard with walnut trees, all located in Volos area, Central Greece, were utilized. The details of these fields are illustrated in Table 2. The area of each field, as well as the number of locations that drones should navigate through in each case, is different.

Table 3 presents a brief description regarding the main features of the five indicative TSP variations, namely:

- Single TSP (sTSP)
- Multiple TSP (mTSP) (without a fixed depot)
- Multiple TSP, single (fixed) depot (mTSPsD)
- Multiple TSP, multiple (fixed) depots (mTSPmD)
- Multiple TSP, multiple (fixed) depots and constrained travelling distance (mTSPmDcT)

Additionally, Table 4 summarizes the results obtained from the implementation of these instances in three different fields. More specifically, the distance covered is presented in Table 4 in conjunction with the computation time required for the implementation of each algorithm. The simulation studies were conducted on a Microsoft Windows 10 with Intel Core i5 7th Gen, 256GB storage, and 4GB memory.

Table 2 Parameters of examined fields

Field	Sampling points	Field Area (ha)
Field 1	50	7.52
Field 2	83	0.65
Field 3	100	9.5

Table 3 Brief description of applied algorithms in this section

Problem instance	Brief description
Single TSP (sTSP)	Single drone moving to the nearest neighbor
Multiple TSP (mTSP)(without a fixed depot)	Multiple drones, multiple start/end points
Multiple TSP, single (fixed) depot (mTSPsD)	Multiple drones, one common fixed start/end point
Multiple TSP, multiple (fixed) depots (mTSPmD)	Multiple drones, multiple fixed start/end points
Multiple TSP, multiple (fixed) depots, constrained distance (mTSPmDcT)	Multiple drones, multiple fixed start/end points, maximum covered distance per drone

Table 4 Overall results obtained from variations of TSP regarding the distance covered by drones and the corresponding computation time for the three representative fields

Field	Instance	Distance covered by drones (km)	Computation time (s)
Field 1	sTSP	1.911	5.44
	mTSP	1.819	39.85
	mTSPsD	2.035	41.64
	mTSPmD	1.905	39.25
	mTSPmDcT	2.117	178.33
Field 2	sTSP	0.905	8.66
	mTSP	0.801	45.19
	mTSPsD	0.885	48.05
	mTSPmD	0.769	42.32
	mTSPmDcT	0.858	196.70
Field 3	sTSP	3.259	9.47
	mTSP	4.092	59.22
	mTSPsD	3.557	55.03
	mTSPmD	3.465	55.14
	mTSPmDcT	3.221	214.90

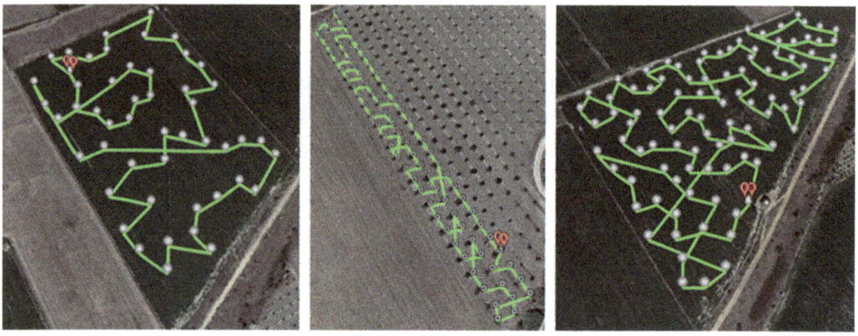

a) Field 1 *b) Field 2* *c) Field 3*

Fig. 17 Drones routes obtained by single TSP (sTSP); Red placemarks denote the initial/final point (Start/End) of each route

6.1 *Single TSP (sTSP)*

As stated previously, drone route is initialized at its take-off position. Then, closest point to the take-off position is visited by the drone. This procedure is repeated until all points are visited exactly once and the drone returns back to its initial point. This approach generates routes with some non-optimal moves, as depicted in Fig. 17, which shows drones routes obtained based on the NN algorithm concerning the Fields 1, 2, and 3 containing 50, 83, and 100 locations, respectively. Overall, this algorithm produces fairly good results, despite its simplicity. As in all figures that follow, red placemarks stand for the start (S) and end (E) points of each route.

6.2 Multiple TSP (mTSP) (Without a Fixed Depot)

There is a distinct probability that one drone will not manage to execute the whole route, due to its limited battery life. Therefore, the overall route can be split into smaller routes with fewer visited points. This approach was implemented by a variation of the previous GA, which has as an input parameter the number of drones. The near-optimal generated routes for the three tested field cases described above are given in Fig. 18. Remarkably, the results concerning the first two fields correspond to shorter routes, in comparison to the sTSP. More specifically, the distances covered using mTSP were approximately 4.8% and 11.5% shorter than utilizing sTSP for the Fields 1 and 2, respectively. Additionally, the computation time for the implementation of mTSP was almost seven and five times longer regarding Fields 1 and 2, respectively. In contrast, as far as Field 3 is concerned, mTSP resulted in 25.6% longer distance and six times increased computation time.

Routes produced by this algorithm can be executed by either exploiting one single drone or a specific number of drones, that is three drones for Field 1 and Field 2 and 4 drones for Field 3, as shown in Fig. 18. In the latter case, navigation issues arise since there are crossed routes in the case of Field 2 and Field 3.

6.3 Multiple TSP, Single (Fixed) Depot (mTSPsD)

An extension of the previous GA is to utilize a predefined depot for all the drones. This technique seems suitable for operations where only one drone is exploited, and its battery needs to be changed after the completion of each route. Moreover, each drone route is restricted to visit a minimum number of points to balance the workload during their missions. The produced routes are presented in Fig. 19. The computation times are roughly equal to the previous approach for all fields. In effect, these

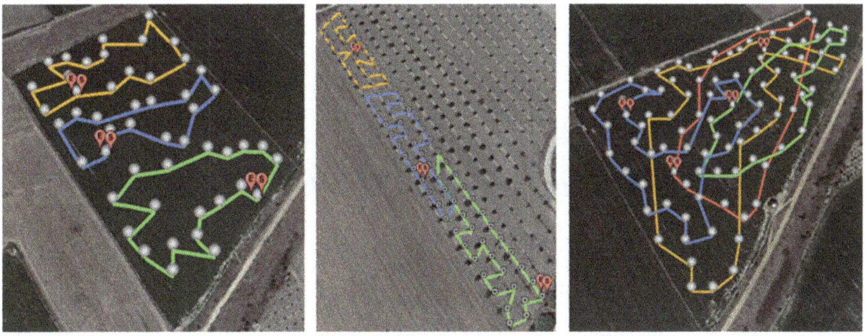

a) Field 1 *b) Field 2* *c) Field 3*

Fig. 18 Routes produced by multiple TSP (mTSP); Each colored line expresses the route of each drone and red placemarks denote the initial/final point of each route

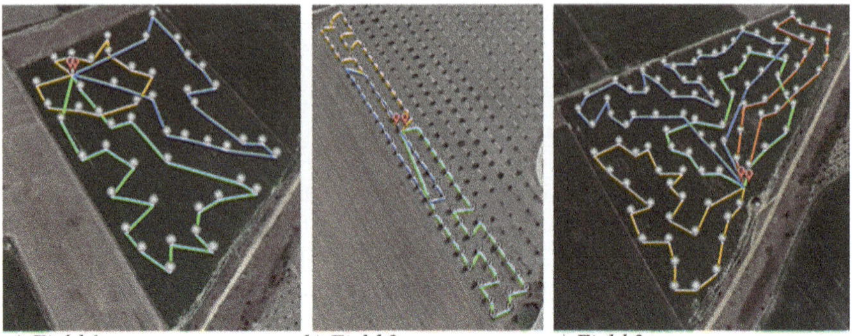

a) Field 1 b) Field 2 c) Field 3

Fig. 19 Routes produced by multiple TSP, single (fixed) depot (mTSPsD); Each colored line expresses the route of each drone and red placemark denotes the predefined depot of each route

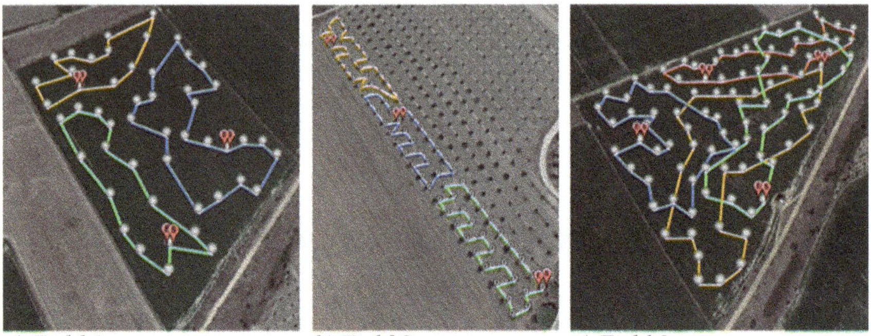

a) Field 1 b) Field 2 c) Field 3

Fig. 20 Routes produced by multiple TSP, multiple (fixed) depots (mTSPmD); Each colored line expresses the route of each drone and red placemarks denote the initial/final point of each route

times are somewhat longer, due to the additional parameters that were introduced. Their covered distances were approximately 11.9% and 10.5% longer for Fields 1 and 2, respectively, while the distance for Field 3 is 13.1% shorter as compared with mTSP.

6.4 Multiple TSP, Multiple (Fixed) Depots (mTSPmD)

One way to further improve the above algorithms is to define different depots for each drone. By applying this extension, each drone is able to navigate through specific regions of the fields and shorten its mission by 6.4%, 13.1%, and 2.6% for the Fields 1, 2, and 3, respectively, compared to the previous GA. Figure 20 illustrates the routes calculated based on this algorithm. Generated routes for the first two fields are not overlapped.

6.5 Multiple TSP, Multiple (Fixed) Depots and Constrained Travelling Distance (mTSPmDcT)

The last GA applied for demonstrating drones routing in agriculture is a combination of the previous ones with one additional constraint, namely drones' maximum travelled distance. Each drone is restricted to start its route at a predefined point, navigate for a minimum number of points and travel up to a maximum distance. This constraint leads to routes focused on specific regions of the fields and adjusted so as drones travel in accordance with the battery limitations. The generated routes are presented in Fig. 21. As expected, this algorithm requires much more computational effort as compared with the other presented algorithms.

7 Conclusions

This chapter pertains to the application of UAVs for sampling operations in agriculture. Firstly, a literature survey was presented on this topic. Subsequently, the definition and the importance of drones' route planning were highlighted. In general, drones should cover a specific area via consuming its available energy in a safe and cost-efficient manner. The main approach for problems of drones routing is TSP. TSP focuses on calculating the optimal route by allowing for only one visit of each point (sampling location) and returning to the initial position. Representative algorithms for solving TSP were briefly presented along with some indicative demonstrations of drones routing for three selected fields consisting of 50, 83, and 100 points of interest.

In a nutshell, UAVs applications in precision agriculture are gaining increasing interest. These applications include, for example, fertilizers usage, weed

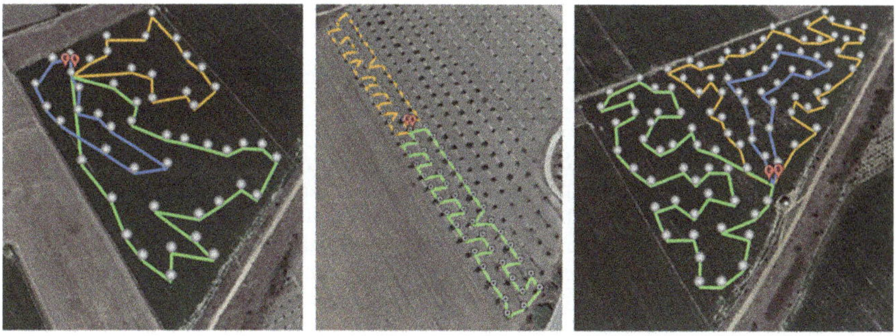

a) Field 1 *b) Field 2* *c) Field 3*

Fig. 21 Routes produced by multiple TSP, multiple (fixed) depots, and constrained travelling distance (mTSPmDcT). Each colored line expresses the route of each drone and red placemarks denote the predefined depot of each route

management, and crop disease inspection [84]. However, UAVs have still some limitations associated mainly with power autonomy, user-friendly features, and scalability that still need to be overcome [85]. Limitations also include aspects of costing and agricultural operations modifications [86, 87]. Thus, UAVs in conjunction with modern technologies, such as Internet of Things (IoT), machine learning, remote sensing, UGVs and agricultural robotics, have the potential of transforming conventional agriculture into smart farming [88].

References

1. Sylvester, G. (2018). *e-Agriculture in action: Drones in agriculture*. Food and Agriculture Organization of the United Nations and International Telecommunication Union, Bangkok.
2. Lampridi, M., Kateris, D., Sørensen, C. G., & Bochtis, D. (2020). Energy footprint of mechanized agricultural operations. *Energies, 13*, 769. https://doi.org/10.3390/en13030769
3. Lampridi, M., Sørensen, C., & Bochtis, D. (2019). Agricultural sustainability: A review of concepts and methods. *Sustainability, 11*, 5120. https://doi.org/10.3390/su11185120
4. Sørensen, C. G., Fountas, S., Nash, E., et al. (2010). Conceptual model of a future farm management information system. *Computers and Electronics in Agriculture, 72*, 37–47. https://doi.org/10.1016/j.compag.2010.02.003
5. Sorensen, C. G., Pesonen, L., Fountas, S., et al. (2010). A user-centric approach for information modelling in arable farming. *Computers and Electronics in Agriculture, 73*, 44–55. https://doi.org/10.1016/j.compag.2010.04.003
6. Radoglou-Grammatikis, P., Sarigiannidis, P., Lagkas, T., & Moscholios, I. (2020). A compilation of UAV applications for precision agriculture. *Computer Networks, 172*, 107148. https://doi.org/10.1016/j.comnet.2020.107148
7. Bochtis, D. D., Sørensen, C. G. C., & Busato, P. (2014). Advances in agricultural machinery management: A review. *Biosystems Engineering, 126*, 69–81. https://doi.org/10.1016/j.biosystemseng.2014.07.012
8. Angelopoulou, T., Tziolas, N., Balafoutis, A., et al. (2019). Remote sensing techniques for soil organic carbon estimation: A review. *Remote Sensing, 11*, 676. https://doi.org/10.3390/rs11060676
9. Angelopoulou, T., Balafoutis, A., Zalidis, G., & Bochtis, D. (2020). From laboratory to proximal sensing spectroscopy for soil organic carbon estimation—A review. *Sustainability, 12*, 443. https://doi.org/10.3390/su12020443
10. Zhang, C., & Kovacs, J. M. (2012). The application of small unmanned aerial systems for precision agriculture: A review. *Precision Agriculture, 13*, 693–712. https://doi.org/10.1007/s11119-012-9274-5
11. Hunt, E. R., & Daughtry, C. S. T. (2018). What good are unmanned aircraft systems for agricultural remote sensing and precision agriculture? *International Journal of Remote Sensing, 39*, 5345–5376. https://doi.org/10.1080/01431161.2017.1410300
12. Rojas Viloria, D., Solano-Charris, E. L., Muñoz-Villamizar, A., & Montoya-Torres, J. R. (2020). Unmanned aerial vehicles/drones in vehicle routing problems: A literature review. *International Transactions in Operational Research, 28*(4), 1626–1657. https://doi.org/10.1111/itor.12783
13. Bochtis, D. D., & Vougioukas, S. G. (2008). Minimising the non-working distance travelled by machines operating in a headland field pattern. *Biosystems Engineering, 101*, 1–12. https://doi.org/10.1016/j.biosystemseng.2008.06.008
14. Bochtis, D. (2008). *Planning and control of a fleet of agricultural machines for optimal management of field operations*. Aristotle University.

15. AUVSI. (2013). *The economic impact of unmanned aircraft systems integrations in the United States*. The Association for Unmanned Vehicle Systems International (AUVSI).
16. Chebrolu, N., Labe, T., & Stachniss, C. (2018). Robust long-term registration of UAV images of crop fields for precision agriculture. *IEEE Robot Autom Lett, 3*, 3097–3104. https://doi.org/10.1109/LRA.2018.2849603
17. Allred, B., Eash, N., Freeland, R., et al. (2018). Effective and efficient agricultural drainage pipe mapping with UAS thermal infrared imagery: A case study. *Agricultural Water Management, 197*, 132–137. https://doi.org/10.1016/j.agwat.2017.11.011
18. Barrero, O., & Perdomo, S. A. (2018). RGB and multispectral UAV image fusion for Gramineae weed detection in rice fields. *Precision Agriculture, 19*, 809–822. https://doi.org/10.1007/s11119-017-9558-x
19. Christiansen, M., Laursen, M., Jørgensen, R., et al. (2017). Designing and testing a UAV mapping system for agricultural field surveying. *Sensors, 17*, 2703. https://doi.org/10.3390/s17122703
20. Verger, A., Vigneau, N., Chéron, C., et al. (2014). Green area index from an unmanned aerial system over wheat and rapeseed crops. *Remote Sensing of Environment, 152*, 654–664. https://doi.org/10.1016/j.rse.2014.06.006
21. Vasudevan, A., Kumar, D. A., & Bhuvaneswari, N. S. (2016). Precision farming using unmanned aerial and ground vehicles. In *Proceedings - 2016 IEEE international conference on technological innovations in ICT for agriculture and rural development, TIAR 2016*. Institute of Electrical and Electronics Engineers Inc., pp. 146–150.
22. Barrientos, A., Colorado, J., del Cerro, J., et al. (2011). Aerial remote sensing in agriculture: A practical approach to area coverage and path planning for fleets of mini aerial robots. *Journal of Field Robotics, 28*, 667–689. https://doi.org/10.1002/rob.20403
23. Faical, B. S., Pessin, G., Filho, G. P. R., et al. (2014). Fine-tuning of UAV control rules for spraying pesticides on crop fields. In *Proceedings - International conference on tools with artificial intelligence, ICTAI* (pp. 527–533). IEEE Computer Society.
24. Kaabi, J., & Harrath, Y. (2019). Permutation rules and genetic algorithm to solve the traveling salesman problem. *Arab Journal of Basic and Applied Sciences, 26*, 283–291. https://doi.org/10.1080/25765299.2019.1615172
25. Bochtis, D. D., & Sørensen, C. G. (2009). The vehicle routing problem in field logistics part I. *Biosystems Engineering, 104*, 447–457.
26. Bochtis, D. D., & Sørensen, C. G. (2010). The vehicle routing problem in field logistics: Part II. *Biosystems Engineering, 105*, 180–188.
27. Orfanou, A., Busato, P., Bochtis, D. D., et al. (2013). Scheduling for machinery fleets in biomass multiple-field operations. *Computers and Electronics in Agriculture, 94*, 12–19. https://doi.org/10.1016/j.compag.2013.03.002
28. Edwards, G., Sørensen, C. G., Bochtis, D. D., & Munkholm, L. J. (2015). Optimised schedules for sequential agricultural operations using a Tabu Search method. *Computers and Electronics in Agriculture, 117*, 102–113. https://doi.org/10.1016/J.COMPAG.2015.07.007
29. Bochtis, D., Griepentrog, H. W., Vougioukas, S., et al. (2015). Route planning for orchard operations. *Computers and Electronics in Agriculture, 113*. https://doi.org/10.1016/j.compag.2014.12.024
30. Bochtis, D., Vougioukas, S., Tsatsarelis, C., & Ampatzidis, Y. (2007). Optimal dynamic motion sequence generation for multiple harvesters. *Agricultural Engineering International: CIGR Journal, 07*, 001.
31. Bochtis, D. D., Vougioukas, S. G., & Griepentrog, H. W. (2009). A mission planner for an autonomous tractor. *Transactions of the ASABE, 52*, 1429–1440.
32. Moysiadis, V., Tsolakis, N., Katikaridis, D., et al. (2020). Mobile robotics in agricultural operations: A narrative review on planning aspects. *Applied Sciences, 10*, 3453. https://doi.org/10.3390/app10103453
33. Abdulkarim, H., Alshammari, I. F., Abdulkarim, H. A., & Alshammari, I. F. (2015). *Comparison of algorithms for solving traveling salesman problem CW laser combined with LED to*

reduce the FWM in SAC-OCDMA networks. View Project Face Regocnition View project Comparison of Algorithms for Solving Traveling Salesman Problem.
34. Király, A., & Abonyi, J. (2015). Redesign of the supply of mobile mechanics based on a novel genetic optimization algorithm using Google Maps API. *Engineering Applications of Artificial Intelligence, 38*, 122–130. https://doi.org/10.1016/j.engappai.2014.10.015
35. Bona, M. *Walk through combinatorics, A: An introduction to enumeration and graph theory* (3Rd ed.). 9789814460002: Amazon.com: Books.
36. University of Waterloo C World Traveling Salesman Problem.
37. Dreyfus, S. (2002). Richard bellman on the birth of dynamic programming. *Operations Research, 50*(1), 48–51. https://doi.org/10.1287/opre.50.1.48.17791
38. Bellman, R. (1962). Dynamic programming treatment of the travelling salesman problem. *Journal of the ACM, 9*, 61–63. https://doi.org/10.1145/321105.321111
39. Held, M., & Karp, R. M. (1962). A dynamic programming approach to sequencing problems. *Journal of the Society for Industrial and Applied Mathematics, 10*, 196–210. https://doi.org/10.1137/0110015
40. Gjorshevski, H., Trivodaliev, K., Kosovic, I. N., et al. (2018). Dynamic programming approach for drone routes planning. In *2018 26th telecommunications forum, TELFOR 2018 - Proceedings*. Institute of Electrical and Electronics Engineers.
41. Little, J. D. C., Murty, K. G., Sweeney, D. W., & Karel, C. (1963). An algorithm for the traveling salesman problem. *Operations Research, 11*, 972–989. https://doi.org/10.1287/opre.11.6.972
42. Lawler, E. L., & Wood, D. E. (1966). Branch-and-bound methods: A survey. *Operations Research, 14*, 699–719. https://doi.org/10.1287/opre.14.4.699
43. Huang, C. Y., Lai, C. Y., & Cheng, K. T. (2009). *Fundamentals of algorithms* (pp. 173–234). Electronic Design Automation. Elsevier.
44. Morrison, D. R., Jacobson, S. H., Sauppe, J. J., & Sewell, E. C. (2016). Branch-and-bound algorithms: A survey of recent advances in searching, branching, and pruning. *Discrete Optimization, 19*, 79–102. https://doi.org/10.1016/j.disopt.2016.01.005
45. Chauhan, C., Gupta, R., & Pathak, K. (2012). Survey of methods of solving TSP along with its implementation using dynamic programming approach. *International Journal of Computer Applications, 52*(4).
46. Lebedev, G., Goncharenko, V., Mikhaylin, D., & Rumakina, A. (2017). Aircraft group coordinated flight route optimization using branch-and-bound procedure in resolving the problem of environmental monitoring. In *ITM Web conference—Seminar on systems analysis*.
47. Kelley, J. E., Jr. (1960). The cutting-plane method for solving convex programs. *Journal of the Society for Industrial and Applied Mathematics, 8*, 703–712. https://doi.org/10.1137/0108053
48. Padberg, M., & Rinaldi, G. (1991). Branch-and-cut algorithm for the resolution of large-scale symmetric traveling salesman problems. *SIAM Review, 33*, 60–100. https://doi.org/10.1137/1033004
49. Schermer, D., Moeini, M., & Wendt, O. (2020). A branch-and-cut approach and alternative formulations for the traveling salesman problem with drone. *Networks, 76*(2), 164–186. https://doi.org/10.1002/net.21958
50. Méndez-Díaz, I., & Zabala, P. (2006). A Branch-and-Cut algorithm for graph coloring. *Discrete Applied Mathematics, 154*, 826–847. https://doi.org/10.1016/j.dam.2005.05.022
51. Applegate, D. L., Bixby, R. E., Chvátal, V., et al. (2009). Certification of an optimal TSP tour through 85,900 cities. *Operations Research Letters, 37*, 11–15. https://doi.org/10.1016/j.orl.2008.09.006
52. Rego, C., Gamboa, D., Glover, F., & Osterman, C. (2011). Traveling salesman problem heuristics: Leading methods, implementations and latest advances. *European Journal of Operational Research, 211*, 427–441. https://doi.org/10.1016/j.ejor.2010.09.010
53. Williamson, D. (2012). *The Design of approximation algorithms*. Cambridge University Press.

54. van Bevern, R., & Slugina, V. A. (2020). A historical note on the 3/2-approximation algorithm for the metric traveling salesman problem. *Historia Mathematica*. https://doi.org/10.1016/j.hm.2020.04.003
55. Christofides, N. (1976). *"T" AD-A025 602 worst-case analysis of a new heuristic for the travelling salesman problem*. Carnegie-Mellon University Prepared for Office of Naval Research.
56. Ben Ghorbel, M., Rodriguez-Duarte, D., Ghazzai, H., et al. (2018). Energy efficient data collection for wireless sensors using drones. In *IEEE vehicular technology conference* (pp. 1–5). Institute of Electrical and Electronics Engineers.
57. Ball, M. O. (2011). Heuristics based on mathematical programming. *Surveys in Operations Research and Management Science, 16*, 21–38.
58. De Sourav, S. D. *Recent advances in hybrid metaheuristics for data clustering*. Google Books.
59. Rosenkrantz, D. J., Stearns, R. E., & Lewis, P. M., II. (1977). An analysis of several heuristics for the traveling salesman problem. *SIAM Journal on Computing, 6*, 563–581. https://doi.org/10.1137/0206041
60. Bentley, J. J. (1992). Fast algorithms for geometric traveling salesman problems. *ORSA Journal on Computing, 4*, 387–411. https://doi.org/10.1287/ijoc.4.4.387
61. Lin, S., & Kernighan, B. W. (1973). Effective heuristic algorithm for the traveling-salesman problem. *Operations Research, 21*, 498–516. https://doi.org/10.1287/opre.21.2.498
62. Gutin, G., Yeo, A., & Zverovich, A. (2002). Traveling salesman should not be greedy: Domination analysis of greedy-type heuristics for the TSP. *Discrete Applied Mathematics, 117*, 81–86. https://doi.org/10.1016/S0166-218X(01)00195-0
63. Bang-Jensen, J., Gutin, G., & Yeo, A. (2004). When the greedy algorithm fails. *Discrete Optimization, 1*, 121–127. https://doi.org/10.1016/j.disopt.2004.03.007
64. Murray, C. C., & Chu, A. G. (2015). The flying sidekick traveling salesman problem: Optimization of drone-assisted parcel delivery. *Transportation Research Part C: Emerging Technologies, 54*, 86–109. https://doi.org/10.1016/j.trc.2015.03.005
65. Johnson DS, Mcgeoch LA (1995) The Traveling Salesman Problem: A Case Study in Local Optimization.
66. Rigas, E. S., Kolios, P., & Ellinas, G. (2020). Extending the multiple traveling salesman problem for scheduling a fleet of drones performing monitoring missions.
67. Helsgaun, K. (2009). General k-opt submoves for the Lin-Kernighan TSP heuristic. *Mathematical Programming Computation, 1*, 119–163. https://doi.org/10.1007/s12532-009-0004-6
68. Anbuudayasankar, S. P., Ganesh, K., Mohapatra, S., et al. (2014). Survey of methodologies for TSP and VRP. In *Models for practical routing problems in logistics* (pp. 11–42). Springer International Publishing.
69. Kaufmann, M., & Park, M. (2014). *Stochastic local search: Foundations and applications*. Holger H. Hoos, Thomas Stützle-Google Books.
70. Cho, D. H., Jang, D. S., & Choi, H. L. (2018) Heterogeneous, multiple depot multi-UAV path planning for remote sensing tasks. In *AIAA Information Systems-AIAA Infotech at Aerospace, 2018*. American Institute of Aeronautics and Astronautics Inc, AIAA.
71. Chandra, B., Karloo, H., & Tovey, C. (1994). New results on the old k-opt algorithm for the TSP. In *SODA '94: Proceedings of the fifth annual ACM-SIAM symposium on Discrete algorithms*, pp. 150–159.
72. Okada, M. (1999). Studies on probabilistic analysis of λ-opt for traveling salesperson problems. https://doi.org/10.11501/3153830.
73. Fink, A., & Voß, S. (1999). Generic metaheuristics application to industrial engineering problems. *Computers and Industrial Engineering, 37*, 281–284. https://doi.org/10.1016/S0360-8352(99)00074-1
74. Thomas, G. (1999). *Stutzle local search algorithms for combinatorial problems—Analysis, improvements, and new applications*.
75. Blum, C., & Roli, A. (2003). Metaheuristics in combinatorial optimization: Overview and conceptual comparison. *ACM Computing Surveys, 35*, 268–308.

76. Holland, J. H. (1984). Genetic algorithms and adaptation. In *Adaptive control of Ill-defined systems* (pp. 317–333). Springer US.
77. Dorigo, M., Colorni, A., & Maniezzo, V. (1991). *Distributed optimization by ant colonies.*
78. Ahmed, F., & Deb, K. (2013). Multi-objective optimal path planning using elitist non-dominated sorting genetic algorithms. *Soft Computing, 17*, 1283–1299. https://doi.org/10.1007/s00500-012-0964-8
79. Singh, A., & Baghel, A. S. (2009). A new grouping genetic algorithm approach to the multiple traveling salesperson problem. *Soft Computing, 13*, 95–101. https://doi.org/10.1007/s00500-008-0312-1
80. Article, R., Gümüşçü, A., Tenekeci, E., & Tabanlıoğlu, A. (2018). The shortest path detection for unmanned aerial vehicles via genetic algorithm on aerial imaging of agricultural lands. *International Advanced Researches and Engineering Journal, 2*(3), 315–319.
81. Dorigo, M., Bonabeau, E., & Theraulaz, G. (2000). Ant algorithms and stigmergy. *Future Generation Computer Systems, 16*, 851–871.
82. Bakhtiari, A. A., Navid, H., Mehri, J., & Bochtis, D. D. (2011). Optimal route planning of agricultural field operations using ant colony optimization. *Agricultural Engineering International: CIGR Journal, 13*, 1–16.
83. Chen, J., Ye, F., & Li, Y. (2017). Travelling salesman problem for UAV path planning with two parallel optimization algorithms. In *Progress in electromagnetics research symposium. electromagnetics academy*, pp. 832–837.
84. Liakos, K., Busato, P., Moshou, D., et al. (2018). Machine learning in agriculture: A review. *Sensors, 18*, 2674. https://doi.org/10.3390/s18082674
85. Boursianis, A. D., Papadopoulou, M. S., Diamantoulakis, P., et al. (2020). Internet of Things (IoT) and Agricultural Unmanned Aerial Vehicles (UAVs) in smart farming: A comprehensive review. *Internet of Things*, 100187. https://doi.org/10.1016/j.iot.2020.100187
86. Lampridi, M. G., Kateris, D., Vasileiadis, G., et al. (2019). A case-based economic assessment of robotics employment in precision arable farming. *Agronomy, 9*, 175. https://doi.org/10.3390/agronomy9040175
87. Marinoudi, V., Sørensen, C. G., Pearson, S., & Bochtis, D. (2019). Robotics and labour in agriculture. A context consideration. *Biosystems Engineering, 184*, 111–121. https://doi.org/10.1016/J.BIOSYSTEMSENG.2019.06.013
88. Benos, L.,Tagarakis, A., Dolias, G., Berruto, R., Kateris, D., & Bochtis, D. (2021). Machine learning in agriculture: A comprehensive updated review. *Sensors, 21*, 3758. https://doi.org/10.3390/s21113758

3D Scenery Construction of Agricultural Environments for Robotics Awareness

Aristotelis Christos Tagarakis, Damianos Kalaitzidis, Evangelia Filippou, Lefteris Benos, and Dionysis Bochtis

1 Introduction

In recent years, depth sensors have been widely used in various applications that include 3D reconstruction of a scene for indoor [1, 2] and outdoor environments [3, 4]. More specifically, 3D modelling supports applications in fields such as object recognition [5, 6], object tracking [7], autonomous navigation [8, 9], and visualization [10]. During the 3D reconstruction process, a scenery is represented by gathering data from a sensor and storing them in a form that is efficient for further processing. A plethora of researchers have approached the 3D reconstruction of a scene by measuring the depth value of the surroundings through image acquisition from RGB-D cameras. This method was firstly applied by Nevatia et al. [11] as a means to present depth relations of objects from image acquiring. Furthermore, retrieving distance values from stereoscopic images, captured by two monocular cameras or a stereo camera, led to the calculation of 3D object's position by using the geometrical derivation method [12, 13]. Subsequently, the technology of structured light was introduced to overcome some of the stereoscopy problems and, later, the time-of-flight technology started being used. In [14], the evaluation of the aforementioned technologies with the use of low-cost depth cameras took place in agricultural applications presenting the advantages and disadvantages of certain RGB-D camera models [14]. In particular, the images collected from depth cameras consist of a numeric digital map that contains, apart from the Red (R), Green (G),

A. C. Tagarakis (✉) · D. Kalaitzidis · E. Filippou · L. Benos · D. Bochtis
Institute for Bio-Economy and Agri-Technology (IBO), Centre for Research and Technology Hellas (CERTH), Thessaloniki, Greece
e-mail: a.tagarakis@certh.gr

D. D. Bochtis et al. (eds.), *Information and Communication Technologies for Agriculture—Theme III: Decision*, Springer Optimization and Its Applications 184, https://doi.org/10.1007/978-3-030-84152-2_6

and Blue (B) digital values, the distances between the sensor and the position of the camera for every pixel that constructs the image. As a result, through digital image processing, it is possible to get information about the color and geometry of the objects of a landscape.

In agricultural fields, 3D reconstruction of plants from various sensors is a trait of interest as plant architecture can be used to understand physiological processes of plant status [15, 16]. In some cases, it is not necessary to fully reconstruct and characterize the parameters of the entire plant structure, but only provide information about the canopy [17–19], given that 3D acquisition can be a laborious and time-consuming procedure. In the study of Chéné et al. [20], an imaging system using a low-cost depth camera was presented for 3D phenotyping of entire plants. In this algorithm, leaf segmentation was accomplished from depth images, without any prior spatial information of the plants, while results between imaging measurements and ground truth measurements were found to be in good agreement [20].

Modern agriculture aims at increasing food production while ensuring low environmental impact. New concepts towards more efficient management of agricultural crops have been emerged during the last decades. One of these concepts, called precision agriculture (or digital agriculture, or more recently agriculture 4.0), utilizes all available technologies for mapping, analyzing, and managing each part of the field of interest according to the real needs. Depth cameras have also been used as part of the available tools in precision agriculture systems. In a recent work [21], depth cameras were used for weed volume estimation with encouraging results indicating that crop and weed biomass can be accurately measured. Consequently, a system using technologically advanced sensors has many possibilities for automating agricultural processes.

1.1 Depth Cameras

Nowadays, there are three available technologies that depth cameras use for image acquisition, namely stereoscopy, structured light and, more recently, time-of-flight (ToF). Usually, one or two of these technologies are utilized by the different depth cameras available on the market. Cameras that use stereoscopic vision (such as Stereolabs ZED; PointGrey Bumblebee; Minoru 3D System) are based on the main stereo view rules that, by triangulation, the depth of an object can be calculated and visualized [22]. The structured light cameras (such as MS Kinect, Asus Xtion Pro, Intel RealSense) can determine the distortions of the sensor emissions of the infrared light (IR) to the object [23]. Cameras that use the ToF technology (such as PMD CamBoard, PMD CamCube, SwissRanger SR4000) work by considering the time delay and estimating it as a depth value.

Interestingly, depth cameras have gained field in 3D mapping, given the combination of characteristics they present such as of low cost, ease of usage, and effectiveness on accurate results comparing to other sensors, such as laser scanners. Lachat et al. [24] evaluated 3D modelling from a depth sensor and a close range laser scanner. The tests from this study showed that using a low-cost depth sensor had sufficient results in depth measurements and the use of depth cameras has a great potential for further applications. Stereoscopic vision technology was implemented in many studies for 3D mapping and 3D object construction in indoor environments such as [25–27]. Furthermore, the ZED stereo camera (Stereolabs, CA, USA) has been used in many different indoor scenarios and applications, including volume determination of objects through image processing [28], crack detection, and measurement on concrete surfaces [29], terrestrial photogrammetric modelling [30], and generation of 3D indoor mapping aiming to have a better understanding in the study of "smart" cities [31]. In another study [32], these various technologies were broadly evaluated in an experiment conducted in indoor environments for estimating bias, precision and lateral noise, under different lighting conditions, targeted objects and sensor setups, by using ten different depth sensors.

Although there has been an extensive usage of depth cameras regarding indoor environments, there is a lack of their implementation in outdoor environments, as they are vulnerable to noise, while demonstrating limitations in viewing angles and restrictions in estimating minimum and maximum distance from the target. However, in recent years, several researchers have evaluated the performance of depth cameras in agricultural fields. For example, Wang and Li [33] proposed a system for the diameter and volume of onions, while Wang and Zhang [34] made a 3D reconstruction of an artificial cherry tree in laboratory environment, by using a structured light depth sensor in both studies. The latter study indicated that bright sunlight conditions in outdoor environments affect the performance of structured light camera providing non-acceptable results. Furthermore, Andujar et al. [35] presented a method for extraction of 3D model structural parameters of cauliflower plants for the purpose of detecting the best harvesting time. In addition, ToF technology has been evaluated in plant structure characterization of peppers [36]. Comparisons of plant leaf imaging between indoor and outdoor conditions by using ToF cameras in plants were made by Kazmi et al. [37]. It was inferred that despite the sensitivity to ambient light and the range limitations of these cameras, there is a sufficient method to implement them in various agricultural applications like plant phenotyping.

In agricultural robotics, depth cameras can be utilized for applications with autonomous UGV and computer vision systems in order to improve field management, yields, and quality [38]. Recently, depth cameras have been implemented in agriculture for fruit localization systems [39–41]. More specifically, fruit detection, disease detection, and plants localization in agricultural fields are among the most important aspects for agricultural robotics applications in tasks such as yield prediction, yield mapping, and automated harvesting [42–46]. Fu et al. [47] presented a review concerning the advantages and disadvantages of depth cameras with different technologies that were implemented for fruit detection and localization in the field. Three types of depth sensors were implemented, namely structured light, ToF and active infrared stereo, with different specifications: resolution, field of view, and capturing parameters. The data acquired from these depth sensing systems were applied in different image processing methods to evaluate the capability of fruit detection and localization in crops. It was deduced that despite their low cost that makes them an attractive option, the structured light cameras were vulnerable to direct sunlight conditions. Although ToF cameras were less affected by sunlight limitations, they proved to be inadequate since they produced low resolution depth images. Additionally, active stereo cameras demonstrated promising results. Finally, in [47], it was concluded that a combination of the two types of depth cameras can give better results.

Furthermore, Condotta et al. [14] presented an evaluation of low-cost depth cameras for agricultural applications. A comparison was presented among 15 depth cameras of three different technologies, while an image processing algorithm was developed. The results indicated that cameras using ToF technology should be used in outdoor environments only in a close range (1–2 m). On the other hand, stereoscopic cameras, such as ZED, provided the most reliable data, even when being implemented under extremely bright conditions. The errors coming from these cameras appeared to be predictable and repeatable, a fact that indicates that a calibration could provide more accurate depth measurements. Moreover, an investigation on the development of a stereovision sensing system for 3D mapping of crop rows structure and tractor guidance took place in [48]. This study showed that the produced 3D map could be used for height and volume measurements of the plants and for accurate guidance of vehicles in the field.

2 Point Cloud Processing and Digitalization

In this section, some of the most common tools for creating 3D models and maps are briefly described. The tools provided by Robot Operating System (ROS) and Stereolabs can be used for almost all robotic applications, namely for both indoor and outdoor ones. Also, the advantages of implementing a simulation environment in robotic applications are described. The convenience provided by the simulations for testing various algorithms and scenarios, with the aim to resolve various issues before operating in the real outdoor field environment, is discussed.

2.1 3D Mapping

The 3D mapping is the procedure of creating a 3D representation of the operations environment. The map is composed of point cloud data that represent real-world models, which have been developed by high resolution depth sensors. The most commonly used tools for 3D reconstruction of point clouds are the Octomap (University of Freiburg, Germany) [49], Real-Time Appearance-Based Mapping (RTABMap; Sherbrooke, Québec, Canada), and Spatial Mapping (Stereolabs Inc.,

CA, USA). The basic operation of these tools is the consecutive capture of a real-world environment and its continuous extraction in a 3D map, in which the scanning area is represented. As far as Octomap and RTABMap are concerned, they are ROS-related 3D reconstruction mapping libraries that are widely used in robotics.

Octomap is an efficient probabilistic 3D mapping tool based on Octrees. The Octree data structure technique uses parent nodes that consist of eight children and store the capturing information about the scanned environment. Connecting these nodes merges all the acquired data and generates compact 3D maps. Moreover, Octomap probabilistic process offers the ability to recognize dynamic objects and changes in the environment [50]. This process leads to a more efficient representation of a 3D environment with less noise from dynamic object or robot position failures. Octomap meets mainly four requirements, namely full 3D modelling (both free and occupied space), while it is updatable (new information can be added at any time), flexible (the map can be dynamically expanded) and compact (the map can be stored in memory and disk). Notable advantages are also the low computational load and memory usage as compared to other 3D mapping tools. The main disadvantage of Octomap, however, is that it incorporates only depth data (not RGB values) to produce 3D maps, which means that only the position information (x, y, and z coordinates) is provided without any information of the scanned object's color.

In contrast, RTABMap generates a complete representation of scanning environment, which contains both depth and RGB data. It has been distinguished among the other 3D mapping tools since it provides additional useful data for robot navigation by using the so-called Simultaneous Localization and Mapping (SLAM) approach. The disadvantages of RTABMap tool include its inability to recognize dynamic objects, thus, resulting in noisy maps, while the required computational load of RTABMap method is increased [50]. The acquired point cloud can be represented in ROS visualization tool (Rviz) and stored in Point Cloud Data (PCD) or Object (OBJ) file formats for further processing.

2.2 *Digital Twin*

Several depth cameras carry additional sensors, such as Inertial Measurement Units (IMUs) that can calculate and record the exact position of the camera as well as detect its kinematics characteristics. This feature is of particular importance for autonomous robotic applications, as the sensors collect information about the environment, which can be later used as input for robot localization and navigation (situation and environment awareness). In robotic applications, simulation techniques are applied in order to imitate the complicated operations environment before real application. This procedure is significant in algorithm evaluation, according to which robots are tested and implemented in tasks during the development stage to improve their performance.

A known simulation technique is the Digital Twin (DT) technology which produces the characteristics, processes, and performance of a real-world physical object or system. This can be achieved by certain operations and techniques resulting in the DT model. This model is a virtual model that simulates the behavior and performance of a physical object. It is worth mentioning that this model is dynamic indicating that it is not only a simulation model of a physical object, but it can also be used for simulation of the physical environment. In addition, it can be continuously upgraded by collecting more data improving its performance [51]. Finally, DT is composed of three basic elements: the physical object, the virtual object and the connection between them, which works as a link that interacts bidirectionally among the two [52].

In recent years, DT technology has been widely used in industrial [53–55] and engineering applications [56, 57] as well as in manufacturing [58–60] for the optimization of products. These applications use DT in the production or the after-production stage. Interestingly, Tao et al. [61] presented a study on exploiting DT technology in the stage of the designing of products. Furthermore, DT technology was used in the field of robotics for the adaptation of robots' behavior in flexible robotic assembly lines. In particular, in [62] a DT system was developed by collecting data from various sensors and robots to improve the interaction between human behavior and dissimilarity of the physical environment.

Over the last years, DT technology has also been applied in agricultural applications [63]. Many researchers have presented ways to improve the agricultural tasks and produce smart farming systems such as water management and development of Internet of Things (IoT) applications [64]. In particular, the creation of digital environment with information about the soil and other data helped farmers to understand and take management decisions in a better and more efficient way. Moreover, the DT technology concept was presented in [65], confirming that it provides farmers the opportunity to act immediately whenever needed. In general, using advanced low-cost depth sensors is an affordable and efficient way to develop outdoor 3D models for various applications.

2.3 Simulation Environments

The visualization of the robot in a simulation environment provides a powerful advantage for testing robotic applications [66–68]. By developing a simulation environment, specific real-world scenarios can be created. Trees, robots, sensors, and humans can participate in order to create a more realistic simulation and test the interaction of the robot with the parts of this ecosystem with safety [69–74]. Some of the most common simulation environments for agricultural robotics applications are Gazebo, Webots, Actin, RoboDK, Morse, OpenRAVE, OpenHrp3, ARGoS, and V-REP [75, 76].

A key role in robotics research has the rapid and reliable testing of new algorithms and tools for making conclusions about robot response without depending on direct robot applications in real environments [77]. Such practices can aid in developing new projects, evaluating robots' performance in various tasks and numerous environments, reducing the cost and the required time for development and testing. The use of simulation environments provides the ability to operate and observe the kinematics of the robot in a 3D world and optimize its behavior before the actual tests in the field [78] while maximizing the safety of the equipment, thus, preventing possible failures in real applications. For a more realistic 3D reconstruction, DT technology can be used for the representation of a real-world model in a simulation environment.

2.4 Aim of This Chapter

In this section, an overview on depth cameras as well as point cloud processing and digitalization and simulation environments was presented. In the following section, the development of a DT model for an outdoor real field environment is presented. The model was developed using real data from a commercial orchard and was applied in Gazebo for simulation. Moreover, the most common software environments are described along with the procedure followed for the analysis. This approach involves point cloud data acquisition, processing, and simulating of the Orchard environment at hand.

3 Demonstrative Scenario: An In-field Application

In this study, an in-field application of depth camera for 3D construction of the environment was performed in a commercial walnut orchard. The camera was installed on a Thorvald (SAGA Robotics SA, Norway) UGV (Fig. 1). The setup was used for navigation in the field, scanning the trees, and collecting the data to construct a 3D point cloud of the orchard. Thus, the 3D model of the real environment was constructed, including all the elements that are present in the field such as trees, boulders, and other obstacles, for use in further robotic applications.

3.1 Point Cloud Data Acquisition

The point cloud data for the 3D construction were collected using the ZED 2 stereo camera. The given dataset was composed of a pixel array with XYZ-RGB values. In order to use the ZED SDK and ROS-related tools, a system was constructed which was composed of an NVIDIA Jetson TX2 development kit with the Ubuntu GNU/Linux 18.04 operating system. In this system, ROS melodic distro was installed which supports ZED 2 camera and provides accessibility to ROS tools.

The Jetson TX2 consisted of 8GB of RAM, 32GB Flash Storage, 2 Denver 64-bit CPUs, and Quad-Core A57 Complex. All GPU cores were set in full performance

Fig. 1 System setup used for the 3D mapping of the trees within the orchard. The ZED 2 camera, the Jetson TX2, and the RTK-GPS were positioned on Thorvald autonomous modular agricultural robot

for ensuring maximum speed and robustness of the system to achieve the best possible results. The point clouds were obtained by scanning samples of walnut trees of different height, volume, and shape. All targeted objects were scanned in direct sunlight conditions close to noon to minimize shading. The 3D reconstruction of trees was performed using the spatial mapping module of Stereolabs Software Development Kit (SDK) tool and RTABMap package of ROS. Moreover, following the Application Programming Interface (API) documentation, several parameters were set accordingly to fit the system to the field conditions. More specifically, the resolution of the camera was set to 720p (1280 × 720 pixels), while the mapping resolution of the point cloud was set to 2 cm to create geometry of high resolution and high density. In addition, the point cloud mapping range, that sets the range of depth data, was set to 5 m maximum and 0.4 m minimum distance from the camera. Thus, it included all objects in the surrounding areas of the scanned trees. This setup is more appropriate for outdoor orchard environments capturing the full range of the tree canopies. The camera has a vertical field of view of 72° and a horizontal field of view of 104° in 720p resolution. As a result, the device effectively scanned all targets in the field in both short and longer distances by adjusting the angle of the camera according to the height and volume of each tree.

The camera readings were acquired at a high frame rate, namely 50 frames per second. Consequently, there were overlapping areas between the different frames in continuous measurements scanning the tree canopy perimetrically. The overlapping sections provide better 3D models, as the SDK tool merges the additional points that are acquired and creates a joined point cloud.

The ZED 2 camera and Jetson TX2 were positioned on the Thorvald robotic vehicle, which was equipped with a high accuracy (RTK) GPS to provide the position of the robot and consequently the true position of the camera (Fig. 1). Thus, georeferencing of the measurements was achieved resulting in the production of georeferenced point cloud. With the intention of properly obtaining the tree images and facilitating the 3D mapping of the actual shapes within the orchard, the ZED camera was oriented towards the direction of the tree of interest and was adjusted in the appropriate height manually. The adjustment of the camera was necessary due to the geometry characteristics of the tree canopies, volume, and height as well as the variations among the trees.

The RGB-D measurements were received by navigating the robot-camera system around each targeted tree for one minute which, according to the acquisition rate, resulted in capturing 3500 frames per sample. After completing a whole round of each tree, the acquired point cloud was stored for further processing. In particular, the OBJ and PCD file formats were chosen for storing the 3D point clouds. These file formats are compatible with various processing software such as Meshlab and CloudCompare.

3.2 Point Cloud Data Processing

The point cloud processing begins with the in-field data capturing and storing, followed by the conversion of the Point Cloud Data format (PCD) to Digital Asset Exchange (DAE) in order to facilitate measurements' extraction for calculating tree height and canopy volume. An illustrative flow chart of the implemented procedure can be seen in Fig. 2.

From the generated point cloud of each walnut tree, the height and volume were calculated, using the cloud processing software CloudCompare and its internal tools. For the volume calculation, the 2.5-dimensions (2.5D) tool, available in the software, was used. This tool is effective in visualization of the density of points among the trees. As expected, the vast amount of the tree volume is concentrated in the canopy (Fig. 3).

Moreover, an automated procedure was established via creating a script in Python programming language, in which volume and height of a multitude of individual trees are calculated and stored (Fig. 4). In this approach, the point clouds were converted in arrays and processed using the appropriate Python libraries for height and volume calculations.

The generated point cloud was georeferenced and imported in Quantum Geographic Information System (QGIS) to reference the points of the tree point cloud from a relative to a real-world coordinate system. The georeferenced tree point cloud was produced by converting the depth point cloud (X, Y, Z) to geographic projection system point cloud (latitude, longitude, altitude) (Fig. 5). For the conversion, it was necessary to record the GPS coordinates of the initial position of the camera, when the image was captured. This information was extracted from the GPS coordinates recorded by the RTK-GPS installed on the agricultural robotic platform. The georeferenced 3D point clouds aim to be used in future tasks for in-field applications, such as spraying and harvest.

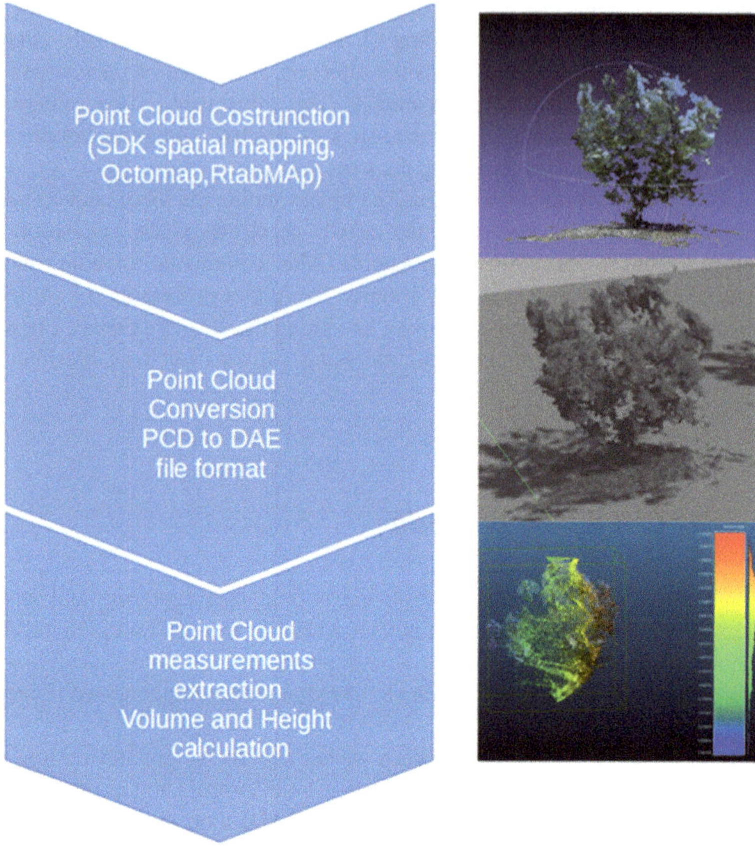

Fig. 2 Flow chart of the procedure for 3D point cloud data analysis acquired by RGB-D camera

3.3 Orchard's Simulation Environment

Among the available simulation environments, Gazebo was selected for our application owing to its effectiveness to accurately reflect complex environments and models where the robot could possibly encounter. Due to the communication between Gazebo and ROS software, a powerful simulation system was available for our study. This system not only provides the ability to visualize a virtual environment, but the user can also introduce in this virtual world the specific robot to be used in the real application in order to interact with the environment and accomplish various operations. The results from Gazebo simulation regarding robot navigation testing and obstacle detection displayed by ROS visualization tool (Rviz) are depicted in Figs. 6a and b, respectively.

The acquired point clouds were converted to mesh file format and imported in Gazebo simulation software for testing the robot's navigation. In this simulation environment, the ability of the robot to overcome obstacles in the field was evaluated

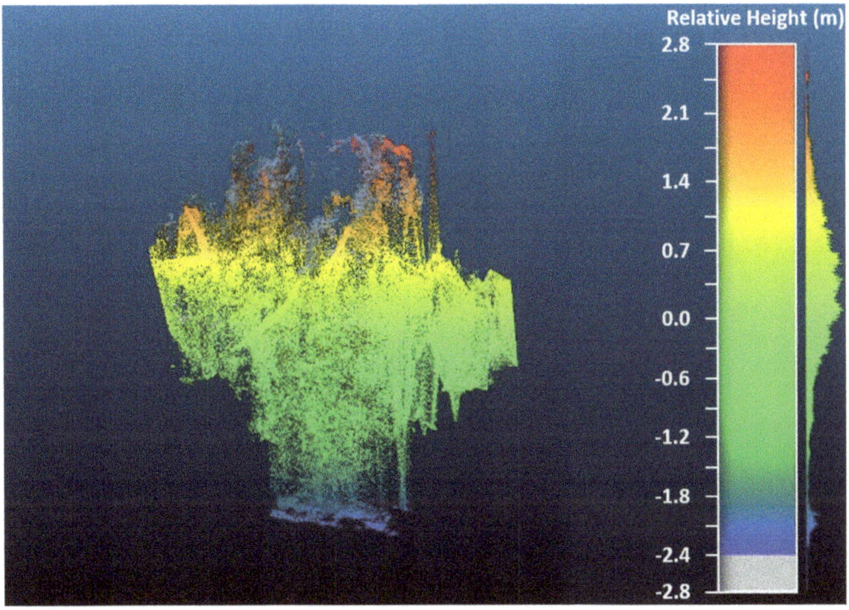

Fig. 3 Analysis of the 3D point cloud dataset to extract the point density for calculating the tree volume and height

Fig. 4 3D point cloud representation of a single tree of the orchard analyzed using script in Python programming for volume and height calculations

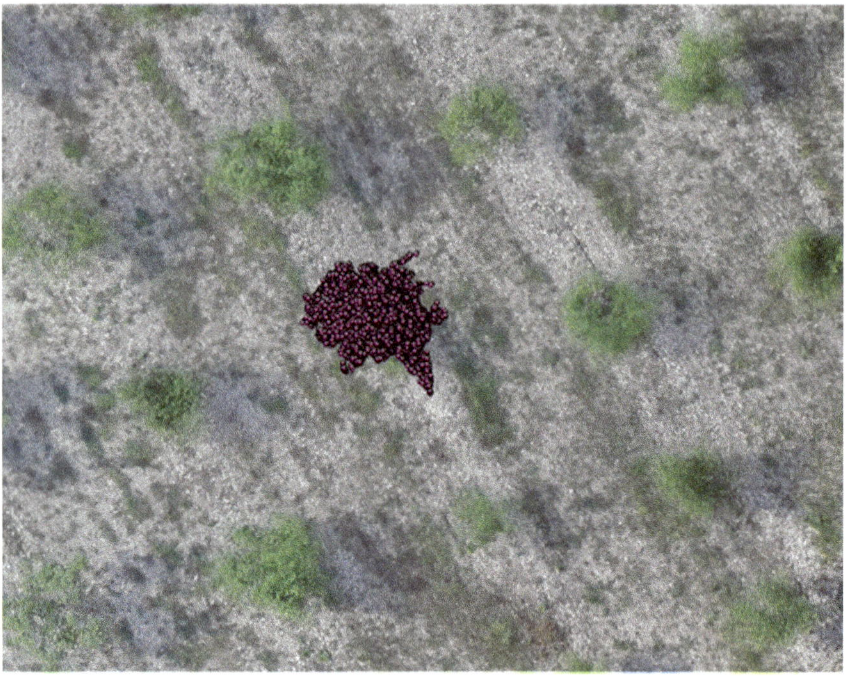

Fig. 5 Georeferenced 3D point cloud using geographic projection system

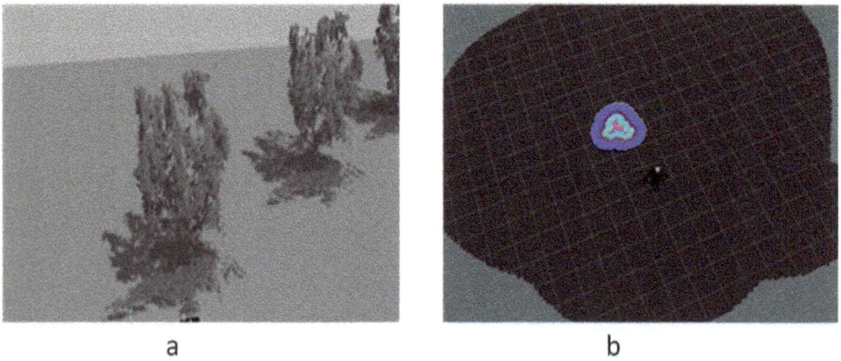

Fig. 6 Models of the scanned trees 3D point clouds imported in Gazebo simulation software for (**a**) robot navigation testing and (**b**) obstacle detection displayed in ROS visualization tool (Rviz)

before operating in real field conditions. This awareness gave us the opportunity to handle the obstacle avoidance procedures and prepare the robot for the actual navigation testing in the field.

Regarding the captured point clouds, it is worth mentioning that the trees had not the required point density to depict details, such as leaves or nuts. Furthermore, there were some blank spaces in the clouds, due to blind spots of the camera view when capturing the trees that changed the real shape of the targets in the visualization. However, despite the errors, the generated point cloud provided a contour of the trees very close to reality. Thus, the setup used in this work provided reasonable results of the volume and the height, while the visualization model that was imported in the Gazebo provided the possibility to run tests in virtual environment that were an accurate representation of the real orchard environment. Changes in the parameterization of the camera and increase of the image frames acquired per tree could optimize the generated 3D model of the tree. Nevertheless, this would cost in processing time and would require higher hardware requirements with considerably larger data size.

4 Conclusions

Over the last years, 3D reconstruction of plants has made great progress in agricultural applications, as plant architecture aims to be used for understanding of plants' processes. In this chapter, the use of low-cost depth cameras in agricultural applications was presented. Preliminary results of in-field application highlighted the potential of RGB-D cameras, such as ZED 2, to be used for 3D mapping and reconstruction of real objects in agricultural environments. The virtual models produced by appropriate analysis of RGB-D-derived 3D point clouds can be applied in simulation techniques to interact with autonomous agricultural robotic systems. Using these virtual objects in simulation environments provides the potential to implement robotic operations in a virtual field and test various tasks before the in-field implementation of them.

In this approach, the realistic point clouds of walnut trees were successfully generated from RGB-D images and the samples were imported in the Gazebo which simulated the outdoor environment. The depth data provided information about the geometry characteristics of the trees and were used for volume and height measurements with sufficient accuracy. As a result, the constructed point clouds showed the ability of depth cameras to provide spatial and color information as well as geometrical information of trees.

The uncertain morphology of the agricultural environment and the variability in the structure of the trees' shapes led to the formation of blind spots in the sensor's field of view and consequently blank spaces in the point cloud. This issue can be resolved by increasing the frames captured per tree to provide more dense point cloud. However, this should be done in accordance with the limitations of the processing unit characteristics and the in-field operation's temporal limitations.

The procedure of collecting and processing the data is not yet fully automated, as there are some certain tasks that require different dedicated software, e.g., for point clouds generation and for the measurements of trees height and volume. Future work includes the use of produced point clouds in supervised learning techniques, such as classification. Training a predictive model according to its spatial and color information could provide classification for the different parts of the tree, such as the leaves or trunks. Certainly, there is still a long way towards establishing an automated procedure for simulating agricultural environments. In this regard, more advanced hardware with higher computing power and more sophisticated programming is required to develop a fully operational automated robotic vision system for agricultural applications.

References

1. Izadi, S., Kim, D., Hilliges, O., Molyneaux, D., Newcombe, R., Kohli, P., Shotton, J., Hodges, S., Freeman, D., Davison, A., et al. (2011). KinectFusion: Real-time 3D reconstruction and interaction using a moving depth camera. In *Proceedings of the UIST'11 - Proceedings of the 24th annual ACM symposium on user interface software and technology*. ACM Press, New York, NY, pp. 559–568.
2. Pan, H., Guan, T., Luo, Y., Duan, L., Tian, Y., Yi, L., Zhao, Y., & Yu, J. (2016). Dense 3D reconstruction combining depth and RGB information. *Neurocomputing, 175*, 644–651. https://doi.org/10.1016/j.neucom.2015.10.104
3. Biskup, B., Scharr, H., Schurr, U., & Rascher, U. (2007). A stereo imaging system for measuring structural parameters of plant canopies. *Plant, Cell and Environment, 30*, 1299–1308. https://doi.org/10.1111/j.1365-3040.2007.01702.x
4. Schöps, T., Sattler, T., Häne, C., & Pollefeys, M. (2017). Large-scale outdoor 3D reconstruction on a mobile device. *Computer Vision and Image Understanding, 157*, 151–166. https://doi.org/10.1016/j.cviu.2016.09.007
5. Li, J., Huang, W., Shao, L., & Allinson, N. (2014). Building recognition in urban environments: A survey of state-of-the-art and future challenges. *Information Sciences (NY), 277*, 406–420. https://doi.org/10.1016/j.ins.2014.02.112
6. Guo, Y., Sohel, F., Bennamoun, M., Wan, J., & Lu, M. (2015). A novel local surface feature for 3D object recognition under clutter and occlusion. *Information Sciences (NY), 293*, 196–213. https://doi.org/10.1016/j.ins.2014.09.015
7. Kim, D. Y., & Jeon, M. (2014). Data fusion of radar and image measurements for multi-object tracking via Kalman filtering. *Information Sciences (NY), 278*, 641–652. https://doi.org/10.1016/j.ins.2014.03.080
8. Yang, D., Liu, Z., Sun, F., Zhang, J., Liu, H., & Wang, S. (2014). Recursive depth parametrization of monocular visual navigation: Observability analysis and performance evaluation. *Information Sciences (NY), 287*, 38–49. https://doi.org/10.1016/j.ins.2014.07.025
9. Viejo, D., Garcia-Rodriguez, J., & Cazorla, M. (2014). Combining visual features and Growing Neural Gas networks for robotic 3D SLAM. *Information Sciences (NY), 276*, 174–185. https://doi.org/10.1016/j.ins.2014.02.053
10. Gálvez, A., & Iglesias, A. (2012). Particle swarm optimization for non-uniform rational B-spline surface reconstruction from clouds of 3D data points. *Information Sciences (NY), 192*, 174–192. https://doi.org/10.1016/j.ins.2010.11.007
11. Nevatia, R. (1976). Depth measurement by motion stereo. *Comput. Graph. Image Process., 5*, 203–214. https://doi.org/10.1016/0146-664X(76)90028-9
12. Mrovlje, J., & Vrani, D. (2008). Distance measuring based on stereoscopic pictures. In *9th Int. PhD Work. Syst. Control Young Gener. Viewp.*
13. Liu, Z., & Chen, T. (2009). Distance measurement system based on binocular stereo vision. In *Proceedings of the IJCAI international joint conference on artificial intelligence*; pp. 456–459.
14. Condotta, I. C. F. S., Brown-Brandl, T. M., Pitla, S. K., Stinn, J. P., & Silva-Miranda, K. O. (2020). Evaluation of low-cost depth cameras for agricultural applications. *Computers and Electronics in Agriculture, 173*, 105394. https://doi.org/10.1016/j.compag.2020.105394

15. Bertheloot, J., Cournède, P.-H., & Andrieu, B. PART OF A SPECIAL ISSUE ON FUNCTIONAL-STRUCTURAL PLANT MODELLING—NEMA, a functional-structural model of nitrogen economy within wheat culms after flowering. I. Model description. *Annals of Botany, 108*, 1085–1096. https://doi.org/10.1093/aob/mcr119
16. Vos, J., Evers, J. B., Buck-Sorlin, G. H., Andrieu, B., Chelle, M., & De Visser, P. H. B. (2010). Functional-structural plant modelling: A new versatile tool in crop science. *Journal of Experimental Botany, 61*, 2101–2115.
17. Zhu, C., Zhang, X., Hu, B., & Jaeger, M. (2008). Reconstruction of tree crown shape from scanned data. In *Proceedings of the lecture notes in computer science (including subseries Lecture Notes in Artificial Intelligence and Lecture Notes in Bioinformatics)*, Springer, Berlin, Heidelberg, Vol. 5093 LNCS, pp. 745–756.
18. Omasa, K., Hosoi, F., & Konishi, A. (2007). 3D lidar imaging for detecting and understanding plant responses and canopy structure. *Journal of Experimental Botany, 58*, 881–898.
19. Binney, J., & Sukhatme, G. S. (2009). 3D tree reconstruction from laser range data. In *Proceedings of the proceedings - IEEE international conference on robotics and automation*, pp. 1321–1326.
20. Chéné, Y., Rousseau, D., Lucidarme, P., Bertheloot, J., Caffier, V., Morel, P., Belin, É., & Chapeau-Blondeau, F. (2012). On the use of depth camera for 3D phenotyping of entire plants. *Computers and Electronics in Agriculture, 82*, 122–127. https://doi.org/10.1016/j.compag.2011.12.007
21. Andújar, D., Dorado, J., Fernández-Quintanilla, C., & Ribeiro, A. (2016). An approach to the use of depth cameras for weed volume estimation. *Sensors, 16*, 972. https://doi.org/10.3390/s16070972
22. Orriordan, A., Newe, T., Dooly, G., & Toal, D. (2019). Stereo vision sensing: Review of existing systems. In *Proceedings of the Proceedings of the international conference on sensing technology, ICST; IEEE computer society*, Vol. 2018-Decem, pp. 178–184.
23. Tran, V. L., & Lin, H. Y. (2018). A structured light RGB-D camera system for accurate depth measurement. *International Journal of Optics, 2018*. https://doi.org/10.1155/2018/8659847
24. Lachat, E., Macher, H., Mittet, M.-A., Landes, T., & Grussenmeyer, P. (2015). FIRST EXPERIENCES WITH KINECT V2 SENSOR FOR CLOSE RANGE 3D MODELLING. In *ISPRS - Int. Arch. Photogramm. Remote Sens. Spat. Inf. Sci.*, XL-5/W4, 93–100. https://doi.org/10.5194/isprsarchives-XL-5-W4-93-2015.
25. Kirsten, E., Inocencio, L. C., Veronez, M. R., Da Silveira, L. G., Bordin, F., & Marson, F. P. (2018). 3D data acquisition using stereo camera. In *Proceedings of the international geoscience and remote sensing symposium (IGARSS)*, Institute of Electrical and Electronics Engineers Inc., Vol. 2018-July, pp. 9214–9217.
26. Marin, G., Agresti, G., Minto, L., & Zanuttigh, P. (2019). A multi-camera dataset for depth estimation in an indoor scenario. *Open Data BR, 27*, 104619. https://doi.org/10.1016/j.dib.2019.104619
27. Endres, F., Hess, J., Sturm, J., Cremers, D., & Burgard, W. (2014). 3-D Mapping with an RGB-D camera. *IEEE Transactions on Robotics, 30*, 177–187. https://doi.org/10.1109/TRO.2013.2279412
28. Tran, T. M., Ta, K. D., Hoang, M., Nguyen, T. V., Nguyen, N. D., & Pham, G. N. (2020). A study on determination of simple objects volume using ZED stereo camera based on 3D-points and segmentation images. *International Journal of Emerging Engineering Research and Technology, 8*, 1990–1995. https://doi.org/10.30534/ijeter/2020/85852020
29. Sarker, M. M., Ali, T. A., Abdelfatah, A., Yehia, S., & Elaksher, A. (2017). A cost-effective method for crack detection and measurement on concrete surface. *ISPRS - International Archives of the Photogrammetry, Remote Sensing and Spatial Information Science, XLII-2/W8*, 237–241. https://doi.org/10.5194/isprs-archives-XLII-2-W8-237-2017
30. Burdziakowski, P. (2018). Low cost real time UAV stereo photogrammetry modelling technique-accuracy considerations. In *Proceedings of the E3S web of conferences, EDP sciences*, Vol. 63, p. 00020.

31. Gupta, T., & Li, H. (2017). Indoor mapping for Smart Cities-An affordable approach: Using kinect sensor and ZED stereo camera. In *Proceedings of the 2017 international conference on indoor positioning and indoor navigation, IPIN 2017*, Institute of Electrical and Electronics Engineers Inc., Vol. 2017-Janua, pp. 1–8.
32. Halmetschlager-Funek, G., Suchi, M., Kampel, M., & Vincze, M. (2019). An empirical evaluation of ten depth cameras: Bias, precision, lateral noise, different lighting conditions and materials, and multiple sensor setups in indoor environments. *IEEE Robotics and Automation Magazine, 26*, 67–77. https://doi.org/10.1109/MRA.2018.2852795
33. Wang, W., & Li, C. (2014). Size estimation of sweet onions using consumer-grade RGB-depth sensor. *Journal of Food Engineering, 142*, 153–162. https://doi.org/10.1016/j.jfoodeng.2014.06.019
34. Wang, Q., & Zhang, Q. (2013). Three-dimensional reconstruction of a dormant tree using RGB-D Cameras. In *Proceedings of the 2013 Kansas City, Missouri, July 21–July 24, 2013*, American Society of Agricultural and Biological Engineers, St. Joseph, MI, Vol. 2, pp. 1341–1350.
35. Andújar, D., Ribeiro, A., Fernández-Quintanilla, C., & Dorado, J. (2016). Using depth cameras to extract structural parameters to assess the growth state and yield of cauliflower crops. *Computers and Electronics in Agriculture, 122*, 67–73. https://doi.org/10.1016/j.compag.2016.01.018
36. van der Heijden, G., Song, Y., Horgan, G., Polder, G., Dieleman, A., Bink, M., Palloix, A., van Eeuwijk, F., & Glasbey, C. (2012). SPICY: Towards automated phenotyping of large pepper plants in the greenhouse. *Functional Plant Biology, 39*, 870. https://doi.org/10.1071/FP12019
37. Kazmi, W., Foix, S., & Alenyà, G. (2012). Plant leaf imaging using time of flight camera under sunlight, shadow and room conditions. In *Proceedings of the 2012 IEEE International Symposium on Robotic and Sensors Environments, ROSE 2012 – Proceedings*, IEEE Computer Society, pp. 192–197.
38. Swain, K. C., Zaman, Q. U., Schumann, A. W., Percival, D. C., & Bochtis, D. D. (2010). Computer vision system for wild blueberry fruit yield mapping. *Biosystems Engineering, 106*. https://doi.org/10.1016/j.biosystemseng.2010.05.001
39. Mai, C., Zheng, L., Sun, H., & Yang, W. (2015). Research on 3D reconstruction of fruit tree and fruit recognition and location method based on RGB-D camera. Nongye Jixie Xuebao/Transactions Chinese Soc. *Agricultural Machinery, 46*, 35–40. https://doi.org/10.6041/j.issn.1000-1298.2015.S0.006
40. Nguyen, T. T., Vandevoorde, K., Wouters, N., Kayacan, E., De Baerdemaeker, J. G., & Saeys, W. (2016). Detection of red and bicoloured apples on tree with an RGB-D camera. *Biosystems Engineering, 146*, 33–44. https://doi.org/10.1016/j.biosystemseng.2016.01.007
41. Tian, Y., Duan, H., Luo, R., Zhang, Y., Jia, W., Lian, J., Zheng, Y., Ruan, C., & Li, C. (2019). Fast recognition and location of target fruit based on depth information. *IEEE Access, 7*, 170553–170563. https://doi.org/10.1109/ACCESS.2019.2955566
42. Gené-Mola, J., Gregorio, E., Auat Cheein, F., Guevara, J., Llorens, J., Sanz-Cortiella, R., Escolà, A., & Rosell-Polo, J. R. (2020). Fruit detection, yield prediction and canopy geometric characterization using LiDAR with forced air flow. *Computers and Electronics in Agriculture, 168*, 105121. https://doi.org/10.1016/j.compag.2019.105121
43. Häni, N., Roy, P., & Isler, V. (2020). A comparative study of fruit detection and counting methods for yield mapping in apple orchards. *Journal of Field Robotics, 37*, 263–282. https://doi.org/10.1002/rob.21902
44. Bochtis, D. D., Sørensen, C. G., Jørgensen, R. N., Nørremark, M., Hameed, I. A., & Swain, K. C. (2011). Robotic weed monitoring. *Acta Agriculturae Scandinavica Section B Soil and Plant Science*, 61. https://doi.org/10.1080/09064711003796428
45. Anagnostis, A., Tagarakis, A. C., Asiminari, G., Papageorgiou, E., Kateris, D., Moshou, D., & Bochtis, D. (2021). A deep learning approach for anthracnose infected trees classification in walnut orchards. *Computers and Electronics in Agriculture, 182*, 105998. https://doi.org/10.1016/j.compag.2021.105998

46. Anagnostis, A., Asiminari, G., Papageorgiou, E., & Bochtis, D. (2020). A convolutional neural networks based method for anthracnose infected walnut tree leaves identification. *Applied Sciences, 10*. https://doi.org/10.3390/app10020469
47. Fu, L., Gao, F., Wu, J., Li, R., Karkee, M., & Zhang, Q. (2020). Application of consumer RGB-D cameras for fruit detection and localization in field: A critical review. *Computers and Electronics in Agriculture, 177*, 105687.
48. Kise, M., & Zhang, Q. (2008). Development of a stereovision sensing system for 3D crop row structure mapping and tractor guidance. *Biosystems Engineering, 101*, 191–198. https://doi.org/10.1016/j.biosystemseng.2008.08.001
49. Hornung, A., Wurm, K. M., Bennewitz, M., Stachniss, C., & Burgard, W. (2013). OctoMap: An efficient probabilistic 3D mapping framework based on octrees. *Autonomous Robots, 34*, 189–206. https://doi.org/10.1007/s10514-012-9321-0
50. De Silva, K. T. D. S., Cooray, B. P. A., Chinthaka, J. I., Kumara, P. P., & Sooriyaarachchi, S. J. (2019). *Comparative analysis of octomap and RTABMap for multi-robot disaster site mapping* (pp. 433–438). Institute of Electrical and Electronics Engineers (IEEE).
51. Wang, Z. (2020). Digital twin technology. In *Industry 4.0 - Impact on Intelligent Logistics and Manufacturing*, IntechOpen.
52. Glaessgen, E. H., & Stargel, D. S. (2012). The digital twin paradigm for future NASA and U.S. Air force vehicles. In *Proceedings of the Collection of Technical Papers - AIAA/ASME/ASCE/AHS/ASC Structures, Structural Dynamics and Materials Conference*.
53. Vachalek, J., Bartalsky, L., Rovny, O., Sismisova, D., Morhac, M., & Loksik, M. (2017). The digital twin of an industrial production line within the industry 4.0 concept. In *Proceedings of the proceedings of the 2017 21st international conference on process control, PC 2017*, Institute of Electrical and Electronics Engineers Inc., pp. 258–262.
54. Knapp, G. L., Mukherjee, T., Zuback, J. S., Wei, H. L., Palmer, T. A., De, A., & DebRoy, T. (2017). Building blocks for a digital twin of additive manufacturing. *Acta Materialia, 135*, 390–399. https://doi.org/10.1016/j.actamat.2017.06.039
55. Söderberg, R., Wärmefjord, K., Carlson, J. S., & Lindkvist, L. (2017). Toward a Digital Twin for real-time geometry assurance in individualized production. *CIRP Annual, 66*, 137–140. https://doi.org/10.1016/j.cirp.2017.04.038
56. Rosen, R., Von Wichert, G., Lo, G., & Bettenhausen, K. D. (2015). About the importance of autonomy and digital twins for the future of manufacturing. *Proceedings of the IFAC-Papers, 28*, 567–572.
57. Gabor, T., Belzner, L., Kiermeier, M., Beck, M. T., & Neitz, A. (2016). A simulation-based architecture for smart cyber-physical systems. In *Proceedings of the Proceedings – 2016 IEEE International Conference on Autonomic Computing, ICAC 2016*, Institute of Electrical and Electronics Engineers Inc., pp. 374–379.
58. Kang, H. S., Lee, J. Y., Choi, S., Kim, H., Park, J. H., Son, J. Y., Kim, B. H., & Noh, S. (2016). Do Smart manufacturing: Past research, present findings, and future directions. *International Journal of Precision Engineering and Manufacturing-Green Technology, 3*, 111–128. https://doi.org/10.1007/s40684-016-0015-5
59. Wang, L., Törngren, M., & Onori, M. (2015). Current status and advancement of cyber-physical systems in manufacturing. *Journal of Manufacturing Systems, 37*, 517–527. https://doi.org/10.1016/j.jmsy.2015.04.008
60. Tao, F., Zhang, L., & Laili, Y. (2015). *CLPS-GA for energy-aware cloud service scheduling* (pp. 191–224).
61. Zhang, M., Sui, F., Liu, A., Tao, F., & Nee, A. Y. C. (2020). Digital twin driven smart product design framework. In *Digital twin driven smart design* (pp. 3–32). Elsevier.
62. Kousi, N., Gkournelos, C., Aivaliotis, S., Giannoulis, C., Michalos, G., & Makris, S. (2019). Digital twin for adaptation of robots' behavior in flexible robotic assembly lines. In *Proceedings of the procedia manufacturing* (Vol. 28, pp. 121–126). Elsevier B.V.

63. Moshrefzadeh, M., Machl, T., Gackstetter, D., Donaubauer, A., & Kolbe, T. H. (2020). Towards a distributed digital twin of the agricultural landscape. *Journal of Digital Landscape Architecture, 5*, 173–118.
64. Alves, R. G., Souza, G., Maia, R. F., Tran, A. L. H., Kamienski, C., Soininen, J. P., Aquino, P. T., & Lima, F. (2019). A digital twin for smart farming. In *Proceedings of the 2019 IEEE global humanitarian technology conference, GHTC 2019*, Institute of Electrical and Electronics Engineers Inc.
65. Verdouw, C., & Kruize, J. W. (2017). Digital twins in farm management: Illustrations from the FIWARE accelerators SmartAgriFood and Fractals. In *Proceedings of the 7th Asian-Australasian conference on precision agriculture*, pp. 1–5.
66. Qian, W., Xia, Z., Xiong, J., Gan, Y., Guo, Y., Weng, S., Deng, H., Hu, Y., & Zhang, J. (2014). Manipulation task simulation using ROS and Gazebo. In *Proceedings of the 2014 IEEE international conference on robotics and biomimetics, IEEE ROBIO 2014*, Institute of Electrical and Electronics Engineers Inc., pp. 2594–2598.
67. Edwards, G., Christiansen, M., Bochtis, D., & Sørensen, C. (2013). A test platform for planned field operations using LEGO mindstorms NXT. *Robotics, 2*, 203–216. https://doi.org/10.3390/robotics2040203
68. Moysiadis, V., Tsolakis, N., Katikaridis, D., Sørensen, C. G., Pearson, S., & Bochtis, D. (2020). Mobile robotics in agricultural operations: A narrative review on planning aspects. *Applied Sciences, 10*, 3453. https://doi.org/10.3390/app10103453
69. Anagnostis, A., Benos, L., Tsaopoulos, D., Tagarakis, A., Tsolakis, N., & Bochtis, D. (2021). Human activity recognition through reccurent neural networks for human-robot interaction. *Applied Sciences*, 11(5), 2188. https://doi.org/10.3390/app11052188
70. Benos, L., Bechar, A., & Bochtis, D. (2020). Safety and ergonomics in human-robot interactive agricultural operations. *Biosystems Engineering, 200*, 55–72. https://doi.org/10.1016/j.biosystemseng.2020.09.009
71. Benos, L., Tsaopoulos, D., & Bochtis, D. (2020). A review on ergonomics in agriculture. Part I: Manual operations. *Applied Sciences, 10*, 1–21.
72. Benos, L., Tsaopoulos, D., & Bochtis, D. (2020). A review on ergonomics in agriculture. part II: Mechanized operations. *Applied Sciences, 10*, 3484.
73. Marinoudi, V., Sørensen, C. G., Pearson, S., & Bochtis, D. (2019). Robotics and labour in agriculture. A context consideration. *Biosystems Engineering, 184*, 111–121. https://doi.org/10.1016/j.biosystemseng.2019.06.013
74. Marinoudi, V., Lampridi, M., Sørensen, C. G., Pearson, S., & Bochtis, D. (2021). The agricultural occupations landscape in view of work automation. In *Bio-economy and agri-production* (pp. 289–348). Elsevier.
75. Shamshiri, R. R., Hameed, I. A., Pitonakova, L., Weltzien, C., Balasundram, S. K., Yule, I. J., Grift, T. E., & Chowdhary, G. (2018). Simulation software and virtual environments for acceleration of agricultural robotics: Features highlights and performance comparison. *International Journal of Agricultural and Biological Engineering, 11*, 15–31. https://doi.org/10.25165/ijabe.v11i4.4032
76. Tsolakis, N., Bechtsis, D., & Bochtis, D. (2019). Agros: A robot operating system based emulation tool for agricultural robotics. *Agronomy*. https://doi.org/10.3390/agronomy9070403
77. Lavrenov, R., Zakiev, A., & Magid, E. (2017). Automatic mapping and filtering tool: From a sensor-based occupancy grid to a 3D Gazebo octomap. In *Proceedings of the 2017 international conference on mechanical, system and control engineering, ICMSC 2017*, Institute of Electrical and Electronics Engineers Inc., pp. 190–195.
78. Guzman, R., Navarro, R., Beneto, M., & Carbonell, D. (2016). Robotnik—Professional service robotics applications with ROS. *Studies in Computational Intelligence, 625*, 253–288. https://doi.org/10.1007/978-3-319-26054-9_10

A Weed Control Unmanned Ground Vehicle Prototype for Precision Farming Activities: The Case of Red Rice

Aristotelis Koulousis, Damianos Kalaitzidis, Dimitrios Bechtsis, Christos Yfoulis, Naoum Tsolakis, and Dionysis Bochtis

1 Introduction

Red rice (*Oryza sativa f. spontanea*) is a challenge for rice farmers around the globe as it constitutes one of the most severe weed infestations in direct seeding rice systems (Fig. 1) [1]. The greatest peril of red rice associates with its inherent competition with commercially cultivated rice over sunlight and nutrients, mainly owing to its: (1) higher growth rates [2] and (2) greater resistance to environmental conditions during winter [3]. To that effect, red rice results in inferior quality, lower yields, and reduced prices of the contaminated cultivated paddy fields [4]. Indicatively, Durand-Morat et al. [5] estimated that red rice infestation in the USA resulted in rice yield losses of about 5.7 million tons with an environmental cost of US$457 million for the period 2002–2014, under a modest infestation scenario.

Notwithstanding the ramifications of red rice, effective field-management strategies and mechanisms against red rice infestations are lacking [6]. The simplest option to tackle red rice is its manual removal from the field. However, this process is arduous and requires significant workforce, while it must be performed at the end of the rice harvesting season where temperatures are typically high. Furthermore, the use of herbicides and other chemicals to control red rice during the production of

A. Koulousis · D. Bechtsis (✉) · C. Yfoulis
Industrial Engineering and Management Department, International Hellenic University (IHU), Thessaloniki, Greece
e-mail: dimbec@ihu.gr

D. Kalaitzidis · N. Tsolakis · D. Bochtis
Institute for Bio-Economy and Agri-Technology (iBO), Centre for Research and Technology Hellas (CERTH), Thessaloniki, Greece

D. D. Bochtis et al. (eds.), *Information and Communication Technologies for Agriculture—Theme III: Decision*, Springer Optimization and Its Applications 184, https://doi.org/10.1007/978-3-030-84152-2_7

Fig. 1 Red rice (Oryza sativa f. spontanea) plant

commercially cultivated rice is not preferable as herbicides typically damage the cultivars [7], considering that the commercially cultivated rice has the same genetic structure with red rice. The development of herbicides' resistance of the weeds further eliminates the potential of the herbicides-based solution [8]. Furthermore, the use of typical mechanical means to control red rice (e.g., moldboard plough, blade harrow) has been proven to be time consuming and less effective compared to chemical treatments [9].

In Greece, the total cultivation of rice in 2018 was 28,374 hectares resulting in the production of 233,361 tons [10]. In particular, in 2018 northern Greece accounted for more than 87% of the national rice production. However, in northern Greece, red rice infestation in commercial paddy fields is documented to result in yield losses of about 46% and 58% for the Ariette and the Thaibonnet cultivars, respectively [11].

The aim of this research is to automate the process of weed control in rice paddies with a special focus on the red rice plant. The concept of the developed process is based on the automation of the manual methods used to control red rice, i.e., the application of herbicides during the period when red rice has a significant height difference from commercially cultivated rice, so that red rice is accessible from above. In particular, a prototype robot for weed control in paddy fields was developed; the robot system consists of an Unmanned Ground Vehicle (UGV) which is further equipped with a specially designed rod mechanism covered with a sponge that is saturated with herbicides and uses a sensor-based control mechanism for applying the herbicides only to the upper part of red rice plants. The rod constantly balances at a predefined height while it remains horizontally aligned to the ground for having a direct contact only with the red rice plants, despite the geomorphology of the terrain. The rod is covered with a sponge that is soaked with herbicides in order to apply it to the upper part of the red rice plants. It should be highlighted that red rice plants have a height difference from the commercial rice plants, and this enables the use of a control system that balances the rod in order to retain a predefined height and be horizontally aligned to the landscape while the autonomous vehicle navigates across the paddy field.

The remainder of this research is organized as follows: Sect. 2 provides a literature review and introduces the need for the developed robot system. Thereafter, Sect. 3 describes the related materials and methods for the prototype robot system design. Section 4 presents the prototype, while Sect. 5 discusses the results generated at both simulation and real-world environments. Finally, conclusions, limitations, and future research recommendations are discussed in Sect. 6.

2 Literature Review

Rice is one of the leading crops in terms of production rates [12] and is providing sustenance and nutrition for billions of people with a global production of about 470 million tons [13]. Technological solutions and autonomous vehicles are being widely used in agricultural robotics [14, 15] in order to increase production and overall efficiency, and this trend is expanded to paddy fields. Mousazadeh [16] presented a review of navigation methods for agricultural autonomous vehicles. The author stated that autonomous vehicles could improve efficiency in agricultural activities and he also considered navigation methods in paddy fields. Unmanned Surface Vehicles (USV) that navigate on the water surface have gained attention as Puddled Transplanted Rice (PTR) methods have been widely used. Liu et al. [17] focused on a Global Positioning System (GPS) airboat navigation in paddy fields. They stated that agricultural machinery could damage the soil and destruct the navigation path due to the narrow spacing between the crops. They designed an airboat vehicle for autonomous fertilizing and herbicides applications and conducted field experiments. Chen et al. [18] developed a machine vision method for a weeding robot in paddy fields. Rice farming is significant in Japan and new methods are proposed for optimizing the cultivation, transplanting, and harvesting using technological tools. A camera was mounted on a small boat and sample images were captured when the rice height was approximately 10–65 cm above the water surface. The camera identified the travel directrix at the paddy field as well as the level of growth of the rice plants. Liu et al. [19] developed an USV for navigating in paddy fields. The USV used a GPS for safely navigating across the paddy fields. The surface vehicle used a predefined map and followed the predetermined route using the GPS.

Over the last years, Direct-Seeded Rice (DSR) is gaining attention over the PTR as it is more environmentally friendly and its cultivation requires reduced water volumes, energy, and even labor sources. However, DSR rice production is prone to yield decreases due to weeds and red rice is considered an emerging constraint to global rice production [20]. Autonomous vehicles are regarded as an effective technological solution that could increase efficiency to agricultural activities. Bechar and Vigneault [21] studied agricultural robots for field operations and compared robot performance and technical progress. They stated that more robust and reliable solutions are needed for addressing real-world field conditions. Han et al. [22] developed an application of a 3D tractor for auto-guided tillage operations. Both computer simulation methods and a test platform were developed for exploring navigation methods.

A range of autonomous vehicles and external robotic mechanisms have been developed and tested by researchers to support operations in rice paddies. Luo and Zhang [23] studied navigation methods for autonomous vehicles and designed a controller for headland turning in rice paddies. Kim et al. [24] built a robotic system for weeding between rice lines in harsh surface environments, while Yin et al. [25] developed an automatically quidded rice transplanter using an automated steering and speed control mechanism and different types of headland turnings in a real-world implementation. A Proportional–Integral–Derivative (PID) controller was used for minimizing the error levels. Nagasaka et al. [26] stated that there is a trend in Japan for diligently working in paddy fields. The authors proposed a high-tech six-row rice transplanter with the use of a real-time kinematic GPS and a fiber optic gyroscope. They also proposed a path planning algorithm that starts to operate near the entrance of the field and covers the field after segmenting it according to the transplanter's width. Finally, Nishiwaki et al. [27] developed a weed sprayer in rice fields that could be embedded to autonomous vehicles.

In this study, a robot system is proposed that comprises of an autonomous vehicle with an embedded rod mechanism for weed control in rice paddies. In particular, a case study is provided that develops: (1) a 3D simulation of the proposed robot prototype and (2) a corresponding real-world testbed for evaluating the performance of the prototype system.

3 Materials and Methods

In order to investigate the pragmatic perspective of research outputs, the object of scrutiny should be the analysis of real-world case studies [28]. The research design and the case study relevant to this research are detailed below.

3.1 Research Design

The objective of this research is to develop an operational prototype technology system that can be used to efficiently tackle the red rice challenge. In this regard, we adopted the systematic assessment view of autonomous vehicles, prior to their implementation in actual working environments that was proposed by Tsolakis et al. [29]. In particular, this research focuses on the simulation and physical testbed application stages so that a proportional engagement to the cyber and physical spaces is achieved to support the proactive analysis of the developed prototype robot system.

In particular, to test the performance of the fully automated weed control UGV, a simulation model was first developed and then a real-world prototype was manufactured and calibrated. At the simulation level, the Gazebo simulation environment was used to test the UGV's navigation and the additional components that were mounted on the UGV. Regarding the prototype robot system, this uses the TurtleBot3 Burger autonomous vehicle and a custom-made rod mechanism that was embedded in the vehicle. For both the simulation modeling effort and the real-world prototype, we used the Robot Operating System (ROS) open source framework and we developed a routing algorithm for navigating the UGV across paddy fields.

3.2 Case Study

The developed technological solution was first tested at a simulation environment and then a corresponding prototype system was developed and piloted in a recreated landscape. In order to autonomously apply herbicides on the top of the red rice foliage, a rotating aluminum roller coated with a porous absorbent material (e.g., sponge) is required to avoid dripping of the liquid. To that end, it is necessary to control two physical quantities: (1) slope and (2) height. This could be possible with the use of a specially designed rod mechanism. Slope control can be performed via a sensor that provides rotation data in degrees, while height control data can be gathered via a laser-type system. The respective sensor-based control algorithms were implemented on a single programmed microcontroller and were used to control the movement of several servomotors that balance the rod mechanism at the desirable position. The calibration of these systems is facile while always delivering consistent results.

4 Robot Prototype Development

The proposed technological prototype integrates a rod mechanism, attached to an autonomous UGV, for implementing a weed control mechanism that could be applied at rice paddy fields. The developed technological solution includes two components, namely: (1) the rod mechanism and (2) the autonomous vehicle.

4.1 Rod Mechanism

The rod mechanism consists of three components, namely: (1) mechanical; (2) electronic; and (3) software. First, the mechanical component includes the rod and two supporting parts that enable the up-down and the left-right movement of the rod (Fig. 2a). Servo motors are mounted to the body of the mechanism and are responsible for the kinematics of the moving and rotating parts. A gear system provides the transmission of the moving power to the rod and controls the movement in great precision (Fig. 2b).

The rod is covered with a porosity sponge which supplies the herbicides through the tubes. A container, positioned in front of the vehicle, provides the proper amount of herbicides to the rod using a pump and two plastic flexible connectors at the edges of the rod. Bearings ensure the firm connection of the tubes to the rod's edges while the rod is moving. Through this operation, herbicides are applied at the upper part of the red rice which is significantly higher than the sowed rice during the specific cultivation stage.

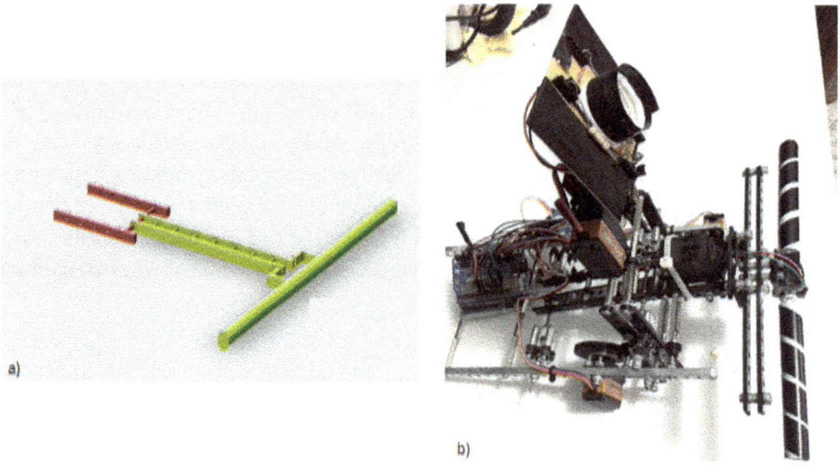

Fig. 2 Mechanical components of the rod mechanism: (**a**) basic concept design; (**b**) prototype technology

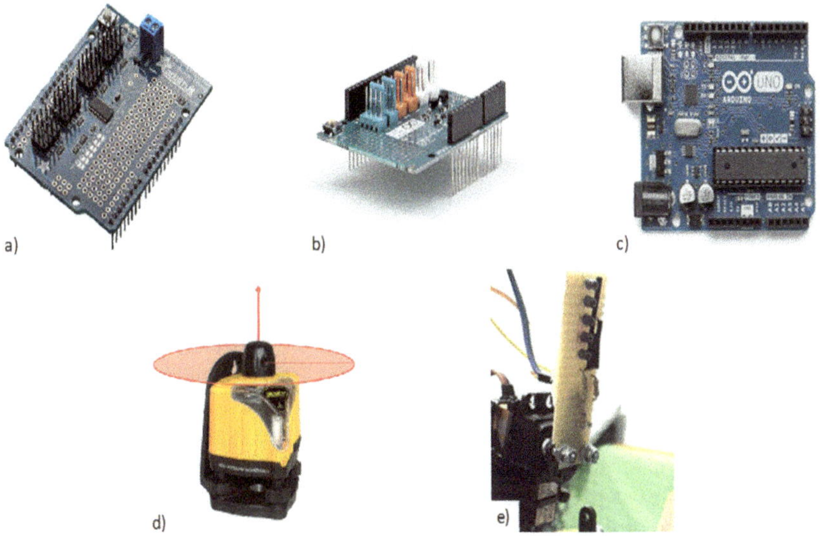

Fig. 3 Electronic components of the rod mechanism: (**a**) servo motor driver; (**b**) gyroscope sensor; (**c**) microcontroller; (**d**) laser transmitter; (**e**) laser receiver

The electronic component of the prototype technology consists of a laser transmitter and a laser receiver, a microcontroller, a gyroscope sensor, and four servo motors (Fig. 3). The laser beam is used in order to provide a universal zero level at the system and thus keep the mechanism at the specified height using a servo motor. Two servo motors are also connected at the microcontroller in order to horizontally balance the rod so that it remains parallel to the ground, while an additional servo motor is used to provide the herbicides at the sponge that is attached to the rod. The electronic part includes two main subsystems that control the rod's movement. The first subsystem uses an independent laser transmitter outside from the field in order to have a reference point and retains the rod mechanism at the specified height in order to affect only the red rice plants. The vehicle moves across the field and reads the laser beam from the receiver mounted on the vehicle. The receiver is located at the highest point of the vehicle, and this is considered as the mechanism's zero level. When the vehicle's receiver is off the zero level (either on a positive or a negative range), a servo motor is used as an actuator to respond to the disturbances in order to retain the rod's height and regulate it at the zero level. The second subsystem keeps the rotating rod stable in a horizontal position while the vehicle slopes to the right or to the left due to the harsh terrain. In order to accomplish this task, a microcontroller receives field data from a gyroscope sensor and responds to real-time disturbances.

The software component implements the necessary regulation algorithms by retrieving real-time data from the sensors and controlling the rod's mechanism. The regulation is performed using three PID controller loops, one for altitude (height) control and two for slope control (covering two different axes, i.e., pitch and roll). The height controller is based on the signal of the laser receiver, while the slope controller reads the gyroscope position. Depending on the deflection-error, the slope controller calculates a rotation value in order to keep the rod in parallel to the ground.

4.2 Autonomous Vehicle

The autonomous vehicle can navigate paddy fields and uses the proposed rod mechanism in an attempt to effectively and efficiently reduce the effects of red rice plants. The rotating rod mechanism is located at the upper part of the vehicle. Due to the specific implementation, only the necessary amount of herbicides is soaked to the rod and is applied directly to the upper part of the red rice plants so that the cultivated rice plants are not affected. In order to assess the performance of the real-world application, a 3D simulation model (Fig. 4a) and a real-world prototype (Fig. 4b) were devised.

At a virtual environment, a UGV was designed in the Gazebo simulation environment [30]. The authors developed a urdf.xacro file, which contains the description of the robot and the rod mechanism, and simulated the exact physical characteristics of the UGV (i.e., dimensions, weight, inertia of individual components). Therefore, it was possible to observe, in a simulation environment, how the physical characteristics of the UGV affect the movement of the vehicle and the rod mechanism (Fig. 5a). Furthermore, the simulation helped to proactively assess the technical and operational feasibility of a respective real-world prototype. The rod mechanism was also designed and integrated to the UGV and several tests were conducted (i.e., straight and curved movement) in a manual mode so that the optimal vehicle's speed was determined. At the optimal speed, the vehicle was expected to have a better response while not being affected by the physical characteristics of the rod mechanism (i.e., weight, dimensions). The rod mechanism was expected to apply the herbicides to the red rice plants while remaining at a certain height and being horizontally aligned to the field's landscape (Fig. 5b). Thereafter, an algorithm

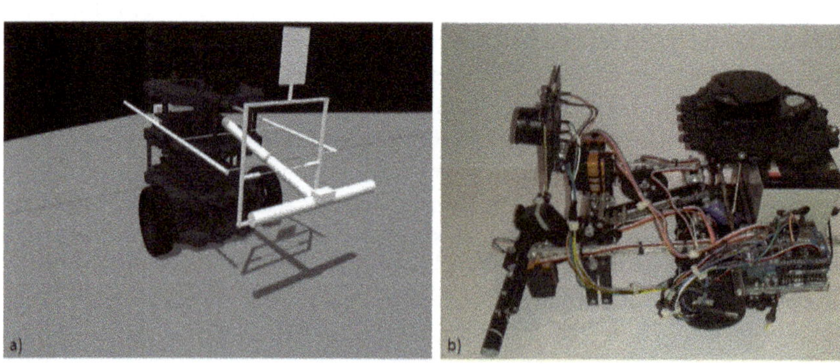

Fig. 4 Autonomous vehicle: (**a**) 3D simulation model; (**b**) real-world prototype

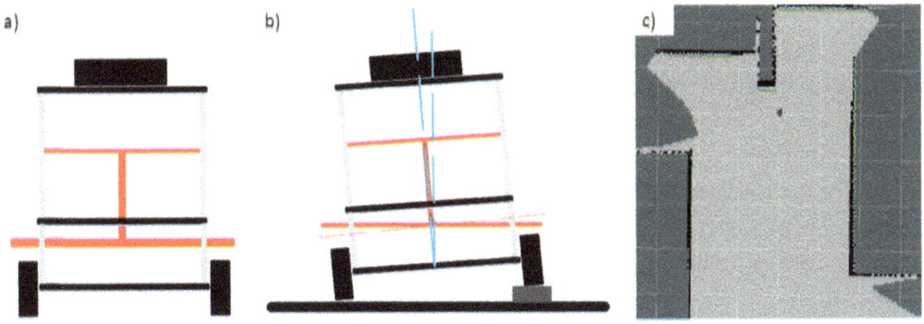

Fig. 5 Simulation analysis of autonomous vehicle: (**a**) prototype technology structure; (**b**) vehicle's movement and balance of rod mechanism; (**c**) vehicle's autonomous navigation, mapping, and obstacle avoidance

was implemented at the ROS Navigation Stack [31] to test the autonomous navigation, mapping, and obstacle avoidance of the vehicle in the simulation environment (Fig. 5c). For the autonomous navigation tests, the dynamic window approach planner was used [32] and customized in order to achieve efficient driving conditions.

Following the tests at the cyber space, a real-world robot prototype was developed in order that the behavior of the proposed robot system was evaluated under physical conditions. In terms of configuration, the TurtleBot3 Burger vehicle was customized to integrate the rod balance mechanism and a battery for power supply. The TurtleBot3 Burger's Lidar sensor was placed at a high point so that the mapping and obstacle avoidance functions were not blocked due to the installed electronic components. It should be stated that the real-world prototype implements all the algorithms that were developed at the simulation environment.

5 Results

In this section, we first insert the simulation modeling results in terms of the examined path planning algorithm. Thereafter, the testbed's application is presented by investigating the UGV's movement across a recreated harsh landscape.

5.1 Simulation Environment

At the Gazebo simulation environment, the movement of the UGV with the embedded rod mechanism was tested, while it was following a predefined path at the simulated agri-field. The simulation model showcased the manner that the physical characteristics of the rod mechanism could affect the autonomous vehicle's navigation, prior to proceeding in real-world tests. The actual rod mechanism uses an external laser beam that stabilizes the rod at a certain height above the rice crop; however, the laser beam was not modeled at the simulation tests and the positioning of the embedded mechanism was manually controlled as the overall movement and the balance of the UGV at the field were tested.

The simulation results highlighted the small size of the robot compared to the weight of the rod mechanism; the rod mechanism deviates from the robot's center of gravity. There are some difficulties in balancing the UGV while it starts and stops, as the articulated structural synthesis of the robot applies several forces to the joints and to the UGV. Such forces are exerted due to the motors torque and the harsh landscape. In order to overcome the abovementioned difficulties, the UGV's speed was significantly reduced and as a result the actuation forces that were generated during the movement were decreased. Furthermore, the addition of extra weight near the UGV's center of gravity significantly reduced the forces during the robot's movement. After the rod's mechanism actual position was finalized, it was experimentally validated that the overall navigation of the UGV and the efficiency of the planners that were developed were not being affected by the addition of the rod mechanism. Minor observed slips and deviations from the actual path were considered insignificant as they were automatically corrected by the navigation tools.

The path planning algorithm implements a point-to-point navigation, as the UGV forms the boustrophedon movement for the coverage of known spaces [33] and constantly changes direction in order to fully cover the agri-field's area. The path is segmented according to the direction change points and the UGV follows the path by traveling from point-to-point. The shortest path at every point-to-point movement is created by using the DWA planner and the UGV travels along a straight line. In case a moving obstacle blocks the movement of the robot system, the local planner is activated in order to bypass the obstacle and return to the initially planned path. Figure 6 presents the boustrophedon-type algorithm along with the actual vertical and horizontal length of the path. The vertical path is equal to the y-axis of the agri-field. The horizontal path equals to half of the rod's mechanism length.

In case of a maneuver to bypass an obstacle, the UGV will not be depositing herbicides to a small part of the agri-field. However, the presence of obstacles across

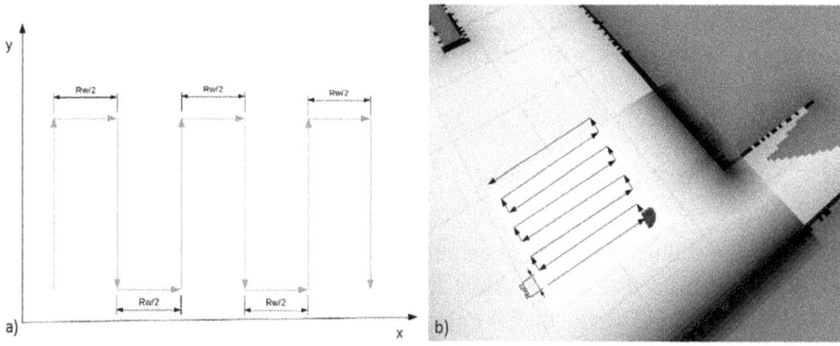

Fig. 6 Path planning: (**a**) point-to-point navigation; (**b**) boustrophedon movement

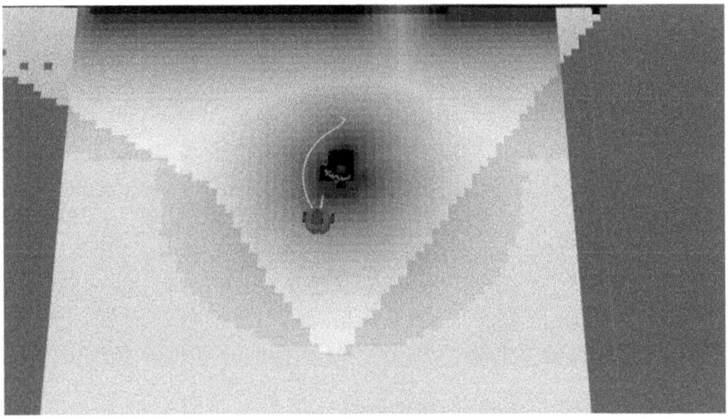

Fig. 7 Robot maneuver to avoid an obstacle within the simulated paddy field

rice paddies is rarely the case. In Fig. 7, this scenario is presented, where pesticides are not applied at a certain agri-field area during the obstacle avoidance movement. The dimensions of the UGV, along with the rod mechanism, define the exact area that would be left untapped.

5.2 *Real-World Environment*

To test the overall performance of the real-world prototype robot system, a boustrophedon-type algorithm was developed as a path planning algorithm. Based on the simulation findings, this could ensure that the rod mechanism would cover a given agri-field to effectively and efficiently apply pesticides to the red rice crop. To test the responsiveness and functionality of the rod mechanism's balancing, an external laser was placed at the appropriate height. It was experimentally observed that the rod mechanism easily retained its horizontal position during the robot's movement, even in case some obstacles were placed on the path (the obstacles had an average height of 1–3 cm) in order to emulate a harsh landscape.

Figure 8 presents snapshots of the UGV's movement at a recreated harsh landscape. At the experimental setting of the testbed, terrain anomalies were introduced in order to observe the system's ability to respond to vehicle's changes in altitude and in tilting. Following a number of tests, we concluded that by decreasing the navigation speed of the vehicle the balancing mechanism has a quicker and smoother response. In particular, it was observed that:

- at high vehicle's velocities, the system could not adequately adjust to the predefined height
- at low UGV's velocities, the robot avoided any obstacles in a smoother way

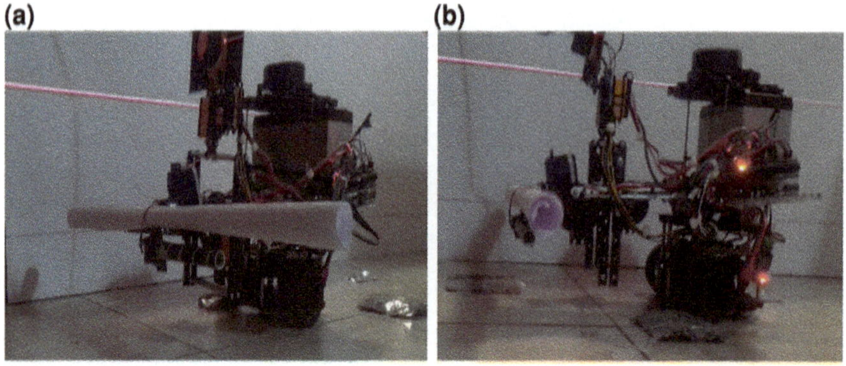

Fig. 8 Experimental tests of the robot's movement

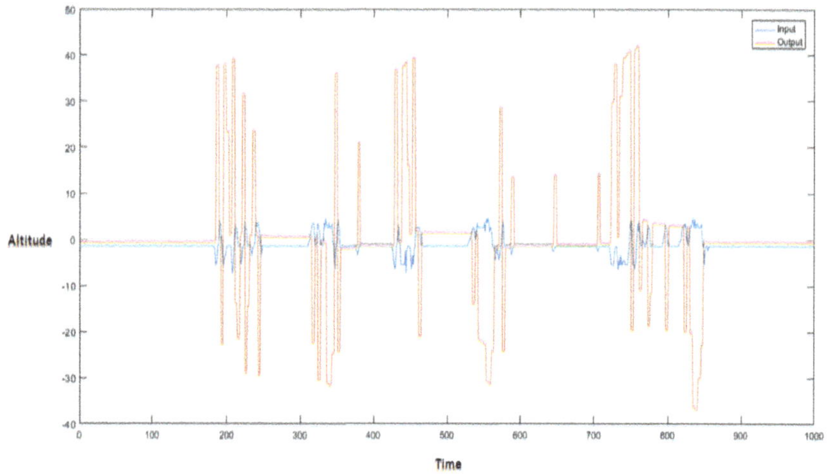

Fig. 9 The height control signals of the prototype rod mechanism

The PID values which were used at the prototype technology system were empirically selected. These values were initially $K_p = 8$, $K_i = 4$, *and* $K_d = 0.2$ for the height controller. In addition, a switching gains method was used to avoid the trembling of the mechanism near the equilibrium position, where the PID values were updated to $K_p = 8$, $K_i = 4$, *and* $K_d = 0$. Similar PID controllers were used for the slope control which were also calibrated empirically.

Figure 9 presents the experimental values related to the altitude control as a function of time (in samples, where each sample corresponds to 10 ms). The altitude of the prototype rod mechanism is the "Input" (in blue), while the control effort is the "Output" (in red), to reflect their role in the PID controller block. In particular:

- the "Input" of the PID controller block is the feedback signal, i.e., the laser beam value returned by the laser receiver.

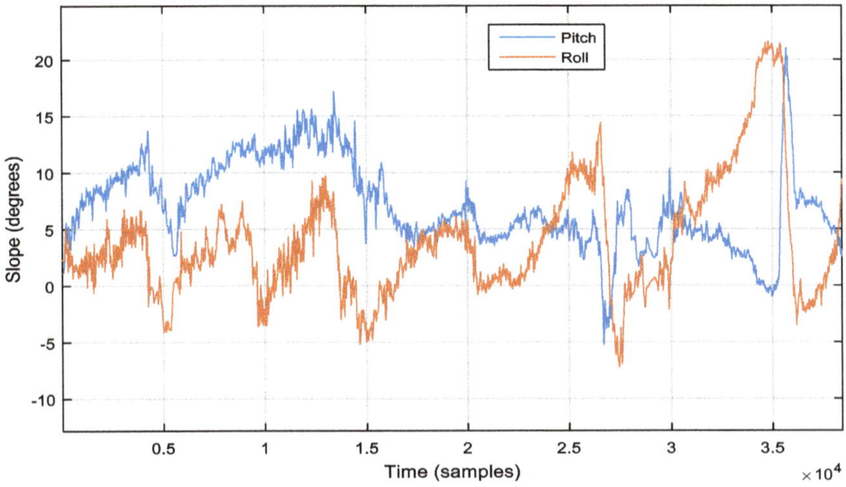

Fig. 10 The slope control signals of the prototype rod mechanism

- the "Output" of the PID controller block is the control signal used to rotate the servo motor and retain the rod mechanism at the desired height.

In Fig. 10, the slope control experimental values are presented as a function of time (in number of samples). The slope values (pitch and roll) returned by the gyroscope are depicted in degrees. The represented data history corresponds to several disturbances introduced by navigation on harsh terrain, while the prototype's slope is regulated in both axes around the desired equilibrium of 0°.

6 Conclusions

Red rice (Oryza sativa f. spontanea) is a wild variety of the commercial rice. Compared to common rice, red rice grows faster and is more resilient to weather conditions. To that end, the population of this hardy plant gradually increases on an annual basis, thus farmers lose most of their yield. Unless farmers adopt effective procedures for removing red rice from the field, the yield production loses will be significant. Although red rice is edible with significant organoleptic properties, the harvest of red rice is practically almost impossible due to its seeds. Red rice seeds are thin and cover entirely the plant, hence prohibiting their collection. Instead when machineries are about to collect them, the seeds are scattered all over the field; this fosters their distribution to the field rather than their collection.

The use of state-of-the-art technology in precision farming activities could boost productivity and provide added value results to daily farming activities. Rice is one of the leading crops in production rates and is providing sustenance and nutrition for billions of people. The implications of red rice on the total production of commercially cultivated rice are widely documented, and it is considered as a major weed challenge as conventional and chemical-based solutions cannot address red rice due to its genetic similarity to the commercial rice.

The presented research developed and experimentally tested a prototype robot system, involving a UGV for weed control in paddy fields, with a particular focus on the balancing of a specially designed rod mechanism used for applying herbicides only to the top of red rice plants. Experimental results both at a cyber and physical space helped to calibrate the control parameters of the automation system that retains the rod mechanism's height at the appropriate height. Nevertheless, at this preliminary stage of development, only the overall UGV's movement on the field was tested along with the balancing of the rod mechanism. In addition, the TurtelBots3 Lidar sensor was placed at a higher point so that the mapping and obstacle avoidance functions were not blocked due to the additional components installed. This is not a viable solution as usually Lidar sensors are installed near the ground in order to avoid collisions with low obstacles.

The commercial application of this system could limit the impact of red rice on the cultivated rice varieties thus ensuring stability in the commodity's supply downstream agri-food value networks.

References

1. Delouche, J. C., Burgos, N. R., Gealy, D. R., de San Martín, G. Z., Labrada, R., Larinde, M., & Rosell, C. (2007). Weedy rices—Origin, biology, ecology and control. In *Rome: FAO Plant Production and Protection Paper 188*. Food and Agriculture Organization of the United Nations.
2. Estorninos, L. E., Gealy, D. R., Gbur, E. E., Talbert, R. E., & McClelland, M. R. (2005). Rice and red rice interference. II. Rice response to population densities of three red rice (Oryza sativa) ecotypes. *Weed Science, 53*, 683–689.
3. Baek, J. S., & Chung, N. J. (2012). Seed wintering and deterioration characteristics between weedy and cultivated rice. *Rice, 5*(1), 21.
4. Fisher, A., & Ramirez, A. (1993). Red rice (Oryza sativa): Competition studies for management decisions. *International Journal of Pest Management, 39*, 133–138.
5. Durand-Morat, A., Nalley, L. L., & Thoma, G. (2018). The implications of red rice on food security. *Global Food Security, 18*, 62–75.
6. Busconi, M., Rossi, D., Lorenzoni, C., Baldi, G., & Fogher, C. (2012). Spread of herbicide-resistant weedy rice (red rice, Oryza sativa L.) after 5 years of Clearfield rice cultivation in Italy. *Plant Biology, 14*, 751–759.
7. Sudianto, E., Beng-Kah, S., Ting-Xiang, N., Saldain, N. E., Scott, R. C., & Burgos, N. R. (2013). Clearfield® rice: Its development, success, and key challenges on a global perspective. *Crop Protection, 49*, 40–51.
8. Owen, M. D. K., & Zelaya, I. A. (2005). Herbicide-resistant crop and weed resistance to herbicides. *Pest Management Science, 61*, 301–311.
9. Ferrero, A., Vidotto, F., Balsari, P., & Airoldi, G. (1999). Mechanical and chemical control of red rice (Oryza sativa L. var. sylvatica) in rice (Oryza sativa L.) pre-planting. *Crop Protection, 18*(4), 245–251.

10. Hellenic Statistical Authority. (2020). Areas and Production, 2018: Table 02a. Cereals for grain. Areas and production, by Region and Regional Unity. Retrieved August 25, 2020, from https://www.statistics.gr/en/statistics/-/publication/SPG06/-.
11. Eleftherohorinos, I., Dhima, K. V., & Vasilakoglou, I. B. (2002). Interference of red rice in rice grown in Greece. *Weed Science, 50*(2), 167–172.
12. Trade and Markets Division, Food and Agriculture Organization of the United Nations. (2016). Rice market monitor 2016, Volume XIX, Issue No. 3. Retrieved August 25, 2020, from http://www.fao.org/fileadmin/templates/est/COMM_MARKETS_MONITORING/Rice/Images/RMM/RMM-Dec16_H.pdf.
13. IRRI, Africa Rice, and C.I.A.T. (2010, November). Global Rice Science Partnership (GRiSP).
14. Bechtsis, D., Moisiadis, V., Tsolakis, N., Vlachos, D., & Bochtis, D. (2019). Unmanned ground vehicles in precision farming services: An integrated emulation modelling approach. In M. Salampasis & T. Bournaris (Eds.), *Information and communication technologies in modern agricultural development. HAICTA 2017. Communications in computer and information science* (Vol. 953, pp. 177–190). Springer.
15. Tsolakis, N., Bechtsis, D., & Bochtis, D. (2019a). Agros: A robot operating system based emulation tool for agricultural robotics, Open Access. *MDPI Agronomy, 9*(7), 403.
16. Mousazadeh, H. (2013). A technical review on navigation systems of agricultural autonomous off-road vehicles. *Journal of Terramechanics, 50*(3), 211–232. https://doi.org/10.1016/j.jterra.2013.03.004
17. Liu, Y., Noguchi, N., & Liang, L. (2019). Development of a positioning system using UAV-based computer vision for an airboat navigation in paddy field. *Computers and Electronics in Agriculture, 162*(April), 126–133. https://doi.org/10.1016/j.compag.2019.04.009
18. Chen, B., Tojo, S., & Watanabe, K. (2003). Machine vision for a micro weeding robot in a paddy field. *Biosystems Engineering, 85*(4), 393–404. https://doi.org/10.1016/S1537-5110(03)00078-3
19. Liu, Y., Noguchi, N., & Yusa, T. (2014). Development of an unmanned surface vehicle platform for autonomous navigation in paddy field. In IFAC *Proceedings Volumes (IFAC-PapersOnline)* (Vol. 19, Issue 3). IFAC. https://doi.org/10.3182/20140824-6-za-1003.00616.
20. Ziska, L. H., Gealy, D. R., Burgos, N., Caicedo, A. L., Gressel, J., Lawton-Rauh, A. L., Avila, L. A., Theisen, G., Norsworthy, J., Ferrero, A., Vidotto, F., Johnson, D. E., Ferreira, F. G., Marchesan, E., Menezes, V., Cohn, M. A., Linscombe, S., Carmona, L., Tang, R., & Merotto, A. (2015). Weedy (Red) rice. An emerging constraint to global rice production. In *Advances in agronomy* (Vol. 129). Elsevier. https://doi.org/10.1016/bs.agron.2014.09.003
21. Bechar, A., & Vigneault, C. (2017). Agricultural robots for field operations. Part 2: Operations and systems. *Biosystems Engineering, 153*, 110–128. https://doi.org/10.1016/j.biosystemseng.2016.11.004
22. Han, X., Kim, H. J., Jeon, C. W., Moon, H. C., Kim, J. H., & Yi, S. Y. (2019). Application of a 3D tractor-driving simulator for slip estimation-based path-tracking control of auto-guided tillage operation. *Biosystems Engineering, 178*, 70–85. https://doi.org/10.1016/j.biosystemseng.2018.11.003
23. Luo, X., & Zhang, Z. (2007). DGPS navigation control system for rice transplanter. In *ASABE annual international meeting, Minnesota*, 17–20 June 2007.
24. Kim, GH, Kim, SC, Hong, YK, Han, KS & Lee, SG 2012, A robot platform for unmanned weeding in a paddy field using sensor fusion in 2012 IEEE International Conference on Automation Science and Engineering: Green Automation Toward a Sustainable Society, CASE 2012., 6386466, IEEE International Conference on Automation Science and Engineering, pp. 904-907.. https://doi.org/10.1109/CoASE.2012.6386466
25. Yin, X., Du, J., Geng, D., & Jin, C. (2018). Development of an automatically guided rice transplanter using RTK-GNSS and IMU. *IFAC-PapersOnLine, 51*(17), 374–378. https://doi.org/10.1016/j.ifacol.2018.08.193

26. Nagasaka, Y., Umeda, N., Kanetai, Y., Taniwaki, K., & Sasaki, Y. (2004). Autonomous guidance for rice transplanting using global positioning and gyroscopes. *Computers and Electronics in Agriculture, 43*(3), 223–234. https://doi.org/10.1016/j.compag.2004.01.005
27. Nishiwaki, K., Amaha, K., & Otani, R. (2004). *Development of nozzle positioning system for precision sprayer*. Paper presented at the Automation Technology for Off-Road Equipment. Kyoto: Japan.
28. Gibbert, M., Ruigrok, W., & Wicki, B. (2008). What passes as a rigorous case study? *Strategic Management Journal, 29*(13), 1465–1474.
29. Tsolakis, N., Bechtsis, D., & Srai, J. S. (2019b). Intelligent autonomous vehicles in digital supply chains: From conceptualisation, to simulation modelling, to real-world operations. *Business Process Management Journal, 25*(3), 414–437.
30. Takaya, K., Asai, T., Kroumov, V., & Smarandache, F. (2016). Simulation environment for mobile robots testing using ROS and Gazebo, 2016 20th International Conference on System Theory, Control and Computing (ICSTCC). *Sinaia, 2016*, 96–101. https://doi.org/10.1109/ICSTCC.2016.7790647
31. ROS Wiki. (2020). Navigation Tutorials. Retrieved from http://wiki.ros.org/navigation/Tutorials/RobotSetup.
32. Fox, D., Burgard, W., & Thrun, S. (1997). The dynamic window approach to collision avoidance. *IEEE Robotics & Automation Magazine, 4*(1), 23–33.
33. Choset, H. (2000). Coverage of known spaces: The boustrophedon cellular decomposition. *Autonomous Robots, 9*, 247–253.

Decision-Making and Decision Support System for a Successful Weed Management

P. Kanatas, I. Travlos, A. Tataridas, and I. Gazoulis

1 Introduction

1.1 The Introduction of Decision Support Systems (DSS) in Agriculture

The rapid technological developments during the last decades have caused radical changes in agriculture. A new era has emerged in which the key to success depends on elaborated decision-making [1]. Decision Support Systems (DSS) are computer-based tools which can aid complex decision-making process [2, 3] in many sectors including agriculture by providing suggestions, and advices, while stakeholders are occasionally allowed to insert their own data, observations, and evaluations [4]. The introduction of DSSs in agriculture targets to optimize farmers' decision-making process by integrating many forms of information and data which are valuable in crop cultivation process [5]. Moreover, the need for a substantial swift of modern agriculture attributes to more sustainable and ecologically based principles, can be achieved by the introduction of DSSs in modern agriculture [6].

According to DSS methodologies, various types of data are collected and analyzed to create crop management strategies which are less harmful to the environment and more profitable as compared to the corresponding strategies included in conventional agriculture [7]. For example, the suggested application rates for pesticides, herbicides, or fertilizers from a DSS can be significantly lower compared to

P. Kanatas (✉)
Department of Crop Science, University of Patras, Mesolonghi, Greece

I. Travlos · A. Tataridas · I. Gazoulis
Department of Crop Science, Agricultural University of Athens, Athens, Greece

D. D. Bochtis et al. (eds.), *Information and Communication Technologies for Agriculture—Theme III: Decision*, Springer Optimization and Its Applications 184, https://doi.org/10.1007/978-3-030-84152-2_8

the recommended label rates [8–11]. In addition, a DSS can forecast future disease outbreaks in croplands, rank lists for pest or weed control treatments and maybe suggest the integration of both chemical and non-chemical-based practices in favor of crop development and productivity [10–14]. However, the success of a DSS is determined by multiple factors. First of all, the cooperative efforts of a team of experts along with technical, technological, and financial resources are required whereas every DSS development process should be directly related to the realistic demands of the farmers [15]. Furthermore, expert knowledge, management models, timely and accurate data capture along with continuous feedback from the farmer after every growing season, pose as key factors affecting DSSs' effectiveness [16, 17].

The first step for the function of a DSS, designed for use in agriculture, includes the collection of data from the environment where the crop is grown. Such data might refer to climatic conditions, crop growth parameters, or the presence of pests and weed flora composition and density. Then the incorporated models in the DSS analyze the data and rank a list of suitable treatments for each case. DSS models can include rules, tips, and schedules of management, mathematical equations, and combinations of decision aids [18].

1.2 The Development of DSSs in Terms of Weed Management

Weeds are considered as the most important threat to agricultural production since they affect the crop production indirectly, by competing with the crop for natural resources, sheltering crop pests, reducing crop yields and quality, and subsequently increasing the cost of processing [19]. There is evidence indicating that weed infestations are the reason for a 5%, 10%, and 25% loss in agricultural production in most countries, less developed countries, and least developed countries [20]. DSS methodologies might be a useful tool in terms of developing more efficient and ecologically friendly weed management strategies in modern days' agriculture.

In general, the function of DSSs is mainly based on complex mathematical models. The models can analyze data regarding the composition of the weed flora in the field, weed density as well as weed and crop growth parameters. Afterwards, estimations for the possible yield losses due to competition can be made, and optimum weed control practices can be suggested [21]. Various DSSs have been developed to help farmers take the appropriate decisions for optimum weed control in their fields, e.g., GESTINF [22], CPO [9], RIM, WeedSOFT® [23], Weed Manager [14], and InfloWeb (Terres [24]). Nowadays, the most common target of DSSs in weed management is to test different herbicide treatments. The adoption of DSS methodologies can reduce herbicide application rates at a significant point and also provide optimum weed control options according to the special needs of each farmer ([9, 12, 25]; Terres [24]). Although in some cases DSSs were also developed to provide information for long-term strategies, either the goals were not fully accomplished or the DSS strategies were focused on the management of specific weed species [14].

The benefits as well as the limitations and the future perspectives of introducing DSSs in weed management practices have been recently reviewed [26]. Developing functional DSSs for weed management might be even more challenging than in other sectors of agriculture. In particular, weed populations are heterogeneous in composition and distribution in the field level and any suggested control practice that does not address this heterogeneity has a degree of inefficiency [27–29]. From a weed science perspective, soil and climatic conditions, weed abundance and ecology, the complex composition of the weed flora and the level of competition between crops and weeds in each agricultural field are noticeable factors affecting DSSs' success in weed management [26]. In addition, DSS's developers should also take into account information regarding farm size and structure, land use, available technologies, biodiversity indices, environmental parameters, and economical aspects [30]. At any case, the communication between farmers, and crop advisors through field and online surveys as well workshops and group meetings has been recognized as the fundamental point of the development of any DSS developed for weed management [31].

The main objective of the present study was to present key factors affecting decision-making process that need to be taken into account before developing a DSS for weed management.

2 Factors Affecting Decision-Making Process in DSSs for Weed Management

2.1 Weed Emergence and Weed Flora Composition in the Field

Most Decision Support Systems (DSSs) designed for weed control use the cost of control treatments as well as their efficacy. Decision-making is based on an optimum economic balance between the economic cost of control options and the yield losses expected for the crop under various cases, including the case where the weeds are left uncontrolled [30]. Considering an agronomic aspect, it is essential to have knowledge of the composition and the abundance of the weed flora in a field and also to have precise information about the growth stage of the crop and the weeds at a specific time for predicting yield losses [26]. Moreover, the majority of farmers lack the knowledge of the biology of weeds, soil seedbank dynamics, the factors affecting weed emergence in the field, how quick do they emerge and grow, or how the yield losses due to competition can be predicted [31]. To achieve making as realistic simulations as possible regarding the impact of either weed infestation or control options on crop's productivity, such agronomic parameters should be taken into account and both farmers and advisers should be provided with such valuable information. The biological models used in DSSs should be relevant to the timing of emergence of the crop and weeds, to weed and crop growth, and to the impact of treatments on weed growth [14].

Capturing data providing knowledge about weed biology as well as weed density in the field is a major priority before adopting any weed management practice. The need for weed control, the selection of weed control method, and the appropriate timing for weed control can be determined by a DSS if such data are captured [32]. DSS models simulate germination, emergence, development, and growth, driven by daily weather data. Capturing weather data is of high importance since they affect weed abundance and emergence in the field and set the level of infestation in an agricultural area. For instance, the effects of climate change and the increase in the average air temperature by 2.5 °C along with control practices on *Lolium multiflorum* (Lam.) have been reported in the study of Pagnoncelli et al. [33]. Furthermore, the study of Sousa-Ortega et al. [34] demonstrated that different emergence models should be developed for different climatic conditions regarding the annual grass species *Lolium rigidum* (Gaud.). Climatic data including maximum and minimum temperatures, radiation, rainfall, and evapotranspiration should be captured as precisely as possible and their effects on weeds and crops should be evaluated since different weed species are characterized by variance regarding their adaptation in different regions. Moreover, climatic conditions can affect yield loss estimations carried out by DSSs models. The experiments of Wright and Wilson [35] showed that yield loss from *Gallium aparine* (L.) competition was not remarkable, while the DSS model overestimated yield losses. These authors noticed the surprisingly low yield loss recorded due to *G. aparine* competition and attributed this result to summer drought. In addition, seed production of annual weeds is influenced by extrinsic factors such as climatic conditions [36, 37]. Predicting weed emergence is vital to make any assessments for crop yield losses due to competition and if the predictions are as realistic as possible, they can increase the efficiency of new introduced smart-farming technologies such as the adoption of a DSS. A crucial point for proper decision-making in weed management is to predict as realistically as possible weed emergence and the composition of the weed flora in the field in order to select the optimum methods for weed control.

The longevity of weed seeds into the soil profiles is attributed to the phenomenon of dormancy that prevents seed germination even under ideal environmental conditions [38]. Dormancy is distinguished into primary and secondary dormancy and the sequence between the two types of the phenomenon is known as dormancy cycling [39, 40]. Seed dormancy along with the environmental factors affecting it, and subsequently seed germination, are key factors to take into account to predict weed emergence under field conditions. Many studies have revealed the effects of environmental factors on weed seed germination and weed emergence (Table 1). Moreover, seed germination and weed emergence are traits varying on species level [49]. For example, the mean base soil temperatures required for germination of summer annuals *Amaranthus albus* (L), *Amaranthus palmeri* (S. Wats.), *Digitaria*

Table 1 Key factors affecting weed emergence, the composition of the weed flora in the field and, consequently, decision-making process for weed management

Factor	Relative weed species	Reference
Air temperature	*L. multiflorum*	Pagnoncelli et al. [33]
Air temperature, soil moisture	*L. rigidum*	Sousa-Ortega et al. [34]
Tillage	*Cirsium arvense, S. arvensis*	Thomas et al. [41]
Fertilization	*A. ludoviciana*	Lack et al. [42]
Fertilization	*A. retroflexus*	Blackshaw and Brandt [43]
Soil pH	*C. canadensis*	Nandula et al. [44]
Soil moisture	*A. myosuroides*	Swain et al. [45]
Soil temperature	*E. crus-galli, S. glauca*	Masin et al. [46]
Tillage	*B. hordeaceus, V. arvensis*	Crawley [47]
Light	*A. fatua, L. amplexicaule*	Milberg et al. [48]
Fertilization, Soil moisture	*Gallium aparine, A. fatua*	Wright and Wilson [35]

sanguinalis (L.), *Echinochloa crus-galli* (L.) P. Beauv., *Portulaca oleracea* (L.), and *Setaria glauca* (L.) was by approximately 40% higher as compared to the value required for the germination of the winter annuals *Hirschfeldia incana* (L.) and *Sonchus oleraceus* (L.) [46]. In addition, there is evidence that the duration as well as the number of diurnal soil temperature fluctuation cycles can exert a great influence on seed dormancy status of several species [50, 51].

The seed germination response to the soil water potential of wild plants is also correlated with the soil water status in their natural habitats [52]. For example, sufficient water potential has been noticed to increase the seed production of noxious weed species such as *Alopecurus myosuroides* (Huds.) seeds [45]. The models which aim to predict weed germination and emergence need to record seed germination in a wide range of water potentials since seeds of various weed species require different values of water potential in order to germinate [53]. Numerous studies have been done to develop emergence prediction models [54–56]. The combined effects of soil temperature and soil moisture content on weed emergence have been used in hydrothermal time models suitable for various weed species [57].

Light is also an environmental factor, affecting seed germination and weed emergence. Lower than 16% germination percentages were recorded for *Avena fatua* (L.) and *Lamium amplexicaule* (L.) in light, whereas seed germination of *Sonchus asper* (L.) Hill reached the level of 99% due to exposure to the light [48]. The range of soil pH values is another crucial factor affecting decision-making process since it influences seed germination and weed emergence. For example, *Eupatorium adenophorum* (Spreng.) can germinate in a range of pH between 5 and 7 whereas *Conyza canadensis* (L.) Cronquist is able to germinate in a pH range from 4 to 10 [45, 58]. There is also evidence showing soil gaseous environment and nitrates' soil content can regulate weed seed dormancy and germination [38, 59, 60].

To optimize the decision-making process, knowledge about the history of an agricultural area is required and farmers should be interviewed very carefully so as to provide researchers with valuable information regarding the dominant weeds of the field and all the agronomic practices applied during the previous years. This is such an important point, given that the biodiversity of weed communities in a cropland can be affected from the adoption of different agronomic practices (Table 1). In a recent study, Travlos et al. [61] highlighted that tillage and fertilization put a significant impact on weed abundance and the composition of the weed flora in the field. Surveys allow researchers to collect users' inputs with a large audience, especially when using semi-open questions that encourage explanations by farmers [62]. Crop management strongly affects the abundance and diversity of weeds and changes in tillage are likely to have a clear effect on the community structure [63]. There is evidence suggesting that timing of tillage interferes with the timing of seed germination and weed emergence acting as a filter of weed communities [64].

Crawley [47] revealed that *Papaver dubium* (L.) and *Viola arvensis* (Murray) emergence was increased by 62.5% and 72%, respectively, due to fall cultivation. In the same study, spring cultivation increased *Bromus hordeaceus* (L.), and *Galinsoga parviflora* (Cav.) emergence by 88% and 92.5%. In the study of Thomas et al. [41], it was observed that the perennial weeds *Cirsium arvense* and *Sonchus arvensis* were associated with reduced- and zero-tillage systems, while annual species were associated with a range of tillage systems. Armengot et al. [65] noticed that weed abundance was higher under reduced tillage systems as compared to conventional tillage systems. Given such evidence, it is well established that tillage affects weed emergence and, thus, the estimations made from models of DSSs. This fact indicates that the tillage practices carried out in a field should be taken into account in decision-making process in weed management.

Fertilization does also affect either weed emergence or abundance and thus, the decision-making process carried out by a DSS model. The kind of fertilization inputs varies between conventionally and organically managed systems and weed species abundance is considered to be favored under organic fertilization [66]. Weed abundance can be negatively affected by increased phosphorus levels whereas species evenness can be negatively correlated with the ratio of organic carbon to total nitrogen in the soil [67]. Fertilization affects growth of weeds and subsequently their competitive ability. For instance, the competitive ability of *Amaranthus retroflexus* was enhanced with higher levels of nitrogen [43]. The added amounts of nitrogen increased the impact of *Avena ludoviciana* Durieu competition on wheat yield performance [42]. Blackshaw and Brandt [43] also mentioned that some species were more competitive under increased phosphorus levels whereas Freyman et al. [68] observed that phosphorus levels did not put any significant impact on the composition of the weed communities observed in the field.

2.2 The Impact of Weed Competition on Crops' Productivity

The seedling emergence pattern should be detailed as well since it informs the timing of weed control treatments with regard to weed growth stage and the necessity of control. Except for the importance of predicting and modeling weed emergence in the field, the efficacy of decisions for weed control such as herbicidal or mechanical treatments are dependent on weeds' growth stages and on the growth stages of the crop. Hence, it is vital for models of DSSs to provide sufficient information regarding weed and crop growth. Zadoks scale is suitable to be used from the models for grass weeds and cereal crops whereas the number of leaves on the entire plant has been used for broadleaf weeds [69, 70]. The next parameter which should be assessed is the interference between the crop and the weeds infesting it, as well as the level of this interference. Plant height and leaf area of both crop and weeds influence daily light competition and the consequent increase in dry matter is partitioned to the different plant organs through photosynthesis. Weeds' along with crop growth estimations are related to the green area per seedling at the time of full hypocotyl or cotyledon expansion and density [69]. There is evidence that it is possible to transform such values predicted regarding the growth of crops and weeds into weed impact indicators to make more simple as well as realistic approaches for the effects of competition on crop's productivity [71, 72].

The levels of weed infestation during crucial crop growth stages, the growth stages of weeds along with the interactions between weeds and other pests are factors affecting the impact of competition on crops' productivity (Table 2). As a result, these factors should be taken into account for making more reliable estimations during decision-making process [31, 73]. According to Kropff and Spitters [75], the estimations of yield losses attributed to competition with a specific weed species are dependent on crop yield under weed free conditions, weed and crop growth, and the damage coefficient of the weed species. Berti et al. [74] suggested that the use of DSSs in areas where both weed composition and densities are known already is more preferable than in areas where weed density is very high or the weed flora is particularly aggressive and difficult to control with the available herbicides (e.g., common lambsquarters or velvetleaf in soybean).

Late-emerging weeds cause less interference with the crop and experience lower reproductive success than early emerging cohorts; however, seed production from late season escapees must be considered when developing effective management tactics [37, 74, 76]. This was also highlighted from the results of the field tests carried out by Rydahl [9] in cereal fields in Denmark where some weed species germinated in large numbers late during the spring season. In the same study, seed production from the late season emerging weeds indicated that supplementary herbicide application was required for their control [9]. Furthermore, weed

Table 2 Key factors affecting the impact of weed competition on crops' growth, yield performance, and, consequently decision-making process for weed management

Factor	Reference
Weed and crop growth	Colbach et al. [71]
Weed density, weed biomass, weed—insect interactions	Mézière et al. [73]
Seed production from late—season emerging weeds	Norsworthy et al. [37]
Daily weed density prediction	Masin et al. [32]
Weed—crop interference, weed and crop growth	Benjamin et al. [69]
Timing of weed emergence, weed and crop growth, weed seedbank dynamics	Parsons et al. [14]
Weed abundance, weed density	Berti et al. [74]
Seed production from late—season emerging weeds	Rydahl [9]
Timing of weed emergence, weed density	Berti and Zanin [22]
Number of seeds per weed plant, weed biomass, fertilization, soil moisture	Wright and Wilson [35]
Crop yield under weed-free conditions, weed and crop growth, weed leaf area, weed density, weed emergence	Kropff and Spitters [75]

population models have proved to be very useful for evaluating weed management strategies and developing decision-making tools [77]. Several authors have used models of weed populations to investigate the economics of weed control [78, 79] and a typical approach is to attempt setting thresholds for intervention [23, 74, 80, 81].

Given that a model of a DSS has made realistic yield loss estimations and has proposed the most appropriate solutions, an attractive option is to create decision trees using the simulatory results of a DSS. This format of presenting the results can offer support to users when deciding which cultural practices to implement and how to combine them in order to achieve their objectives [82]. It is obvious that presenting the possible outcome of the suggested practices in this way has the advantage of illustrating the results in a familiar format to what is already used for guiding decision-making in weed management. Suitable presentation of the solutions would help dealing with the major challenge to convince farmers to be familiarized with DSSs [83]. DSSs are still not broadly accepted and used by farmers [74, 84, 85]. For providing complete information to farmers, emergence models often fail in their predictions regarding the actual weed density under real-field conditions. The survey process of DSSs could be used to relate the actual weed density with a specific emergence percentage and predict the daily weed emergence by using only a proportion for making estimations. The emergence models which are able to predict the daily density of weeds with a single survey, they can result in saving time and money and providing specific information to farmers [32]. These methods can be improved when weed mortality during the crop cycle is considered since not all the weeds that emerge survive until crop harvest [86].

3 Factors Affecting Decision-Making Either in the Short- or in the Long-Term Period and Future Challenges of DSSs Developed for Weed Management

Most of the existing DSSs for weed control are based on the chemical approach, aiming for an optimization of the total herbicide use [22, 80]. To provide the information required for chemical weed control within a growing season, a typical DSS contains models which can estimate crop yield losses. The estimations are carried out according to the growth of both crops and weeds and the effect of competition on crop's yield performance. In other words, the most common method to optimize herbicide application models is to estimate first the need for weed control, where the efficacy of the applications depends on the field-specific weed population since each of the present weeds causes a certain magnitude of yield loss. The information provided by such models is the percentage of actually emerged weeds out of the total number of plants that might emerge during the cropping cycle [32]. The proposed treatments are evaluated not only for their efficacy but also for their application cost and their economic beneficial effect on crop's yield [22, 80]. The effects of pre- and post-drilling cultivations on weed emergence should be separated because then the impact of herbicides on crop's yield can be evaluated very quickly [14].

An example of a DSS used for optimal herbicide applications is the Danish Crop Protection Online (CPO) which provides information about the need for weed control and the optimal herbicide combination at the optimal rates [9, 87, 88]. The herbicidal treatments as proposed from CPO are estimated according to herbicide dose–response curves, which are parameterized with experimental data from many years of herbicide testing under field conditions and data as derived from the manufacturers of the herbicides [89]. The impact of the DSS technology on reducing herbicide use up to 40%, as compared to reference herbicide treatments, has been reported by Jørgensen et al. [12]. Efficient weed control has been recorded under 30% reduced herbicide use in cereal crops while in sugar beet, the advice from a DSS decreased the herbicide use in a similar way [9]. Moreover, herbicide inputs can be decreased up to 60% in cereals, if spraying is carried out according to the estimations of a DSS model [90]. Other studies also reported that potential savings in the number of herbicides sprayed to control broadleaf and grass weed species ranged between 60 and 77% [91].

The frequency of herbicide treatments can be calculated using either the standard or the maximum registered application rates [89]. The effects of herbicide applications can be estimated by reducing weed growth attributes according to the sensitivity of each weed to the herbicide, and data provided by the manufacturers. The efficacy of the herbicides used can be evaluated at each of the timings within each application period to find out which application timing resulted in the minimum yield loss for the crop [14]. Within a growing season, the observed differences regarding the effects of the herbicidal treatments in cost, treatment frequency, crop's yield, and weed coverage at harvest can be analyzed using linear mixed models (R Development Core [92]). Pairwise differences between the means of the treatments can be evaluated using post hoc t-tests with adjustment for multiplicity [93]. Herbicide mixtures can provide the level of control needed on all weeds in the cases where a single herbicide cannot. DSSs models should also assess if the synergistic and antagonistic effects between the mixture components are significant [94]. All the permitted mixtures should be evaluated in the same way as any single herbicide. A single herbicide or mixture that is not superior to a cheaper one should be eliminated [9]. The final outcome can be a list for each period containing applicable herbicides and mixtures with their best timings. Linear optimization should be used to find the least expensive combination of herbicides and application rates that accomplishes the target of sufficient and broad spectrum weed control. This combination of bio-economic DSSs and emergence models can help in optimizing the timing of weed control operations, improving efficacy of control strategies and reducing herbicide use through implementation of Integrated Weed Management [95].

The adoption of DSSs can result in reduced herbicide inputs in agriculture [9, 12]. However, only a few models have been developed to assist in a range of management practices within the cropping system [96, 97]. Nowadays, the development of DSSs should target, in the long-term period, to enhance weed management with strategies involving lower reliance on the use of herbicides. This is quite a challenging issue given the complexity of the weed flora composition under the real-field conditions and also the different biology of each species infesting a crop [98]. Under this context, the validation of a DSS should be based upon evaluation of various case studies in different temporal and spatial scale, taking into consideration the specific characteristics of each farming system such as the farming resources, the land use, and the cropping scenario [99]. The output of DSSs could be characterized as: (1) short-term, which consists of preventive and reactive measures in the crop production and weed management, aiming to provide solutions for specific problems in the short-run; (2) mid-term, that includes single and more complex tactics; and (3) long-term, which is analytical frameworks and strategies that also depend on environmental and economic components at the farm level [30].

Theoretical analyses of weeding strategies have demonstrated the need to consider the long-term consequences over at least 10 years of the decisions rules established to optimize sustainable systems [100]. Long-term weed management requires a good knowledge of weed growth and interactions between crops and weeds according to ecological and agronomical components, and this is something that is often underestimated by DSS developers (Fig. 1). For weed species difficult to control in crops where they cause damage, management should target in reducing the seed bank of these species during previous crops [101]. Another approach is to design or modify a crop rotation system in line with weed control targets [30].

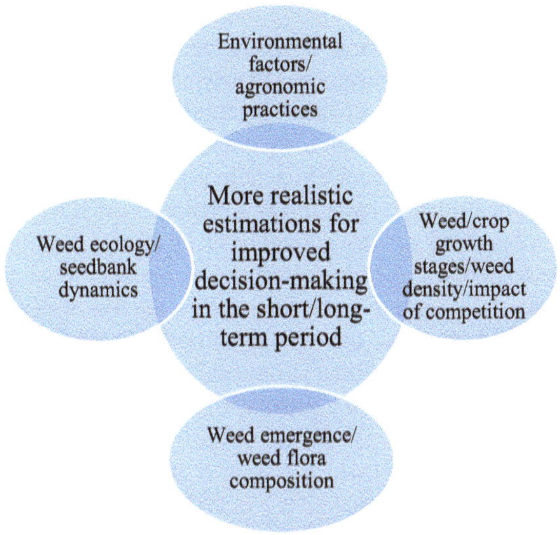

Fig. 1 This figure represents a holistic approach of the key parameters determining the reliability of estimations for improved decision-making in terms of weed management both in the short- and the long-term period

Among the fundamentals of developing more efficient weed management systems is to estimate the soil seedbank dynamics in the field [102] (Fig. 1). Seedbank dynamics could be evaluated given the expected weed density in the absence of weed control measures [69]. The numbers of seeds either in the shallow or in the deeper layers at the start of the growing season as well as the number of seeds in the shallow layer buried to the deeper layers have been used in equations to simulate the soil seedbank dynamics of an agricultural area. The number of germinable seeds, the number of seeds per plant, the proportion of seeds lost by herbivory and the proportion of ungerminated seeds that die in the soil per season are among the parameters used to estimate the number of seeds either in the shallow or in the deeper soil profiles at the beginning of a season [96].

Any weed species that reached the seed set stage and propagated should be monitored the next year more precisely to prevent unsuitable applications or detrimental yield losses due to new infestations. Under this context, a DSS that deals with weed control can provide valuable data in the decision-making process that are supplementary to the farmers' experience and not substitutes [30]. The same authors also suggested that farmers might reap the benefits of the adoption of DSSs that provide cost-effective long-term strategies for weed management and have multidisciplinary participatory characteristics. Moreover, estimating the effects of environmental factors affecting the seed germination and emergence for the dominant weed species of a cultivated area can give the ability to the researchers to predict weed infestation in a field and make the appropriate decisions required for adequate weed management in the field. Such variables influence both weeds and crops since pre-emergence stages are governed by soil structure, temperature, and water potential whereas post-emergence processes are regulated by light availability and air temperature [31]. Adopting biological models to estimate seedbank dynamics can optimize the choice of sowing date, type of cultivation, and control options to find out the most effective weed management strategies throughout a crop rotation which will result in the greatest net returns [14]. The same authors have reported that decision variables should be the sowing time, the type of cultivation and the cost of herbicide or mechanical operations proposed for weed control but not crops' selection in a rotation since this is a choice of the user [14].

Another issue that should be taken into account in decision-making for the long-term period is herbicide resistance. Farmers have a high interest in investigating risky or uncertain events [103]. With the rise of herbicide-resistant weeds in global cropping systems, the need to communicate integrated, diverse weed control practices to ensure cropping sustainability is increasingly important [37]. To mitigate the risks of increasing the populations of herbicide-resistant weeds, each weed can be classified as eliminated, susceptible, moderately susceptible, moderately resistant, or resistant to each herbicide applied at each of weed growth stages [14]. "Ryegrass Integrated Management" is a DSS for testing the performance of *L. rigidum* control practices, a weed notorious for evolving multiple resistance patterns [104, 105]. The model of this DSS estimates grain yield of the crop, the cost of *L. rigidum* management, and treatments' efficacies to develop management strategies for a decade, allowing the comparison of the control options by both agronomic and economic aspects [106]. The formulation of this model is based on the principle where the current weed population is equal to the product of the previous population size multiplied by rates reflecting intrinsic and extrinsic factors [21]. However, developing an actual model assessing for the risk of evolving herbicide resistance patterns would require much further modeling effort and the reintroduction of herbicide modes of actions, which were purposefully removed [107].

At any case, environmental factors, agronomic practices, elements of weed ecology, weed seedbank dynamics into the soil, weed growth as well as the impact of weed competition during crucial growth stages of the crop need to be taken into account to make more realistic estimations (Fig. 1).

DSSs designed for weed management are dynamic programs that are altered occasionally depending on the specific characteristics and needs at the farm level. Although their use in terms of weed management has great potential, strong challenges need to be addressed in the future (Fig. 2). In particular, given the extreme complexity of the weed flora and its variability between field to field, capturing accurate and representative data is quite a difficult and challenging task. Another important issue is the development of models with the ability to predict weed emergence and soil seed bank dynamics under real-field conditions for developing management strategies in the long-term period. The creation of such models might be complex since the effects of both environmental factors and agronomic practices on weed biology should be evaluated through extended experimentation. In addition, before implying DSS methodologies for weed management, the interference among

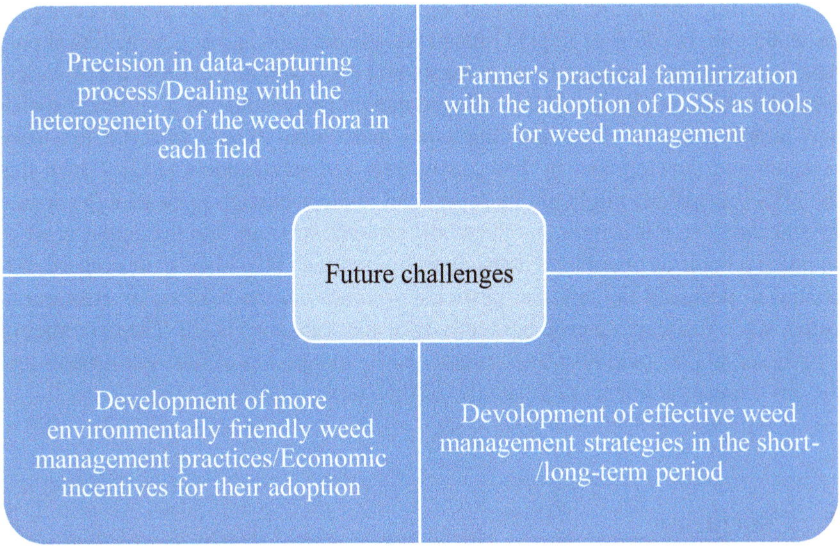

Fig. 2 Future challenges that need to be addressed in order to optimize the adoption of DSSs in the field of weed management

each crop and various noxious weed species should be assessed because some weed species can be more harmful for a specific crop than others. For instance, *A. sterilis* when recorded at density of 120 panicles per square meter has been indicated as more harmful for barley crop as compared to grass species *Phalaris minor* (Retz.) when recorded at density of 400 panicles per square meter, respectively [108]. Moreover, the impact of competition is directly related to the growth stages of weeds and crops. For instance, the critical time of competition in soybean is placed during V2 and V4 growth stages [109]. Defining the importance of such factors affecting the impact of competition on crops' productivity is a crucial point that needs to be further investigated in various crops and under different soil and climatic conditions.

Another important issue that needs to be addressed in the near future is that the majority of farmers might be frustrated at the rate of change in technology, which can discourage the adoption of the technology as a whole. This is not unusual with new, evolving technologies; however, it has been widely recognized as a major limitation in the adoption of remote sensing and DSSs [110]. In one of the most recent studies, Colas et al. [31] highlighted the importance of carrying out field surveys, hosting workshops, and group meetings in order to communicate with farmers and help them familiarize with the adoption of DSS methodologies for weed management. Jørgensen et al. [12] did also emphasize the need to give economic incentives to the farmers in order to persuade them use such new technologies instead of their usual agronomic practices.

Persuading farmers to use DSSs in their crop production processes is a complex issue, though. Booltink et al. [111] stated that the decision and policy making should have a participatory approach and a transparent open structure in order that low- and high-tech technologies are well developed and properly validated in different spatial and temporal scales. The modeling input and output should undergo extensive research and development by both developers and stakeholders [112]. Given that the DSS suggestions might not be robust and realistic enough to be widely adopted by stakeholders, it is certain that the whole systems structure and the output must be as much reliable as possible to assist decision-making [113, 114]. In this regard, it is crucial to persuade farmers that pesticide volumes can be reduced or even eliminated, while maintaining crop productivity at adequate level [115]. The outcomes of web-based DSSs could also be strengthened by integration of risk assessments and the environmental impact of pesticides [116].

4 Conclusion

To conclude, there are several factors that need to be taken into account before developing an effective DSS for weed management. Attention should be paid to the effects of environmental factors and agronomic practices on weed emergence and the composition of the weed flora in an agricultural field. To develop any weed management strategy, it is essential to have a deep understanding of weed biology and ecology. The biological traits of weeds, weed growth, the impact of weed competition during crucial growth stages for the crop should be estimated in order to optimize decision-making process. A better understanding of weed seedbank dynamics into the soil would help experts develop DSSs able to provide management strategies also in the long-term period. These objectives need to be addressed in the future along with communicating with farmers to get them familiarized with the adoption of DSS methodologies. Further research and extended experimentation are needed in order to develop effective DSSs in terms of weed management under different soil and climatic conditions, always according to the special needs of each farmer.

References

1. Fountas, S., Carli, G., Sørensen, C. G., et al. (2015). Farm management information systems: Current situation and future perspectives. *Computers and Electronics in Agriculture, 115*, 40–50. https://doi.org/10.1016/j.compag.2015.05.011
2. Marakas, G. M. (2003). *Decision support systems in the 21st century* (Vol. 134, 2nd ed.). Prentice Hall.
3. Sprague RH, Carlson ED (1982) Building effective decision support systems. Prentice Hall, .

4. Rupnik, R., Kukar, M., Vračar, P., Košir, D., Pevec, D., & Bosnić, Z. (2018). AgroDSS: A decision support system for agriculture and farming. *Computers and Electronics in Agriculture, 161*, 260–271. https://doi.org/10.1016/j.compag.2018.04.001
5. Adebayo, S., Ogunti, E. O., Akingbade, F. K., & Oladimeji, O. A. (2018). Review of decision support system using mobile applications in the provision of day to day information about farm status for improved crop yield. *Periodicals of Engineering and Natural Sciences, 6*, 89–99. https://doi.org/10.21533/pen.v6i2.183
6. Rains, G. C., Olson, D. M., & Lewis, W. J. (2011). Redirecting technology to support sustainable farm management practices. *Agricultural Systems, 104*, 365–370. https://doi.org/10.1016/j.agsy.2010.12.008
7. Agrios, G. N. (2005). *Plant pathology* (5th ed.). Academic Press.
8. Goffart, J. P., Olivier, M., & Destain, J. P. (2005). Presentation of a Decision Support System (DSS) for nitrogen management in potato production to improve the use of resources. In A. J. Haverkort & P. C. Struik (Eds.), *Potato in progress: Science meets practice* (pp. 134–142). Wageningen Academic Publishers.
9. Rydahl, P. (2003). A web-based decision support system for integrated management of weeds in cereals and sugarbeet. *EPPO Bull, 33*, 455–460. https://doi.org/10.1111/j.1365-2338.2003.00679.x
10. Sciarretta, A., Tabilio, M. R., Amore, A., et al. (2019). Defining and evaluating a Decision Support System (DSS) for the precise pest management of the Mediterranean fruit fly, *Ceratitis capitata*, at the farm level. *Agronomy, 9*, 608. https://doi.org/10.3390/agronomy9100608
11. Small, I. M., Joseph, L., & Fry, W. E. (2015). Development and implementation of the BlightPro decision support system for potato and tomato late blight management. *Computers and Electronics in Agriculture, 115*, 57–65. https://doi.org/10.1016/j.compag.2015.05.010
12. Jørgensen, L. N., Noe, E., Langvad, A. M., Jensen, J. E., Ørum, J. E., & Rydahl, P. (2007). Decision support systems: Barriers and farmers' need for support. *EPPO Bull, 37*, 374–377. https://doi.org/10.1111/j.1365-2338.2007.01145.x
13. Kuflik, T., Pertot, I., Moskovitch, R., Zasso, R., Pellegrini, E., & Gessler, C. (2008). Optimization of Fire blight scouting with a decision support system based on infection risk. *Computers and Electronics in Agriculture, 62*, 118–127. https://doi.org/10.1016/j.compag.2007.12.003
14. Parsons, D. J., Benjamin, L. R., Clarke, J., Ginsburg, D., Mayes, A., Milne, A. E., & Wilkinson, D. J. (2009). Weed Manager—A model-based decision support system for weed management in arable crops. *Computers and Electronics in Agriculture, 65*, 155–167. https://doi.org/10.1016/j.compag.2008.08.007
15. Travis, J. W., & Latin, R. X. (1991). Development, implementation and adoption of expert systems. *Annual Review of Phytopathology, 29*, 343–360.
16. Keating, B. A., & McCown, R. L. (2001). Advances in farming systems analysis and intervention. *Agricultural Systems, 70*(2–3), 555–579. https://doi.org/10.1016/S0308-521X(01)00059-2
17. Sonka, S. T., Bauer, M. E., Cherry E. T., et al. (1997) Precision agriculture in the 21st century. In *Geospatial and information technologies in crop management. Committee on assessing crop yield: Site-specific farming, information systems, and research opportunities, board of agriculture*, National Research Council. National Academy Press, Washington.
18. Seem, R. C., & Russo, J. M. (1984). Simple decision aids for practical control of pests. *Plant Disease, 68*, 656–660.
19. Zimdahl, R. L. (2018). *Fundamentals of weed science* (5th ed.). Academic Press.
20. Oerke, E. C., & Dehne, H. W. (2004). Safeguarding production-losses in major crops and the role of crop protection. *Crop Protection, 23*, 275–285. https://doi.org/10.1016/j.cropro.2003.10.001

21. Holst, N., Rasmussen, I. A., & Bastiaans, L. (2007). Field weed population dynamics: A review of model approaches and applications. *Weed Research, 47*, 1–14. https://doi.org/10.1111/j.1365-3180.2007.00534.x
22. Berti, A., & Zanin, G. (1997). GESTINF: A decision model for post-emergence weed management in soybean (*Glycine max* (L.) Merr.). *Crop Protection, 16*, 109–116. https://doi.org/10.1016/S0261-2194(96)00088-9
23. Neeser, C., Dille, J. A., Krishnan, G., Mortensen, D. A., Rawlinson, J. T., Martin, A. R., & Bills, L. B. (2004). WeedSOFT®: A weed management decision support system. *Weed Science, 52*, 115–122. https://doi.org/10.1614/P2002-154
24. Inovia, T., Acta, Agrosup Dijon, Arvalis, Fnams, Inra, Itab, Itb. (2020). Infloweb - Connaître et gérer la flore adventice. Retrieved October 13, 2020, from http://www.infloweb.fr/.
25. Rydahl, P., Bojer, O. M., Jørgensen, R. N., Dyrmann, M., Andersen, P., Jensen, N. P., & Sorensen, M. (2018). Spatial variability of optimized herbicide mixtures and dosages. In: *Proceedings of the 14th international conference on precision agriculture. international society of precision agriculture*, Montreal, QC, 24–27 June 2018.
26. Kanatas, P., Travlos, I. S., Gazoulis, I., Tataridas, A., Tsekoura, A., & Antonopoulos, N. (2020). Benefits and Limitations of Decision Support Systems (DSS) with a Special Emphasis on Weeds. *Agron, 10*, 548. https://doi.org/10.3390/agronomy10040548
27. Brown, R. B., & Noble, S. D. (2005). Site-specific weed management: Sensing requirements—What do we need to see? *Weed Science, 53*, 252–258. https://doi.org/10.1614/WS-04-068R1
28. Brown RB, Steckler PGA, Anderson GW (1994) Remote sensing for identification of weeds in no-till corn. Trans Am Soc Agric Eng 37:297–302. doi:https://doi.org/10.13031/2013.28084.
29. Lamb, D. W., & Weedon, M. (1998). Evaluating the accuracy of mapping weeds in fallow fields using airborne digital imaging. *Panicum effusum* in oilseed rape stubble. *Weed Research, 38*, 443–451. https://doi.org/10.1046/j.1365-3180.1998.00112.x
30. Macé, K., Morlon, P., Munier-Jolain, N., & Quéré, L. (2007). Time scales as a factor in decision-making by French farmers on weed management in annual crops. *Agricultural Systems, 93*, 115–142.
31. Colas, F., Cordeau, S., Granger, S., et al. (2020). Co-development of a decision support system for integrated weed management: Contribution from future users. *European Journal of Agronomy, 114*, 126010. https://doi.org/10.1016/j.eja.2020.126010
32. Masin, R., Vasileiadis, V. P., Loddo, D., Otto, S., & Zanin, G. (2011). A single-time survey method to predict the daily weed density for weed control decision-making. *Weed Science, 59*, 270–275. https://doi.org/10.1614/ws-d-10-00148.1
33. Pagnoncelli, F. D. B., Trezzi, M. M., & Gonzalez-Andujar, J. L. (2020). Modeling the Population dynamics and management of Italian ryegrass under two climatic scenarios in Brazil. *Plants, 9*, 325. https://doi.org/10.3390/plants9030325
34. Sousa-Ortega, C., Chamber, E., Urbano, J. M., et al. (2020). Should emergence models for Lolium rigidum be changed throughout climatic conditions? The case of Spain. *Crop Protection, 128*, 105012. https://doi.org/10.1016/j.cropro.2019.105012
35. Wright, K. J., & Wilson, B. J. (1992). Effects of nitrogen fertiliser on competition and seed production of *Avena fatua* and *Galium aparine* in winter wheat Nitrate and Farming Systems. *Aspects of Applied Biology, 30*, 381–386. https://www.cabdirect.org/cabdirect/abstract/19932328159.
36. Goggin, D. E., Powles, S. B., & Steadman, K. J. (2012). Understanding Lolium rigidum seeds: The key to managing a problem weed? *Agron, 2*, 222–239. https://doi.org/10.3390/agronomy2030222
37. Norsworthy, J. K., Ward, S. M., Shaw, D. R., et al. (2012). Reducing the risks of herbicide resistance: Best management practices and recommendations. *Weed Science, 60*, 31–62. https://doi.org/10.1614/WS-D-11-00155.1

38. Benech-Arnold, R. L., Sánchez, R. A., Forcella, F., Kruk, B. C., & Ghersa, C. M. (2000). Environmental control of dormancy in weed seed soil banks. *Field Crops Research, 67*, 105–122. https://doi.org/10.1016/S0378-4290(00)00087-3
39. Baskin, C. C., & Baskin, J. M. (1998). *Seeds: Ecology, biogeography, and, evolution of dormancy and germination*. Academic Press.
40. Karssen, C. M. (1982). Seasonal patterns of dormancy in weed seeds. In A. Khan (Ed.), *The physiology and biochemistry of seed development, dormancy and germination* (pp. 243–270). Elsevier Biomedical Press.
41. Thomas, A. G., Derksen, D. A., Blackshaw, R. E., et al. (2017). A multistudy approach to understanding weed population shifts in medium- to long-term tillage systems. *Weed Science, 52*, 874–880. https://doi.org/10.1614/WS-04-010R1
42. Lack, S., Parchami, P., & Modhej, A. (2011). Study the effects of nitrogen levels and wild oat (*Avena ludoviciana* L.) densities on grain yield and agronomic nitrogen efficiency of wheat (*Triticum aestivum* L.). *Advances in Environmental Biology, 5*, 2445–2451. http://www.aensionline.com/aeb/2011/2445-2450.pdf/
43. Blackshaw, R. E., & Brandt, R. N. (2009). Phosphorus fertilizer effects on the competition between wheat and several weed species. *Weed Biology and Management, 9*, 46–53. https://doi.org/10.1111/j.1445-6664.2008.00317.x
44. Nandula, V. K., Eubank, T. W., Poston, D. H., Koger, C. H., & Reddy, K. N. (2006). Factors affecting germination of horseweed (*Conyza canadensis*). *Weed Science, 54*, 898–902. https://doi.org/10.1614/WS-06-006R2.1
45. Swain, A. J., Hughes, Z. S., Cook, S. K., & Moss, S. R. (2006). Quantifying the dormancy of *Alopecurus myosuroides* seeds produced by plants exposed to different soil moisture and temperature regimes. *Weed Research, 46*, 470–479. https://doi.org/10.1111/j.1365-3180.2006.00532.x
46. Masin, R., Zuin, M. C., Archer, D. W., Forcella, F., & Zanin, G. (2005). WeedTurf: A predictive model to aid control of annual summer weeds in turf. *Weed Science, 53*, 193–201. https://doi.org/10.1614/WS-04-066R1
47. Crawley, M. J. (2004). Timing of disturbance and coexistence in a species rich ruderal plant community. *Ecology, 85*, 3277–3288. https://doi.org/10.1890/03-0804
48. Milberg, P., Andersson, L., & Thompson, K. (2000). Large-seeded spices are less dependent on light for germination than small-seeded ones. *Seed Science Research, 10*, 99–104. https://doi.org/10.1017/S0960258500000118
49. Forcella, F. (1998). Real-time assessment of seed dormancy and seedling growth for weed management. *Seed Science Research, 8*, 201–210. https://doi.org/10.1017/S0960258500004116
50. Benech-Arnold, R. L., Ghersa, C. M., Sánchez, R. A., & Insausti, P. (1990). A mathematical model to predict *Sorghum halepense* (L.) Pers. seedling emergence in relation to soil temperature. *Weed Research, 30*, 91–99. https://doi.org/10.1111/j.1365-3180.1990.tb01691.x
51. Murdoch, A. J., Roberts, E. H., & Goedert, C. O. (1989). A model for germination responses to alternating temperatures. *Annals of Botany, 63*, 97–111. https://doi.org/10.1093/oxfordjournals.aob.a087733
52. Evans, C. E., & Etherington, J. R. (1990). The effect of soil water potential on seed germination of some British plants. *The New Phytologist, 115*, 539–548. https://doi.org/10.1111/j.1469-8137.1990.tb00482.x
53. Travlos, I., Gazoulis, I., Kanatas, P., Tsekoura, A., Zannopoulos, S., & Papastylianou, P. (2020). Key factors affecting weed seeds' germination, weed emergence and their possible role for the efficacy of false seedbed technique as weed management practice. *Frontiers in Agronomy, 2*, 1. https://doi.org/10.3389/fagro.2020.00001
54. Forcella, F., Benech Arnold, R. L., Sanchez, R., & Ghersa, C. M. (2000). Modeling seedling emergence. *Field Crops Res, 67*, 123–139. https://doi.org/10.1016/S0378-4290(00)00088-5

55. Grundy, A. C. (2003). Predicting weed emergence: A review of approaches and future challenges. *Weed Research, 43*, 1–11. https://doi.org/10.1046/j.1365-3180.2003.00317.x
56. Masin, R., Cacciatori, G., Zuin, M. C., & Zanin, G. (2010). AlertInf: Emergence predictive model for weed control in maize in Veneto. *Italian Journal of Agrometeorology, 1*, 5–9. http://agrometeorologia.it/documenti/Rivista2010_1/AIAM%201-2010_pag5.pdf
57. Finch-Savage, W. E., Steckel, J. R. A., & Phelps, K. (1998). Germination and post-germination growth to carrot seedling emergence: Predictive threshold models and sources of variation between sowing occasions. *The New Phytologist, 139*, 505–516. https://doi.org/10.1046/j.1469-8137.1998.00208.x
58. Lu, P., Sang, W., & Ma, K. (2006). Effects of environmental factors on germination and emergence of Crofton weed (*Eupatorium adenophorum*). *Weed Science, 54*, 452–457. https://doi.org/10.1614/WS-05-174R1.1
59. Boyd, N., & Van Acker, R. (2004). Seed germination of common weed species as affected by oxygen concentration, light, and osmotic potential. *Weed Science, 52*, 589–596. https://doi.org/10.1614/WS-03-15R2
60. Derkx, M. P. M., & Karssen, C. M. (1994). Are seasonal dormancy patterns in Arabidopsis thaliana regulated by changes in seed sensitivity to light, nitrate and gibberellin? *Annals of Botany, 73*, 129–136. https://doi.org/10.1006/anbo.1994.1015
61. Travlos, I. S., Cheimona, N., Roussis, I., & Bilalis, D. J. (2018). Weed-species abundance and diversity indices in relation to tillage systems and fertilization. *Frontiers in Environmental Science, 6*, 11. https://doi.org/10.3389/fenvs.2018.00011
62. Merot, A., Bergez, J. E., Capillon, A., & Wery, J. (2008). Analysing farming practices to develop a numerical, operational model of farmers' decision-making processes: An irrigated hay cropping system in France. *Agricultural Systems, 98*, 108–118. https://doi.org/10.1016/j.agsy.2008.05.001
63. Nichols, V., Verhulst, N., Cox, R., & Govaerts, B. (2015). Weed dynamics and conservation agriculture principles: A review. *Field Crops Research, 183*, 56–68. https://doi.org/10.1016/j.fcr.2015.07.012
64. Smith, R. G. (2006). Timing of tillage is an important filter on the assembly of weed communities. *Weed Science, 54*, 705–712. https://doi.org/10.1614/WS-05-177R1.1
65. Armengot, L., Blanco-Moreno, J., Bàrberi, P., et al. (2016). Tillage as a driver of change in weed communities: A functional perspective. *Agriculture, Ecosystems and Environment, 222*, 276–285. https://doi.org/10.1016/j.agee.2016.02.021
66. Hyvönen, T., Ketoja, E., Salonen, J., Jalli, H., & Tiainen, J. (2003). Weed species diversity and community composition in organic and conventional cropping of spring cereals. *Agriculture, Ecosystems and Environment, 97*, 131–149. https://doi.org/10.1016/S0167-8809(03)00117-8
67. Ma, M. (2005). Species richness vs evenness: Independent relationship and different responses to edaphic factors. *Oikos, 111*, 192–198. https://doi.org/10.1111/j.0030-1299.2005.13049.x
68. Freyman, S., Kowalenko, C., & Hall, J. (1989). Effect of nitrogen, phosphorus and potassium on weed emergence and subsequent weed communities in south coastal British Columbia. *Canadian Journal of Plant Science, 69*, 1001–1010. https://doi.org/10.4141/cjps89-121
69. Benjamin, L. R., Milne, A. E., Parsons, D. J., & Lutman, P. J. (2010). A model to simulate yield losses in winter wheat caused by weeds, for use in a weed management decision support system. *Crop Protection, 29*, 1264–1273. https://doi.org/10.1016/j.cropro.2010.07.015
70. Zadoks, J. C., Chang, T. T., & Konzak, C. F. (1974). A decimal code for the growth stages of cereals. *Weed Research, 14*, 415–421. https://doi.org/10.1111/j.1365-3180.1974.tb01084.x
71. Colbach, N., Colas, F., Pointurier, O., Queyrel, W., & Villerd, J. (2017). A methodology for multi-objective cropping system design based on simulations. Application to weed management. *European Journal of Agronomy, 87*, 59–73. https://doi.org/10.1016/j.eja.2017.04.005
72. Mézière, D., Colbach, N., Dessaint, F., & Granger, S. (2015a). Which cropping systems to reconcile weed-related biodiversity and crop production in arable crops? An approach with simulation-based indicators. *European Journal of Agronomy, 68*, 22–37. https://doi.org/10.1016/j.eja.2015.04.004

73. Mézière, D., Petit, S., Granger, S., Biju-Duval, L., & Colbach, N. (2015b). Developing a set of simulation-based indicators to assess harmfulness and contribution to biodiversity of weed communities in cropping systems. *Ecological Indicators, 48*, 157–170. https://doi.org/10.1016/j.ecolind.2014.07.028
74. Berti, A., Bravin, F., & Zanin, G. (2003). Application of decision-support software for postemergence weed control. *Weed Science, 51*, 618–627. https://doi.org/10.1614/0043-1745(2003)051[0618:AODSFP]2.0.CO;2
75. Kropff, M. J., & Spitters, C. J. T. (1991). A simple model of crop loss by weed competition from early observations on relative leaf area of the weeds. *Weed Research, 31*, 97–105. https://doi.org/10.1111/j.1365-3180.1991.tb01748.x
76. Bagavathiannan, M. V., & Norsworthy, J. K. (2012). Late-season seed production in arable weed communities: Management implications. *Weed Science, 60*, 325–334. https://doi.org/10.1614/WS-D-11-00222.1
77. Gonzalez-Andujar, J. L. (2008). Weed control models. In S. Jørgensen & B. Fath (Eds.), *Population dynamics* (Encyclopedia of ecology) (Vol. 5, pp. 3776–3780). Elsevier.
78. Cousens, R., Doyle, C. J., Wilson, B. J., & Cussans, G. W. (1986). Modelling the economics of controlling *Avena fatua* in winter wheat. *Journal of Pest Science, 17*, 1–12. https://doi.org/10.1002/ps.2780170102
79. Doyle, C. J., Cousens, R., & Moss, S. R. (1986). A model of the economics of controlling Alopecurus myosuroides Huds. in winter wheat. *Crop Protection, 5*, 143–150. https://doi.org/10.1016/0261-2194(86)90096-7
80. Bennett, A. C., Price, A. J., Sturgill, M. C., Buol, G. S., & Wilkerson, G. G. (2003). HADSS, Pocket HERB, and WebHADSS: Decision aids for field crops. *Weed Technology, 17*, 412–420. https://doi.org/10.1614/0890-037X(2003)017[0412:HPHAWD]2.0.CO;2
81. Coble, H. D., & Mortensen, D. A. (1992). The threshold concept and its application to weed science. *Weed Technology, 6*, 191–195. https://doi.org/10.1017/S0890037X00034552
82. Colbach, N., & Cordeau, S. (2018). Reduced herbicide use does not increase crop yield loss if it is compensated by alternative preventive and curative measures. *European Journal of Agronomy, 94*, 67–78. https://doi.org/10.1016/j.eja.2017.12.008
83. McCown, R. L., Hochman, Z., & Carberry, P. S. (2002). Probing the enigma of the decision support system for farmers: Learning from experience and from theory. *Agricultural Systems, 74*, 1–10. https://doi.org/10.1016/S0308-521X(02)00021-5
84. Swanton, C. J., Mahoney, K. J., Chandler, K., & Gulden, R. H. (2008). Integrated weed management: Knowledge-based weed management systems. *Weed Science, 56*, 168–172. https://doi.org/10.1614/WS-07-126.1
85. Wilkerson, G. G., Wiles, L. J., & Bennett, A. C. (2002). Weed management decision models: Pitfalls, perceptions, and possibilities of the economic threshold approach. *Weed Science, 50*, 411–424. https://doi.org/10.1614/0043-1745(2002)050[0411:WMDMPP]2.0.CO;2
86. Zanin, G., & Sattin, M. (1988). Threshold level and seed production of velvetleaf (*Abutilon theophrasti* Medicus) in maize. *Weed Research, 28*, 347–352. https://doi.org/10.1111/j.1365-3180.1988.tb00813.x
87. Kudsk, P. (2008). Optimizing herbicide dose: A straightforward approach to reduce the risk of side effects of herbicides. *Environmentalist, 28*, 49–55. https://doi.org/10.1007/s10669-007-9041-8
88. Kudsk, P., Mathiassen, S. K., & Rydahl, P. (2014). Decision support system for optimized herbicide dose in spring barley. *Weed Technology, 28*, 19–27. https://doi.org/10.1614/WT-D-13-00085.1
89. Sønderskov, M., Fritzsche, R., de Mol, F., et al. (2015). DSSHerbicide: Weed control in winter wheat with a decision support system in three South Baltic regions–Field experimental results. *Crop Protection, 76*, 15–23. https://doi.org/10.1016/j.cropro.2015.06.009
90. Jensen, H. G., Jacobsen, L. B., Pedersen, S. M., & Tavella, E. (2012). Socioeconomic impact of widespread adoption of precision farming and controlled traffic systems in Denmark. *Precision Agriculture, 13*, 661–677. https://doi.org/10.1007/s11119-012-9276-3

91. Gutjahr, C., & Gerhards, R. (2010). Decision rules for site-specific weed management. In E. C. Oerke, R. Gerhards, G. Menz, & R. A. Sikora (Eds.), *Precision crop protection—The challenge and use of heterogeneity* (pp. 223–239). Springer.
92. Team, R. C. (2013). R: A language and environment for statistical computing. Retrieved from http://finzi.psych.upenn.edu/R/library/dplR/doc/intro-dplR.pdf
93. Hothorn, T., Bretz, F., & Westfall, P. (2008). Simultaneous inference in general parametric models. *Biometrical Journal, 50*, 346–363. https://doi.org/10.1002/bimj.200810425
94. Streibig, J. C., Kudsk, P., & Jensen, J. E. (1998). A general joint action model for herbicide mixtures. *Journal of Pest Science, 53*, 21–28. https://doi.org/10.1002/(SICI)1096-9063(199805)53:1%3C21::AID-PS748%3E3.0.CO;2-L
95. Buhler, D. D., Liebman, M., & Obrycki, J. J. (2000). Theoretical and practice challenges to an IPM approach to weed management. *Weed Science, 48*, 274–280. https://doi.org/10.1614/0043-1745(2000)048[0274:TAPCTA]2.0.CO;2
96. Benjamin, L. R., Milne, A. E., Parsons, D. J., Cussans, J., & Lutman, P. J. W. (2009). Using stochastic dynamic programming to support weed management decisions over a rotation. *Weed Research, 49*, 207–216. https://doi.org/10.1111/j.1365-3180.2008.00678.x
97. Monjardino, M., Pannell, D. J., & Powles, S. B. (2003). Multispecies resistance and integrated management: A bioeconomic model for integrated management of rigid ryegrass (*Lolium rigidum*) and wild radish (*Raphanus raphanistrum*). *Weed Science, 51*, 798–809. https://doi.org/10.1614/P2002-118
98. Naudin, K., Husson, O., Scopel, E., Auzoux, S., Giner, S., & Giller, K. E. (2015). PRACT (Prototyping Rotation and Association with Cover crop and no Till)–a tool for designing conservation agriculture systems. *European Journal of Agronomy, 69*, 21–31. https://doi.org/10.1016/j.eja.2015.05.003
99. Castelán-Ortega, O. A., Fawcett, R. H., Arriaga-Jordán, C., & Herrero, M. (2003). A decision support system for smallholder campesino maize–cattle production systems of the Toluca Valley in Central Mexico. Part II—Emulating the farming system. *Agricultural Systems, 75*, 23–46. https://doi.org/10.1016/S0308-521X(01)00110-X
100. Munier-Jolain, N. M., Chavvel, B., & Gasquez, J. (2002). Long-term modelling of weed control strategies: Analysis of threshold-based options for weed species with contrasted competitive abilities. *Weed Research, 42*, 107–122. https://doi.org/10.1046/j.1365-3180.2002.00267.x
101. Melander, B., & Rasmussen, K. (2000). Reducing intrarow weed numbers in row crops by means of a biennial cultivation system. *Weed Research, 40*, 205–218. https://doi.org/10.1046/j.1365-3180.2000.00183.x
102. Buhler, D. D., Hartzler, R. G., & Forcella, F. (1997). Implications of weed seedbank dynamics to weed management. *Weed Science, 45*, 329–336. https://doi.org/10.1017/S0043174500092948
103. Hochman, Z., & Carberry, P. S. (2011). Emerging consensus on desirable characteristics of tools to support farmers' management of climate risk in Australia. *Agricultural Systems, 104*, 441–450. https://doi.org/10.1016/j.agsy.2011.03.001
104. Boutsalis, P., Gill, G. S., & Preston, C. (2012). Incidence of herbicide resistance in rigid ryegrass (*Lolium rigidum*) across southeastern Australia. *Weed Technology, 26*, 391–398. https://doi.org/10.1614/WT-D-11-00150.1
105. Travlos, I., Tabaxi, I., Papadimitriou, D., Bilalis, D., & Chachalis, D. (2016). *Lolium rigidum* Gaud. biotypes from Greece with resistance to glyphosate and other herbicides. *Bulletin UASVM Horticulture, 73*, 1–2. https://doi.org/10.15835/buasvmcn-hort:11772
106. Lacoste, M., & Powles, S. (2015). RIM: Anatomy of a weed management Decision Support System for adaptation and wider application. *Weed Science, 63*, 676–689. https://doi.org/10.1614/ws-d-14-00163.1
107. Lacoste, M., & Powles, S. B. (2014). Upgrading the RIM model for improved support of integrated weed management extension efforts in cropping systems. *Weed Technology, 28*, 703–720. https://doi.org/10.1614/WT-D-14-00020.1

108. Dhima, K. V., Eleftherohorinos, I. G., & Vasilakoglou, I. B. (2000). Interference between *Avena sterilis*, *Phalaris minor* and five barley cultivars. *Weed Research, 40*, 549–559. https://doi.org/10.1046/j.1365-3180.2000.00213.x
109. Knezevic, S. Z., & Datta, A. (2015). The critical period for weed control: Revisiting data analysis. *Weed Science, 63*, 188–202. https://doi.org/10.1614/WS-D-14-00035.1
110. Johannsen, C. J., Carter, P. G., Morris, D. K., Ross, K., & Erickson, B. (2000). The real applications of remote sensing to agriculture. In *Proceedings of the second international conference on geospatial information in agriculture and forestry*, vol. 1. Lake Buena Vista, FL, January 10–12 2000, pp. 1–5.
111. Booltink, H. W. G., Van Alphen, B. J., Batchelor, W. D., Paz, J. O., Stoorvogel, J. J., & Vargas, R. (2001). Tools for optimizing management of spatially-variable fields. *Agricultural Systems, 70*(2–3), 445–476. https://doi.org/10.1016/S0308-521X(01)00055-5
112. Kipling, R. P., Bannink, A., Bellocchi, G., et al. (2016). Modeling European ruminant production systems: Facing the challenges of climate change. *Agricultural Systems, 147*, 24–37. https://doi.org/10.1016/j.agsy.2016.05.007
113. Li, H., Zhao, Y., & Zheng, F. (2020). The framework of an agricultural land-use decision support system based on ecological environmental constraints. *Science of the Total Environment, 717*, 137149. https://doi.org/10.1016/j.scitotenv.2020.137149
114. Starke, S. D., & Baber, C. (2020). The effect of known decision support reliability on outcome quality and visual information foraging in joint decision making. *Applied Ergonomics, 86*, 103102. https://doi.org/10.1016/j.apergo.2020.103102
115. Kristensen, K., & Rasmussen, I. A. (2002). The use of a Bayesian network in the design of a decision support system for growing malting barley without use of pesticides. *Computers and Electronics in Agriculture, 33*, 197–217. https://doi.org/10.1016/S0168-1699(02)00007-8
116. Zhan, Y., & Zhang, M. (2012). PURE: A web-based decision support system to evaluate pesticide environmental risk for sustainable pest management practices in California. *Ecotoxicology and Environmental Safety, 82*, 104–113. https://doi.org/10.1016/j.ecoenv.2012.05.019

Zephyrus: Grain Aeration Strategy Based on the Prediction of Temperature and Moisture Fronts

D. C. Lopes and A. J. Steidle Neto

1 Introduction

The aeration process is conducted to preserve the physical conditions of the stored grain and is considered one of the most important practices of the postharvest sector in terms of food security for the world population [1]. During aeration air is forced through the grain bulk, with the main purpose of cooling the stored product. Other objectives of aeration are to equalize grain temperature throughout the bulk, promote limited drying, and remove fumigant residues and odors [2].

When compared with other technologies, such as modified atmosphere and irradiation, the aeration is the safest, most economic and efficient preventive practice [3]. Since aeration modifies the bulk microclimate during storage, it has been considered essential for controlling mycotoxin production and the localized hotspots that can result from fungi, mites, and insects [4, 5]. This technology is particularly important for food industries, agricultural cooperatives, and farms due to the regulatory constraints on the use of chemicals for pest control, as well as the problems associated with insect and fungal resistance [6].

The variations in weather from region to region, and among cropping seasons, can lead to situations where ambient air conditions are not suitable to cool grain, requiring efficient control strategies and equipment that guarantee the beneficial effects of aeration [4, 7, 8]. Under certain climatic conditions, mainly in tropical and subtropical regions, aeration with air ambient should not achieve grain temperatures below 21 °C, capable of minimizing insect development [9, 10]. In these cases, chilled aeration is recommended for commodities that justify the added expense of

D. C. Lopes (✉) · A. J. Steidle Neto
Federal University of São João del-Rei, Sete Lagoas, Brazil
e-mail: danielalopes@ufsj.edu.br

D. D. Bochtis et al. (eds.), *Information and Communication Technologies for Agriculture—Theme III: Decision*, Springer Optimization and Its Applications 184, https://doi.org/10.1007/978-3-030-84152-2_9

refrigeration unit [11, 12]. However, ambient aeration systems are still dominant for temperature management of stored grains worldwide [1, 13, 14].

Typical aeration systems are composed by a silo with perforated in-floor or on-floor ducts, which are connected to one or more fans to force the air through the grain [2]. These systems also require roof vents for exhaust or air intake, and sensors for monitoring grain temperatures and external air conditions [8, 15, 16]. Additionally, good design and well-defined control strategies of aeration systems are essential for efficient storage management. Specifically, ineffective control strategies can result in high electrical energy consumption, as well as greater grain spoilage.

The simplest and less efficient control strategy is based on mechanical time controllers, which are programmed for turning the aeration fan on during specific and fixed periods of the day [3]. More complex electro-mechanical controllers include thermostats or combined thermostats and humidistats, in which the aeration system operation occurs when the temperature or the combination of temperature and relative humidity of the ambient air is lower than the maximum set on the equipment [7]. Temperature difference, wet-bulb temperature and proportional time controllers are other options for identifying the best moments for operating the aeration systems [2].

Modern control strategies are implemented in microprocessors or computers, which decisions are based on grain aeration simulations [18–20]. These controllers combine the aeration process predictions with the real-time data measurements, relating several important storage parameters, such as grain moisture content, temperature and dry matter loss, as well as temperature and relative humidity of ambient air. For this, they are interesting tools for automatically updating the aeration control set points, adjusting them according to different geographic locations, weather conditions, and aeration systems. Computer-based controllers depend on complex programming codes, limiting their implementation for professionals who work in the postharvest sector and resulting in longer computing times. This complexity is mainly associated with the auxiliary equations required by the models and the additional parameters, including physical and thermal properties of grain and specific data of the region.

This chapter presents an effective, economical, and simple control strategy to manage grain aeration systems based on the prediction of speeds and changes of temperature and moisture fronts while air is passed through a grain bulk.

2 Methodology

2.1 Theory Basis of Zephyrus Control Strategy

Zephyrus control strategy is based on relatively simple approximate equations for speeds and changes of temperature and moisture fronts, which move within the grain bulk as a result of the aeration air movement. The theory basis of the strategy is

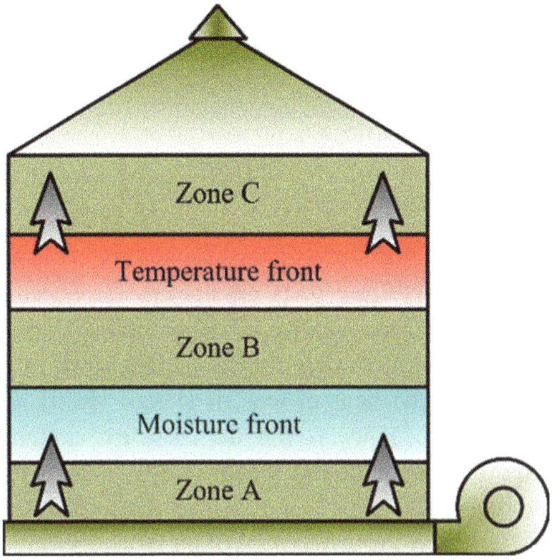

Fig. 1 Zones and fronts formed during the grain aeration

derived from the knowledge that when air is forced through a grain bulk, areas (zones) with different temperatures and moisture contents are formed and separated by change fronts (Fig. 1). Fronts and zones move through the grain bulk in waves, which are limited by leading and trailing edges [21]. Before aeration the grain bulk has not been affected by change fronts and is in an area called zone C, maintaining its original moisture content and temperature [3]. During the process, both moisture and temperature fronts move towards the bulk surface, but the temperature front is more rapidly than the moisture one, and after some time of aeration it moved completely through the grain bulk. Since aeration fans should be operated until the temperature front travels through the entire grain mass, the effects of moisture front are limited to the grain near the entrance of the aeration air. This makes that after aeration zone C disappears and most of the grain bulk is located in an intermediate zone, affected only by the temperature front, and called zone B [22]. These grains present moisture content close to the initial value and temperature in equilibrium with the grain bulk at its initial moisture content after the air has cooled or heated along a wet-bulb temperature. A small amount of grain, located near the air input and affected by moisture and temperature fronts, tends to reach the equilibrium with the temperature and relative humidity of the aeration air. These grains are considered in an area called zone A [23].

The main equations used to simulate the speeds and changes on temperature and moisture fronts were presented by Hunter [23]. They were based on a method in which the grain conditions after the passing of the temperature and moisture fronts can be estimated without recourse to numerical integration:

$$\mathrm{VTF} = v_a \left(\frac{\varepsilon \, \rho_a \, C_a}{\rho_g \, C_g} \left(1 + \frac{w_a h^2}{C_a \, R \, T^2} \right) + \frac{\varepsilon \, \rho_a \, w_a \, h^2 D}{\rho_g \, C_a \, R \, T^2 \left(1 + \frac{w_a h^2}{C_a \, R \, T^2} \right)} \right) \quad (1)$$

$$\mathrm{VMF} = v_a \left(\frac{\varepsilon \, \rho_a \, D}{\rho_g \left(1 + \frac{w_a \, h^2}{C_a \, R \, T^2} \right)} \right) \quad (2)$$

Where VTF and VMF are the speeds of temperature and moisture fronts (m s^{-1}), v_a is the mean velocity of interstitial air (m s^{-1}), ε is the grain bulk porosity (decimal), ρ_a is the air density (1.169 kg m^{-3}), ρ_g is the bulk density (kg m^{-3}), C_a is the specific heat of wet air (J kg^{-1} K^{-1}), C_g is the specific heat of wet grain (J kg^{-1} K^{-1}), w_a is the air moisture content (g g^{-1}), h is the latent heat of vaporization of bound water (J kg^{-1}), R is the gas constant for water equation vapor (461.52 J kg^{-1} K^{-1}), T is the maximum grain temperature before aeration (K), and D is the differential of air moisture content as function of dry basis moisture content of grain.

The differential of air moisture content as function of dry basis moisture content of grain (D) was estimated based on procedure reported by Hunter [23], considering the moisture contents known at the state where the derivative is required (conditions of zone B) and at a state on either side of this state, all at the same temperature. The specific heats of wet air and grain (C_a and C_g) were calculated according to Sutherland et al. [24]. Auxiliary equations and iterative procedures presented by Lopes et al. [25], Melo et al. [26], and Navarro and Noyes [3] were also applied.

The temperature of the grain in zone A is that of the aeration air. The grain moisture content can be estimated from an appropriate isotherm equation. In this work, the Modified Henderson model was used [3].

This same equation was rearranged and applied to predict the equilibrium relative humidity required by the control strategy when estimating grain temperature in zone B. This was performed by using an iterative method, which consists in following the adiabatic line by keeping the enthalpy constant until finding a value for the equilibrium temperature at which relative humidity is equal to the equilibrium relative humidity considering the aeration air and the initial grain moisture content. Thus, a numerical method was used, by calculating and recalculating the relative humidity in zone B as a function of air moisture content and equilibrium temperature by using the psychrometric relationships, until it converges [25, 26]. The air moisture content was incremented or decremented accordingly by a predefined step value of 0.0001 g g^{-1} in each iteration. The grain moisture content in zone B was estimated taking into account that its initial value decreases 1% w.b. for every 28 °C of cooling [3]:

$$\mathrm{WB} = \frac{\mathrm{WI} - (\mathrm{TG} - \mathrm{TB})}{28} \quad (3)$$

The estimate of grain temperature and moisture content in zone B was refined by repeating the calculations and using the grain moisture content obtained in the first estimate as the initial value for estimating temperature in zone B.

The thickness of zone A was obtained by multiplying the speed of the moisture front by the time required by the temperature front to completely move through the grain bulk. It is important to consider this parameter mainly in regions where the aeration potential is limited and in situations where the grain moisture content is not the indicated for safe storage. In these cases, it is possible to aerate the grain bulk even with non-ideal weather conditions if zone A results in a relatively thin layer. The time required for the temperature front leaves the grain bulk was found by dividing the total grain bulk height by the speed of the temperature front.

2.2 Description of Zephyrus Control Strategy

The proposed strategy was called Zephyrus and was implemented in a software package written in Pascal (Turbo Pascal, Borland International, Scotts Valley, USA). This programming language was used since it is a high-level, compiled, and strongly typed language that supports structured and object-oriented projects. Furthermore, many routines previously implemented and validated for data acquisition were required during experimental tests in this work. The resulting codes were easy to read, quick to compile, and allowed a modular programming that associated the data acquisition values with the predicted ones, leading to an integrated decision-making, with automatic activating of the aeration fan, and the recording of monitored and control results.

The goal of Zephyrus was turning the aeration fan on only if this procedure results in safe grain temperatures and moisture contents both in zones A and B. Safe storage conditions include grain moisture contents in equilibrium with air at a relative humidity from 65% to 70%, as well as grain temperature gradients smaller than 3 °C and grain temperatures preferentially smaller than 21 °C [27–29].

For this, two conditions (C1 and C2) were evaluated by the control strategy (Fig. 2). Both conditions presented fixed values, which were used in the comparisons of the control variables. These values were set based on previous simulation tests considering different grains, regions, and times of the year, where they varied, and the performance of the control strategy was evaluated in terms of aeration time, grain bulk cooling, and grain moisture content variations. Each fixed value range was tested based on technical recommendations aiming at limiting the aeration hours with energy saving purposes, but without prejudicing the storage quality [8, 19, 30].

The first condition evaluated by Zephyrus was related to temperature control inside the bin. Since the main objective of aeration is cooling the grain, the temperature in zone B must be smaller than the maximum temperature measured into the grain bulk (Tmax). For energy saving and cooling efficiency purposes a difference of 3.5 °C between these temperatures was considered, following recommendations of Martins et al. [31) and Mrema et al. (32]. This condition also allows

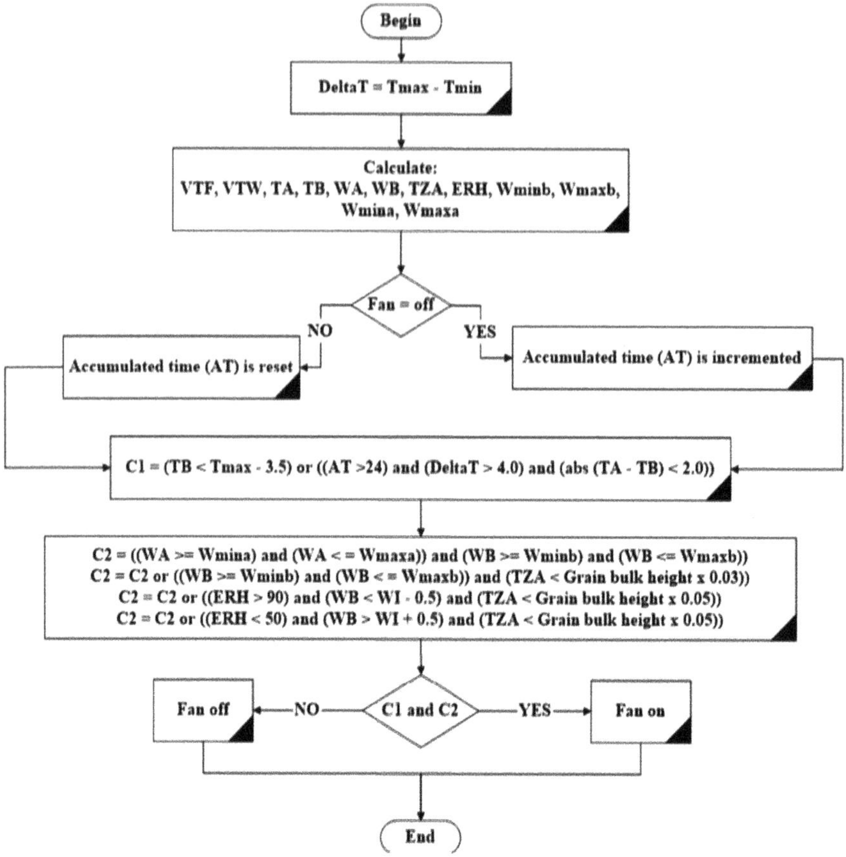

Fig. 2 Simplified flow diagram of Zephyrus control strategy

equalizing the grain temperatures, which is another important objective of the grain aeration as part of integrated pest management practices, mainly in tropical and subtropical regions. In this case, it may be required to turn the fan on if the difference between maximum and minimum temperatures of the grain bulk (DeltaT) is greater than 4.0 °C. According to results of previous tests, this procedure will be economically viable only if the simulated difference between the temperatures in zones A and B is smaller than 2.0 °C and if the fan is off for more than 24 h. The previous tests were conducted when developing the control strategy, considering different grain, regions, and times of the year. These parameters varied, simulations were performed, and the efficiency of the control strategy was evaluated in terms of aeration time, grain bulk cooling, and grain moisture content variations, also comparing results with other works [8, 19, 30].

The objective of the second condition evaluated by Zephyrus was to maintain the moisture content of dry grain at a safe level. For this, safe moisture limits were established within which the aeration fan must operate depending on the grain bulk and aeration air conditions. Four situations can occur when evaluating grain moisture content condition. The first one is that in which the aeration air is suitable and the grain bulk was adequately dried, resulting in safe moisture contents both in zones A and B. Another situation can occur when aeration results in safe moisture content only in zone B, but the thickness of zone A is sufficiently small for not prejudicing the storage quality, considering that short periods of unfavorable aeration conditions will be compensated in the bottom of the silo by later aeration under better conditions. In this study, the thickness threshold was 3 or 5% of the grain bulk height, depending on the combination of the other parameters. Finally, if grain is drier or humid, aeration can be applied in order to increase or decrease the grain moisture content if this procedure tends to reach the safe value.

2.3 *Experimental Evaluation of Zephyrus Control Strategy*

Zephyrus was used for monitoring and controlling an aeration system located at Sete Lagoas city, Minas Gerais State, Brazil, with the following geographical coordinates: 19°27′S latitude, 44°14′W longitude, and 767 m elevation.

The aeration system consisted of a metallic silo prototype with flat bottom and 0.8 m in diameter. It was equipped with a totally perforated floor, a plenum with height of 0.1 m, and roof vents for exhaust air. The bin was filled with bean (*Phaseolus vulgaris*) of 13.0% w.b. to a depth of 1.0 m. A centrifugal fan (model BM5115-04 W-B30, Minibea, USA) was used to force the air through the grain, operating with an airflow rate equal to 6 $m^3\ h^{-1}\ t^{-1}$ from February 10, 2018 to August 31, 2018 (Fig. 3).

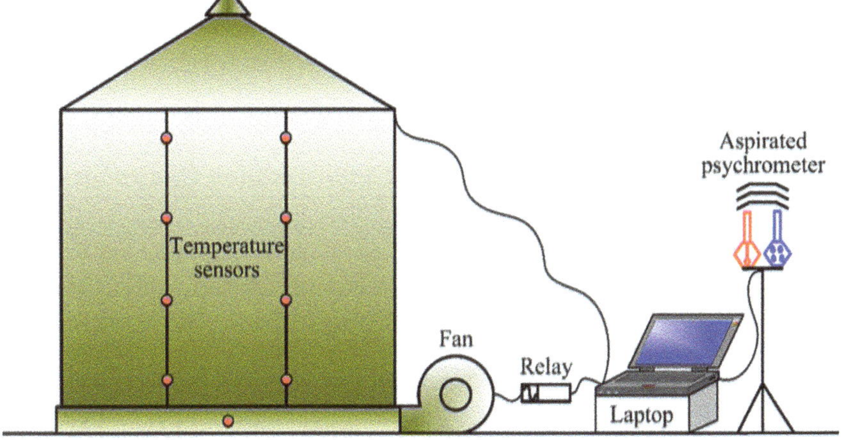

Fig. 3 Schematic diagram of the aeration system used in this work

A thermistor-based system was employed for data acquisition during the field trial [8]. For this, a sensor was located at the plenum of the silo and eight temperature sensors were located in the grain bulk, divided into two cables equally spaced at an interval of 0.3 m. In the vertical direction, the distances between sensors were 0.25 m. The cables with the temperature sensors were housed inside protective cable jackets, which were fastened to the roof and floor of the silo. A digital aspirated psychrometer and a relay circuit for fan operation completed the apparatus (Fig. 3).

Temperatures inside the silo were recorded every 1 min, as well as the temperature and relative humidity of the air, and the control strategy was performed every 30 min. Furthermore, during the field trial grain moisture contents were measured weekly by a capacitive meter (model 999-ES, Motomco, Brazil). Samples were collected with a probe at 0.0, 0.5, and 1.0 m, from the perforated floor. The profiles of environmental data, grain moisture, and temperature, as well as fan operation, were analyzed using electronic spreadsheets by means of tables and graphs relating the variables to time.

Other technologies can be used to perform the environmental measurements required to Zephyrus operation. The in-built equipment necessary to implement the proposed control strategy must include a digital psychrometer for monitoring the ambient air conditions, temperature sensors located inside the grain bulk, a temperature sensor located at the plenum of the bin for monitoring the temperature of aeration air and an electronic switch (or similar equipment) to automatically turn the fan on or off. The number of sensors located inside the grain mass will depend on the dimensions of the structure and on the available budget. Many sensors will perform a better monitoring but can increase the project costs. According to Jian and Jayas [33] and Mrema et al. [32] is 5 m, leading to a reduced cost of equipment and a monitoring radius of approximately 2.5 m per sensor. However, since temperature differences among locations in the grain bulk and over time are common during the storage period, best monitoring result is achieved if the distance between sensors is kept around 1.2–1.8 m or less [16]. These recommendations are based on the high insulating capacity of grain bulks, due to their low thermal conductivity [3], which requires that many temperature sensors are adequately distributed in the silo to ensure timely detection of any slowly developing localized hotspot.

2.4 Comparison of Zephyrus with Other Aeration Controllers

In a second phase, simulations were carried out in order to compare the Zephyrus performance with two other efficient aeration control strategies (AERO and Thermostat combined with humidistats). The aeration process was simulated by using the one-dimensional model proposed by Thorpe [34] and presented in details by Lopes et al. [35] and Thorpe [36], which was implemented in a software written in the Pascal programming language (Turbo Pascal, Borland International, Scotts Valley, USA). This model was based on differential equations and solved by the finite difference numerical method.

The AERO controller was developed with the objectives of maintaining grain quality with minimal energy input, automatically adjusting its set points according to different climates and storage systems based on real-time data acquisition and on simulations of the aeration process [19]. The first condition evaluated by AERO indicates turning the fan on if the dew point temperature of the aeration air is lower than the minimum grain temperature. According to another condition, the fan can be turned on if the difference between the simulated grain moisture content and the safe one is less than 0.5% points or is smaller than the actual one. Temperature management is based on two additional conditions. To equalize grain temperatures, the fan can be turned on if a difference greater than 5 °C is observed inside the bin and if the simulated difference is smaller than the real one. Finally, the fan can also be turned on if the simulated temperature for one of the grain sections is smaller than the real one equivalent to it or if the simulated average grain temperature is smaller than the real one, assuring cooling whenever it is possible.

The aeration control based on thermostat combined with humidistats (THH) recommends turning the fan on when the aeration temperature is lower than a maximum set value and the aeration relative humidity is within a default range. The temperature and humidity limits must be carefully chosen to provide effective aeration operation results. The use of very restrictive set points can limit the amount of usable cooling time, while wide ranges can result in excessive aeration with increasing electrical energy cost [27, 37]. Generally, this controller requires the updating of its set points during the storage period, as well as for systems located at different regions. In this study and after previous tests, the THH controller was set to turn the fan on whenever the aeration temperature is 5 °C below the maximum grain temperature and the aeration relative humidity is between 60 and 80%.

During the tests, each controller operation was simulated considering soybean, wheat, and corn stored into different harvest periods in the Southeast, South and Center-West Brazilian regions. Different storage capacities and initial grain conditions were also applied during simulations, resulting in different storage scenarios (Table 1).

The simulation locations were represented by the cities of Sete Lagoas (Southeast), Diamantino (Center-West), and Cruz Alta (South), located at the Brazilian States of Minas Gerais, Mato Grosso, and Rio Grande do Sul, respectively. These locations were selected based on their grain production [38], and on the availability of weather data [39]. The three selected regions (Southeast, Center-West and South) represented a humid subtropical climate with dry winter and hot summer; a tropical climate with dry winter; and a humid subtropical climate without dry season and hot summer, respectively [40].

Five years (2012–2017) of dry bulb temperatures and relative humidity at hourly scale were obtained from the Brazilian National Institute of Meteorology [39] for the three studied regions and used to simulate the conditions of ambient aeration air. A consistency analysis was performed on the meteorological data with electronic spreadsheet functions to remove all inconsistent values. Additionally, visual analysis of graphs relating the variables to time complemented the data evaluation. Temperature and humidity gaps of 1 to 3 h were filled with the mean values of the time

Table 1 Simulation conditions considered for comparing the Zephyrus controller with the AERO and thermostat combined with humidistats

Treatment	1	2	3	4	5	6	7	8	9
Grain	Corn	Corn	Corn	Wheat	Wheat	Wheat	Soybean	Soybean	Soybean
City	Sete Lagoas	Sete Lagoas	Sete Lagoas	Cruz Alta	Cruz Alta	Cruz Alta	Diamantino	Diamantino	Diamantino
Storage capacity (t)	15.82	15.82	15.82	632.25	632.25	632.25	1669.9	1669.9	1669.9
Airflow (m^3 min^{-1} t^{-1})	0.1	0.1	0.1	0.05	0.05	0.05	0.2	0.2	0.2
Initial grain Moisture (% w.b)	13	14	12	13	14	11	12	13	11
Initial grain temperature (°C)	32	28	36	33	35	31	35	30	36
Storage period	Feb May	Jun Sep	Apr Jul	Oct Jan	Nov Feb	Dec Mar	Jan Apr	Mar Jun	Apr Jul

before and after. Finally, meteorological data relating to the 5 years were averaged and the model equations were applied iteratively considering their hourly variations.

Simulations were performed always considering the months when each crop is harvested and stored in the simulated Brazilian regions. For corn (Southeast), this corresponds to aeration from late summer to late fall, from late fall to early spring and from middle fall to middle winter. For wheat (South) harvests occur from middle spring to middle summer, from late spring to late summer and from early summer to early fall. Finally, for soybean (Center-West), the aeration cycles were simulated from middle summer to early fall, from early fall to early winter and from late fall to late winter [38].

To demonstrate grain cooling and energy requirement variations, simulated data were submitted to variance analysis and averages compared by the F and Scott-Knott tests ($P < 0.05$) through SISVAR software [41].

3 Results and Discussion

3.1 Experimental Evaluation of Zephyrus Control Strategy

The Zephyrus control strategy was evaluated regarding its ability to minimize the potential for spoilage, fungi activity, and insect development into the grain bulk maintaining electrical energy saving. These were represented by moisture content control, grain temperature uniformity, grain cooling, and electrical energy requirements.

The beneficial effects of the proposed control strategy on the stored grain ecosystem along the field trial are presented in Fig. 4. The trial started in February 2018, when the first-crop bean is harvested in Minas Gerais State. That is, the first aeration events occurred during late summer, which resulted in larger frequencies with medium intervals of fan operating due to the high grain temperatures and relative humidity of aeration air below 80% most of the time. From late March through mid-April, the fan was turned on for longer intervals with small frequencies, while after this period the fan operation was reduced and characterized by lager frequencies and small intervals. This was expected since Autumn started in March, and air temperature dropped. Further, Sete Lagoas city is characterized by a dry climate, which contributed to air conditions suitable for safe grain cooling, except for few periods with relative humidity greater than 80%. Sometimes, the fan was turned on even with aeration temperatures greater than the grain one, probably due to the low relative humidity and the grain response to the evaporative cooling effect of the warm dry air, as reported by Steidle Neto and Lopes [8] and Navarro and Noyes [3]. A more pronounced cooling was observed from May, probably due to the early winter, when lower aeration temperatures and relative humidity close to 60% were verified. Considering the aeration air, greater thermal and humidity amplitudes were observed from June to August. The grain cooling achieved before was maintained by the appropriate fan operation and due the insulating properties of the grains.

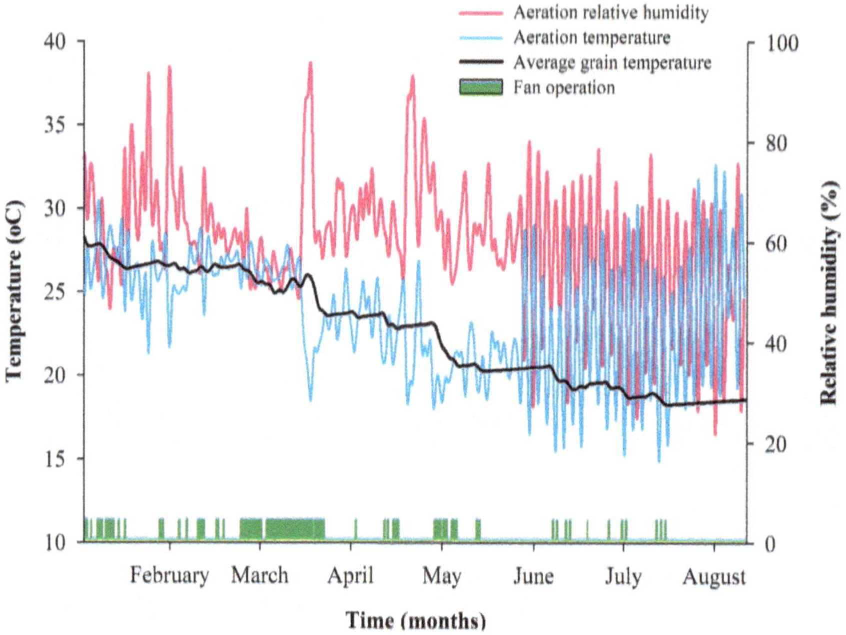

Fig. 4 Effects of Zephyrus strategy on bean stored for 6 months in Sete Lagoas (Brazil)

Average grain temperature dropped from 30 to 18.6 °C, maintaining an average temperature gradient of 2.6 °C throughout the grain bulk. Along the 6 months of study, the total aeration time was 845 h and the energy required for cooling the 0.42 t of grain was 0.06 kWh t^{-1} $°C^{-1}$ (0.30 kWh t^{-1}). Steidle Neto and Lopes [8] reported values of 0.18 kWh t^{-1} $°C^{-1}$ (0.46 kWh t^{-1}) when aerating 0.42 t of corn from June to August (57 days) in the same region of this study, but using the AERO control strategy and reaching a cooling of 2.6 °C. Both trials were carried out using a prototype equipped with a well-designed aeration system, efficient fan, and thermometry technology, as well as employing appropriated control strategies. The lower energy requirement verified when applying the Zephyrus controller can be explained by the longer period of operation compared to the other experiment, which might have contributed to a larger grain cooling in this study. Also, the variations in weather from year to year, storage period and initial grain conditions may have affected the operation of the systems. Regardless of these variations, the Zephyrus controller was effective for reducing grain temperatures to a level that inhibit insect activity, also maintaining the temperature homogeneity inside the grain bulk and grain moisture contents under safe limits, with electrical energy saving.

The results in this trial agree with the recommended by Navarro et al. [2] and Taruvinga et al. [29] who reported that grain temperatures from 16° to 21 °C are considered safe for insect management. Navarro et al. [2] also reinforced the importance of grain aeration for maintaining low temperature gradients throughout the grain bulk with the objective of preventing moisture migration, mainly in tropical and subtropical climates.

During the field trial, the average grain moisture content varied from 13% to 12.4% w.b., which can be explained by sampling variation and moisture meter precision (±0.25%), but mainly by the aeration effects. Collins and Conyers [42], Sanderson et al. [43] and Foster [44] showed that typical moisture variations during aeration processes are from 0.2 to 1.0% w.b., depending on the silo dimensions and on the aeration air conditions. These changes can be explained by the relative humidity of the aeration air which is commonly different than the interstitial grain bulk relative humidity, resulting in heat and moisture transfers. The final moisture content corresponded to equilibrium relative humidity of 60% when considering temperatures around 20 °C, which is considered safe for grain storage [2, 45].

3.2 Comparison of Zephyrus with Other Aeration Controllers

Good results were also obtained when comparing the Zephyrus control strategy with the AERO and THH ones by simulation, as shown by Tables 2 and 3 where it is possible to verify the effects of these controllers on the grain cooling and on the electrical energy requirements, respectively.

Differences in weather and initial grain conditions, as well as the storage capacity, aeration airflow, and stored grain, affected the aeration effects and resulted in significant differences in average grain cooling between some of the simulated conditions and the three evaluated control strategies. On the other hand, despite some control strategies and/or simulated conditions presented small differences in terms of electrical energy requirements, these were no significant.

The effects of initial grain conditions, airflow rates, and bin sizes on aeration results were also reported by Jayas and White [37], Collins and Conyers [42], and Antunes et al. [46]. Generally, higher initial grain temperatures, small airflow rates and large bin sizes tend to increase the required aeration time, which can result in

Table 2 Comparison results of Zephyrus, AERO, and TTH controllers for simulated grain cooling (°C)

Scenario	AERO	THH	ZEPHYRUS
1	8.9 bA	11.4 bA	11.3 bA
2	3.2 cA	5.4 cA	6.9 cA
3	16.4 aA	16.2 aA	16.0 aA
4	8.2 bA	3.6 dB	11.3 bA
5	7.1 bA	3.7 dB	8.5 cA
6	3.2 cA	1.8 dA	6.3 cB
7	9.1 bA	6.9 cA	7.8 cA
8	6.6 bA	4.0 dA	6.6 cA
9	6.4 bA	1.1 dB	4.9 cA

Averages followed by the same lower case letters in a column and capital letters on the lines do not differ significantly by the Scott-Knott test ($p < 0.05$)

Table 3 Comparison results of Zephyrus, AERO, and TTH controllers for simulated aeration energy requirement (kWh t^{-1} °C^{-1})

Scenario	AERO	THH	ZEPHYRUS
1	0.140 aA	0.184 aA	0.133 aA
2	0.292 aA	0.396 aA	0.348 aA
3	0.064 aA	0.098 aA	0.069 aA
4	0.005 aA	0.017 aA	0.008 aA
5	0.007 aA	0.021 aA	0.007 aA
6	0.016 aA	0.023 aA	0.015 aA
7	0.016 aA	0.011 aA	0.026 aA
8	0.036 aA	0.036 aA	0.033 aA
9	0.030 aA	0.023 aA	0.032 aA

Averages followed by the same lower case letters in a column and capital letters on the lines do not differ significantly by the Scott-Knott test ($p < 0.05$)

lower grain cooling depending on the suitable daily aeration hours available in each region. On the other hand, higher airflows and larger storage capacities increase the system energy consumption. Finally, aeration of grain at safe moisture content, without fines and foreign material, tends to be more efficient.

Zephyrus resulted in higher grain cooling for 56% of simulated scenarios, followed by AERO (33%) and THH (11%). When considering the power consumption, each one of Zephyrus and AERO required lower electrical energy per mass of cooled grain in 44.5% of the simulated scenarios, while THH resulted in lower values in 11% of the simulated scenarios. This indicated that the control strategies based on simulations were more efficient in harnessing the aeration potential of different regions and storage conditions.

In addition to the greater efficiency, the proposed controller is more flexible regarding the aeration and thermometry designs when compared with the THH control strategy, also automatically adjusting its settings according to different regions and seasons due to its capacity of simulating the aeration process and relating it to the real-time data acquisition. Regarding the AERO controller, Zephyrus was capable of reaching higher grain cooling with similar electrical energy saving, also maintaining safe grain moisture contents during the storage period. Further, the proposed control strategy requires less storage parameters for its decision-making, as well as a simpler mathematical model. This results in less complex programming codes and computing times, facilitating its implementation for professionals who work in the postharvest sector.

Considering the aeration energy requirements, scenario 2 was the only one which presented significant difference between the other ones, when comparing the average of the three controllers. This scenario resulted in larger electrical energy requirements for the three controllers (Table 3), while achieved relative lower grain cooling (Table 2). Probably, this occurred due to the simulated aeration in small storage, which cooled most rapidly in the late fall and warmed most rapidly in the early spring.

As expected, scenario 3, which considered a small silo aerated from late fall to early winter in a dry region, presented the most efficient grain cooling, with the three control strategies reaching grain temperature reductions close to 16 °C (Table 2). This scenario also resulted in small electrical energy consumptions for the three controllers, when compared with scenarios 1 and 2, that represented similar aeration systems. For a same region, storage capacity, aeration airflow, and stored grain, lower grain cooling was observed when aeration was performed during early spring or summer (scenarios 2, 6, and 7). Significant differences were noted among the temperature reductions of the three control strategies for these scenarios and the 4 and 5 ones, probably due to the aeration period which presented limited number of suitable aeration hours. Additionally, scenarios 4, 5, and 6 simulated the aeration with low airflow in a wet region. These results evidenced the importance of control strategies capable of automatically adjusting their set points according to different aeration systems and weather conditions, in order of taking maximum advantage of the aeration potential of each region, season, and system design.

4 Conclusions

In this work, an effective, economical, and simple control strategy to manage grain aeration was presented, based on the prediction of speeds and changes of temperature and moisture fronts while air is passed through the grain bulk. The experimental tests with Zephyrus showed that this controller has potential for cooling grain bulks, also maintaining safe moisture content and requiring lower electrical energy consumption. These effects contribute for reducing insect and mold activities, minimizing grain deterioration, and increasing the safe storage time. Results of simulations and comparisons between Zephyrus and other efficient controllers confirmed the potential of the proposed strategy for meeting the needs of optimum grain storage and electrical energy saving, automatically adjusting its set points to different regions, grain types, initial grain conditions, and seasons. Additionally, the proposed control strategy can be easily adapted to different data acquisition systems and aeration designs as well, contributing to optimizing the use of aeration and preventing storage losses by using a chemical-free technology. Studies using the proposed controller in real aeration systems located at different world regions and storing different grain types are encouraged. Additionally, future works include the development of applications based on Zephyrus strategy for monitoring and controlling aeration systems from mobile devices and wireless networks. This leads to some challenges, such as the specific requirements of this kind of software regarding external devices like sensors and actuators, and security constraints involving data privacy and information sharing.

Acknowledgments The authors would like to acknowledge the FAPEMIG/Brazil (CAG—APQ-01389-15) for the financial support.

References

1. Serna-Saldivar, S. O., & García-Lara, S. (2016). Cereals: Storage. *Encyclopedia of Food and Health*, 712–717.
2. Navarro, S., Noyes, R. T., Casada, M., & Arthir, F. H. (2012). Grain aeration. In D. W. Hagstrum et al. (Eds.), *Stored product protection* (pp. 121–134). Kansas State University.
3. Navarro, S., & Noyes, R. T. (2001). *The mechanics and physics of modern grain aeration management*. CRC Press.
4. Edde, P. A. (2012). A review of the biology and control of *Rhyzopertha dominica*(F.) the lesser grain borer. *Journal of Stored Products Research, 48*, 1–18. https://doi.org/10.1016/j.jspr.2011.08.007
5. Neme, K., & Mohammed, A. (2017). Mycotoxin occurrence in grains and the role of postharvest management as a mitigation strategies. A review. *Food Control, 78*, 412–425. https://doi.org/10.1016/j.foodcont.2017.03.012
6. Beckett, S. J. (2011). Insect and mite control by manipulating temperature and moisture before and during chemical-free storage. *Journal of Stored Products Research, 47*, 284–292. https://doi.org/10.1016/j.jspr.2011.08.002
7. Nawi, N. M., Chen, G., & Zare, D. (2010). Economics of using aerated storage to minimize the impact of weather damage during wheat harvesting. *Biosystems Engineering, 105*(3), 323–331. https://doi.org/10.1016/j.biosystemseng.2009.12.001
8. Steidle Neto, A. J., & Lopes, D. C. (2015). Thermistor based system for grain aeration monitoring and control. *Computers and Electronics in Agriculture, 116*, 45–54. https://doi.org/10.1016/j.compag.2015.06.004
9. Bradna, J., Šimon, J., Hájek, D., & Vejchar, D. (2018). The impact of weather conditions on microclimate in storage facilities. *Agronomy Research, 16*(4), 1580–1589. https://doi.org/10.15159/ar.18.178
10. Coradi, P. C., Oliveira, M. B., Oliveira Carneiro, L., Souza, G. A. C., Elias, M. C., Brackmann, A., & Teodoro, P. E. (2020). Technological and sustainable strategies for reducing losses and maintaining the quality of soybean grains in real production scale storage units. *Journal of Stored Products Research, 87*, 101624. https://doi.org/10.1016/j.jspr.2020.101624
11. Lopes, D. C., & Steidle Neto, A. J. (2019). Effects of climate change on the aeration of stored beans in Minas Gerais State, Brazil. *Biosystems Engineering, 188*, 155–164.
12. Morales-Quiro, A., Campabadal, C. A., Maier, D. E., Lazzari, S., Lazzari, F., Cook, S., & Phillips, T. W. (2016). Chilling aeration to control pests and maintain grain quality during in-bin storage of wheat in Kansas. In *ASABE annual international meeting, American Society of Agricultural and Biological Engineers*, pp 1–9.
13. Arthur, F. H., Morrison, W. R., & Trdan, S. (2020). Feasibility of using aeration to cool wheat stored in Slovenia: A predictive modeling approach using historical weather data. *Applied Sciences, 10*(17), 6066. https://doi.org/10.3390/app10176066. www.mdpi.com/journal/applsci
14. Li, X., Han, Z., Lin, Q., Wu, Z., Chen, L., & Zhang, Q. (2020). Smart cooling-aeration guided by aeration window model for paddy stored in concrete silos in a depot of Guangzhou, China. *Computers and Electronics in Agriculture, 173*, 105452. https://doi.org/10.1016/j.compag.2020.105452
15. Onibonoje, M. O., Kehinde, L. O., & Owolarafe, O. K. (2015). A wireless sensor network for controlling the effect of the moisture content in stored maize grains. *International Journal of Engineering Research & Technology, 4*(10), 141–147.
16. Singh, C. B., & Fielke, J. M. (2017). Recent developments in stored grain sensors, monitoring and management technology. *IEEE Instrumentation & Measurement Magazine*, 1094–6969.
17. Kaliyan, N., Morey, R. V., Wilcke, W. F., Carrilo, M. A., & Cannon, C. A. (2007). Low-temperature aeration to control Indianmeal moth, *Plodia interpunctella* (Hubner), in stored grain in twelve locations in the United States: A simulation study. *Journal of Stored Products Research, 43*, 177–192. https://doi.org/10.1016/j.jspr.2006.04.004

18. Charles, A. (2006). *Ambient air temperature aeration controller*. Doctoral dissertation, University of Southern Queensland, Australia.
19. Lopes, D. C., Martins, J. H., Lacerda Filho, A. F., Melo, E. C., Monteiro, P. M. B., & Queiroz, D. M. (2008). Aeration strategy for controlling grain storage based on simulation and on real data acquisition. *Computers and Electronics in Agriculture, 63*, 140–146. https://doi.org/10.1016/j.compag.2008.02.002
20. Sun, L., & Zhu, Z. S. (2013). A Bayesian network model for aeration management of stored grain. In *Applied mechanics and materials* (pp. 4751–4756). Trans Tech Publications.
21. Thorpe, G. R. (2008). The application of computational fluid dynamics codes to simulate heat and moisture transfer in stored grains. *Journal of Stored Products Research, 44*, 21–31. https://doi.org/10.1016/j.jspr.2007.07.001
22. Nwaizu, C. C. (2013). *Characterizing airflow paths in grain bulks*. Dissertation. University of Manitoba, Canada.
23. Hunter, A. (1988). Temperature and moisture front movement in an aerated seed bulk. *Journal of Agricultural Engineering Research, 40*, 113–127.
24. Sutherland, J. W., Banks, P. J., & Griffith, H. J. (1971). Equilibrium heat and moisture transfer in air flow through grain. *Journal of Agricultural Engineering Research, 16*(4), 368–386.
25. Lopes, D. C., Steidle Neto, A. J., & Santiago, J. K. (2014). Comparison of equilibrium and logarithmic models for grain drying. *Biosystems Engineering, 118*, 105–114.
26. Melo, E. C., Lopes, D. C., & Corrêa, P. C. (2004). GRAPSI: Programa computacional para o cálculo das propriedades psicrométricas do ar. *Engenharia na Agricultura, 12*(2), 154–162.
27. Lopes, D. C., Martins, J. H., Monteiro, P. M. B., & Lacerda Filho, A. F. (2010). Effects of different control strategies on grain storage in tropical and subtropical regions. *Ceres, 57*(2), 157–167.
28. Mohapatra, D., Kumar, S., Kotwaliwale, N., & Singh, K. K. (2017). Critical factors responsible for fungi growth in stored food grains and non-Chemical approaches for their control. *Industrial Crops and Products, 108*, 162–182. https://doi.org/10.1016/j.indcrop.2017.06.039
29. Taruvinga, C., Mejia, D., & Alvarez, J. S. (2014). *Appropriate seed and grain storage systems for small-scale farmers: Key practices for DRR implements*. FAO. Retrieved Accessed Dec 8, 2016, from http://www.fao.org.br.
30. Navarro, S. (2012). Advanced grain storage methods for quality preservation and insect control based on aerated or hermetic storage and IPM. *Journal of Agricultural Engineering, 49*(1), 13–20.
31. Martins, J. H., Mota, A. M., & Fonseca, J. A. (2001). Simulation of an automatic controller for stored grain aeration systems. *Engenharia na Agricultura, 9*(1), 55–70.
32. Mrema, G. C., Gumbe, L. O., Chepete, H. J., & Agullo, J. O. (2011). Grain crop drying, handling and storage. In *Rural structures in the tropics: Design and development* (pp. 363–411). FAO.
33. Jian, F., & Jayas, D. S. (2012). Temperature monitoring. In D. W. Hagstrum et al. (Eds.), *Stored product protection* (pp. 271–280). Kansas State University.
34. Thorpe, G. R. (1997). Modelling ecosystems in ventilated conical bottomed farm grain silos. *Ecological Modelling, 94*, 255–286.
35. Lopes, D. C., Martins, J. H., Melo, E. C., & Monteiro, P. M. B. (2006). Aeration simulation of stored grain under variable air ambient conditions. *Postharvest Biology and Technology, 42*, 115–120. https://doi.org/10.1016/j.postharvbio.2006.05.007
36. Thorpe, G. R. (2001). Physical basis of aeration. In S. Navarro & R. T. Noyes (Eds.), *The mechanics and physics of modern grain aeration management* (pp. 125–194). CRC Press.
37. Jayas, D. S., & White, N. D. G. (2003). Storage and drying of grain in Canada: Low cost approaches. *Food Control, 14*, 255–261. https://doi.org/10.1016/S0956-7135(03)00014-8
38. Conab. (2018). *Companhia Nacional de Abastecimento. Acompanhamento da safra brasileira de grãos*, vol 6, Safra 2018/19, Brazil.
39. Inmet. (2018). Banco de dados meteorológicos para ensino e pequisa (BDMEP). Reftrieved October 9, 2018, from http://www.inmet.gov.br/projetos/rede/pesquisa/.

40. Alvares, C. A., Stape, J. L., Sentelhas, P. C., Gonçalves, J. L. M., & Sparovek, G. (2013). Köppen's climate classification map for Brazil. *Meteorologische Zeitschrift, 22*(6), 711–728. https://doi.org/10.1127/0941-2948/2013/0507
41. Ferreira, D. F. (2011). Sisvar: A computer statistical analysis system. *Ciên Agrotec, 35*(6), 1039–1042.
42. Collins, L. E., & Conyers, S. T. (2009). Moisture content gradient and ventilation in stored wheat affect movement and distribution of Oryzaephillus surinamensis and have implications for pest monitoring. *Journal of Stored Products Research, 45*, 32–39. https://doi.org/10.1016/j.jspr.2008.07.003
43. Sanderson, D. B., Muir, W. E., & Sinhá, R. N. (1988). Moisture contents within bulks of wheat ventilated with near-ambient air: Experimental results. *Journal of Agricultural Engineering Research, 40*, 45–55.
44. Foster, G. H. (1967). Moisture changes during aeration of grain. *Transactions ASAE, 10*(3), 344–351.
45. Weinberg, Z. G., Yan, Y., Chen, Y., Finkelman, S., Ashbell, G., & Navarro, S. (2008). The effect of moisture level on high-moisture maize (Zea mays L.) under hermetic storage conditions—In vitro studies. *Journal of Stored Products Research, 44*, 136–144. https://doi.org/10.1016/j.jspr.2007.08.006
46. Antunes, A. M., Devilla, I. A., Borges Neto, A. C., Alves, B. G. X., Alves, G. R., & Santos, M. M. (2016). Development of an automated system of aeration for grain storage. *African Journal of Agricultural Research, 11*(43), 4293–4303. https://doi.org/10.5897/AJAR2016.11538

Decision-Making Applications on Smart Livestock Farming

Irenilza de Alencar Nääs and Jair Minoro Abe

1 Smart Livestock Farming

According to the Food and Agriculture Organization of the United Nations [1], there has been an increase in demand for animal protein products in the last years, such as eggs and milk, although maintaining a steady meat demand level by the consumers (Fig. 1). On the other hand, a large rural younger population has moved to urban areas in recent years, leaving a gap in the rural labor force in developed countries. Although this movement is noticeable in high-income countries, it is also a trend in developing countries [2]. Therefore, farmers need to rely on fewer workers and a more qualified labor force to carry on livestock production tasks. Such a scenario led to applications of information technology (IT) in the decision-making approach in rural activities, including livestock farming.

Another consequence of this shift in rural areas, associated with the increasing demand for animal products, is the decline in the number of farms [3]. As a noticeable effect, the number of stock units on livestock farms will continue to rise. Farmers also benefit from raising more animals and reaching a viable income. As a result, modern farmers have less time to attend to individual animals, making it more challenging to manage the animals properly [4].

Besides the difficulty of multiple tasks, new goals related to animal welfare standards and the reduction of environmental impact are striking livestock farmers. Raising profitable livestock, applying proper husbandry, has become a complex process of connecting growth, health, welfare, and environmental issues. Farmer's managerial decisions are typically based on the personal experience of the stockman, who often estimates each effect of a controlled action over time. A prescriptive

I. de Alencar Nääs (✉) · J. M. Abe
Graduate Program in Production Engineering, Universidade Paulista, São Paulo, Brazil
e-mail: irenilza.naas@docente.unip.br

D. D. Bochtis et al. (eds.), *Information and Communication Technologies for Agriculture—Theme III: Decision*, Springer Optimization and Its Applications 184, https://doi.org/10.1007/978-3-030-84152-2_10

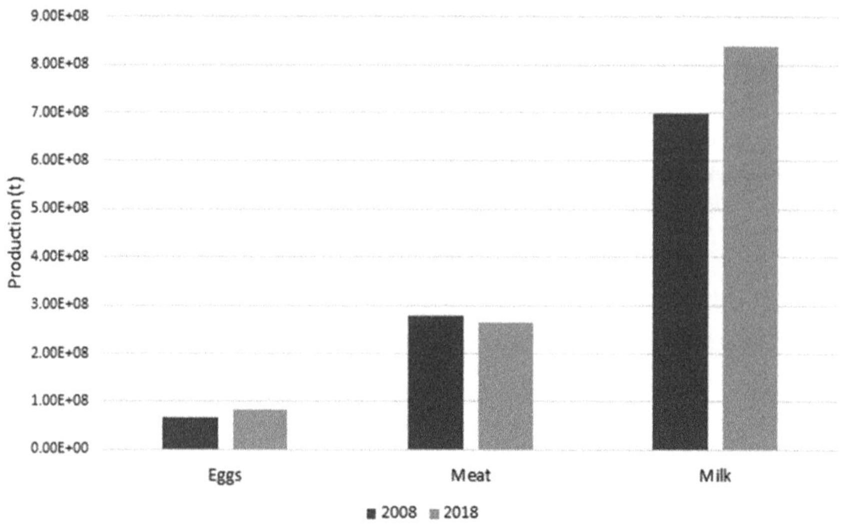

Fig. 1 The scenario of the world's animal protein production. Adapted from [1]

nutrition regime is applied to reach the best results for the specific genetic and management (input). It is expected to produce a required performance based on the growth model [5].

The systems approach to a modern livestock farm integrates the processes inflows looking for the best outputs. The procedure needs to be optimized by reducing the losses [6]. Banhazi and Black (2009) [7] point out the critical steps that need to be combined within a well-planned and controlled production process:

- Integrating the automatic data assessment systems into the production chain
- Developing the protocols for data-combination and automatic data analysis to identify errors in the processes and improve decision-making [8]
- Transferring the results from data analysis as inputs into automatic processes and activating management actions
- Triggering the control systems into standard operating procedures
- Incorporating the procedures to monitor the outcome of control actions [3]

The progress in sensor, computer, and analytical tools technology has made the collection and analysis of large amounts of data highly viable. An on-farm data collection system reduces the requirement for manual data recording and presents the producer with management data in a cost-effective form [7, 9, 10]. Such initiatives make it possible to transfer details of animal performance, labor input, environmental efficiency, and other crucial information to a central data warehouse, where it could be stored, processed, and sent to a control system on the farm.

Such a vision has been implemented before. Parsons et al. (2007) [5] describe the developing and testing of a prototype closed-loop, model-based, real-time system for the integrated control of pig growth. It starts with the identification of the animals, enabling herd traceability, and follows with the recording of all management procedures during growth and fattening [11]. Berckmans (2014) [12] presents multiple applications on smart livestock farming from the identification of the animals to actions such as the traditional environmental measures, extended with direct measures of animal responses utilizing feed intake sensors, growth monitors, cameras, and microphones. In this context, the animal is used as a sensor, and algorithms translate the measured animal responses into key indicators for optimal performance, improved animal welfare, and farm sustainability. The author also infers that in a future stage, the applications will support the farmer in helping management decisions and generate early warnings when something is going wrong in the production process.

Researchers have developed further actions to improve hardware, associated with smart software to extract information from a wide range of data resulting in several possibilities of applications [7, 10, 12]. For instance, Moura et al. (2008) [10] propose software that performs an analysis of piglet vocalizations in stressful and non-stressful situations, presenting different results on the characterization of the sounds emitted during both conditions. In this case, the recording of sounds associated with the software might help the real-time decisions on management and the assessment of welfare. Another use of precision livestock farming is real-time pig weight detection, proposed by Banhazi and Black (2009) [7]. The automated weight system uses image analysis to acquire sufficient data points at adequate time intervals and, with enough accuracy, as animals visit a feeder, and for identifying fatness and carcass yield. The information on the weights of animals can be incorporated into computer simulation models to calculate the daily ration needed to meet specified growth trajectories based on the animal's genotype and the environmental housing conditions.

Consistent development in smart technologies applications has been made to livestock farming in the last years [13]. However, commercial systems need a combination of hardware, observing technical and safety standards, associated with the software and a good user interface to meet farmers' needs. According to Berckmans (2014) [12], results and potential technology are mostly unknown to animal scientists, veterinarians, and ethologists, due to a lack of collaboration among the professionals. Therefore, a multidisciplinary approach is needed to provide useful smart interfaces that allow the interaction among livestock, farmers, stockmen, and the environment based on appropriate technical solutions.

1.1 Concepts and Fundamentals

The concept of smart livestock farming has been quickly evolving since the beginning of the twenty-first century. One of the first initiatives, by the end of the twentieth century, included using the data extracted from accelerometers in dairy cows at the beginning of the information technology (IT) era [14]. Data from dairy cows' activities led to mathematical model developments [15] to forecast several health assessment behaviors and mating. From that time on, the use of IT in farming progressed rapidly. First in agriculture, leading to the development of precision agricultural production, and then it moved further to livestock farming.

Smart livestock farming (SLF) describes real-time monitoring technologies that manage the smallest manageable production unit's temporal variability, known as "the per animal approach" [4]. In 2004, Berckmans [3] defined such initiative as precision livestock farming (PLF). After the development of several models [3, 5, 9, 10, 16] during the first decade of the twenty-first century, Wathes et al. (2008) [17] reviewed the applications of precision livestock farming from the point of the engineering, the farm animal, and the farmer. The authors concluded that the approach was an embryonic technology with great promise, but it requires considerable research and development before stakeholders' acceptance.

Berckmans (2014) [12] states that PLF helps to manage a farm through the monitoring and recording of automatic, real-time measurements of animal-related aspects of production. Borchers and Bewley (2015) [18] suggest that smart livestock farming is related to technologies that enable assessing various animals' signs to optimize herd/flock performance. Halachmi and Guarino (2016) [4] define the PLF actions as the real-time monitoring technologies to manage the smallest manageable production unit (the concept of "the *per* animal approach").

From all those definitions, we can presume that smart livestock farming allows the interconnection of coordinated actions using sensors to assess animal-related vital signals and production variables during the production process in-farm environment and actuators to infer decisions [19]. Such advancement is a step ahead of precision livestock farming, as previously described, leading to the concept of smart livestock farming (SLF). The input of the system is animal-related data combined with environmental data, and the output is decision-making actions.

The new advancement of electronic devices, improved wireless communication, and internet access improved the development of new resources for farmers. Nowadays, farmers have access to new services such as individual feeding, health monitoring, and animal localization to operate their livestock or flock more efficiently, ensuring productivity, animal well-being, and economic benefits.

The architecture of the system can be described as the action enhanced by the farmers, based on assessment and computing of data. The decision-making process is aligned with the available data from resources (herd/flock monitoring, food and feed resources, and environmental recording) as inputs for the farmers' actions (Fig. 2).

During the process, it is crucial to use IT tools to exchange, store, and transform the database into vital knowledge to optimize the activities. Ultimately, robots will replace some stockman's tasks helping the farmers to control larger herds or flocks.

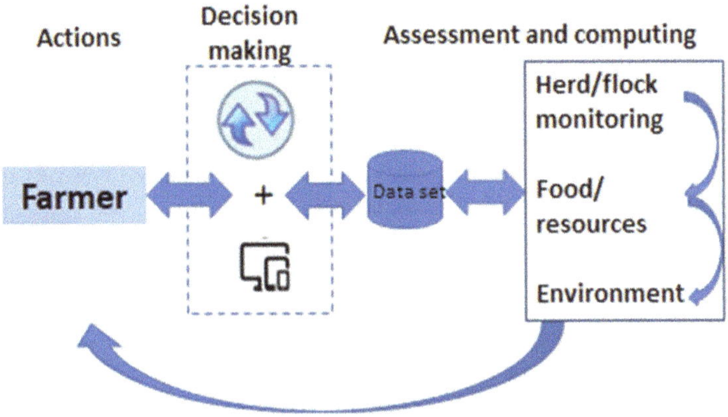

Fig. 2 The general architecture of a smart farming solution. Adapted from [20]

1.2 *Smart Livestock Farming Models Implemented On-farm Actions*

Early on, the stockmen tasks on livestock farms involve herd/flock's follow-up, aiming high efficiency of the production of milk, eggs, beef, pig, and broiler meat. Since the implementation of SLF, developed models and interconnected solutions for optimizing on-site diagnostics in real-time, the electronic and software industry, made available important decision-making tools for workers and farmers.

Identifying the essential tasks for specific livestock processes can be the main challenge, requiring a good deal of creative thinking and analysis. Another way is to undertake strategic planning, where experts across the production value chain identify potential areas where change would considerably affect output. Some are relatively clear, such as removing humans from the dairy cow milking process. Others require more elaborated solutions, such as assessing animal welfare. Here, we present some models applied to pig, poultry, and dairy production.

Pig Production

Pig are homeotherms, and their performance depends on the heat balance between the barn and the ambient temperature as well as the relative humidity. Therefore, swine farms need to rely on proper environmental temperature and relative humidity control to reach optimal performance. Although the ideal housing temperature and other weather characteristics in pig housing are known, to develop a model that expresses all relationships between the environmental conditions and the animal characteristics (growth and reproduction) is a difficult task. A study by Ramirez et al. (2018) [21] describes an approach for evaluating the thermal environment in growing-finishing pig housing. The method uses thermal environmental data as an input from a sensor array associated with estimated body weight. The approach describes a mechanistic thermal balance model and uses the model results to derive a housed swine heat stress index. The developed model analyzes the spatiotemporal thermal environment data indicating the feasibility of the on-field application. The application of such an index provides preliminary insight for using it to assess the thermal housing environment. The index proposal is a significant innovation necessary to improve the housing design, assessment, and control.

The advancements of tools such as Computational Fluid Dynamic (CFD) simulation allowed researchers [22] to evaluate the influence of airflow speed, turbulence intensity, and animal orientation on the convective heat transfer to implement ventilation control algorithms to provide an improved production environment. The study was done using CFD simulations in two geometry models, a pig model with actual size and a simplified cylinder model, in a virtual wind tunnel. The impacts of airspeed, inlet turbulence, air incidence angles, and pigskin modeling on convective heat transfer coefficients were studied. The convection heat transfer of a pig model has a similar trend with the cylindrical model following the increase of airspeed. A correlation was found between the geometry of the actual pig model and the cylinder model. The orientation and turbulence intensity significantly alter the convective heat transfer coefficient of a pig, and the results indicate that the proposed model can be used with limitations in the process of assessing the convective heat transfer of pig.

Several studies present the possibility of predicting weight gain in piggeries [3, 23] using image analysis. The initiative allows farmers to follow up on the herd to target the market both in real-time and using data interpretation. Berckmans (2014) [12] presents a camera-based system positioned in a top-down perspective inside the ceiling of a pig house and generates a floor area visualization. After registering the animals' data, an analysis' software converts the acquired images into indexes of weight prediction, spatial distribution, and activity [24]. These indexes are a measure of the pig's position and movement inside the raising area, and subsequently, make the inference of the animal well-being.

Another way of assessing animal welfare is by using pig vocalization. Vocalization is the generation of sounds to communicate a state of an animal that may occur naturally or from an external event. Most vocalizations are initially recorded in a husbandry system or behavioral framework (such as feeding, social conflict). The recorded calls are subsequently open to the analysis to define specific acoustic parameters. Distinct vocalization means a different degree of excitement on pigs, or their degree of fear or social separation [25]. Support of the understanding of vocal sounds with behavioral parameters allows predicting animal well-being and health status. In this approach, the animal is used as a sensor, and algorithms translate the measured animal responses into key signs for the best performance or improved animal welfare. Such data availability allows the farmer to take management decisions and generate early alerts when something is going wrong in the production process [26], such as the presence of cough, as early warnings of respiratory diseases [27].

Poultry Production

Mathematical models have been used to estimate poultry welfare using pattern recognition. Dawkins et al. (2009, 2012) [28, 29] used a combination of camera and statistical analysis of optical flow patterns to evaluate the behavior of commercial broiler chicken flocks. The movements of the individual birds affect the total activity of the flock, and these were significantly correlated with key welfare measures such as mortality and the numbers of birds with abnormal walking behavior identified in broilers 15–20 days old. Optical flow patterns provided an interesting approach for assessing the welfare of commercial broiler chickens. On the other hand, behavioral models have been used to evaluate welfare status in broiler chickens and presented the evolution of behavioral models applied to animal welfare [30]. The alerts produced by changes in animal behavior can be registered in dashboards on PC or mobile gadgets. Farmers are directly driven to the location of the issue, thus saving time and avoiding production losses. Since most farmers typically can only achieve about 70% of the genetic potential of their herds, the already commercial system indicates that the early warning functionality can result in an improvement in the performance.

Other noninvasive methods have been developed to assess broilers' lameness [31, 32] automatically. Aydin (2017) [31] used some feature variables (speed, step frequency, step length, and the lateral body oscillation) of broilers to predict lameness in the birds. The results showed that all explored feature variables were useful to detect lameness in broilers when the condition was already severe. Such models enabled the estimation of birds' walking ability using pattern recognition techniques.

The displacement of the birds' head is directly related to how the animal eats with direct consequences on its nutrition. A study was carried out to analyze this displacement automatically, by tracking the bird's eye position in two dimensions (2D) in the image, and then to calculate the total movement during each mandibulation [33]. The distance between upper and lower beak tips was determined automatically from the frame. A machine vision procedure was used to carry out the calculations without physical markers attached to the birds' body. The technique consisted of eye detection (as a reference point) to determine the position of the chick's head. The head area was extracted to remove unnecessary background information (part of the picture that is of no interest). The procedure was followed by beak tips detection to analyze the maximum beak gape. Finally, the "removal" of the feed particle pixels was assumed to prevent errors during the beak tips' detection. Four kinematic variables were studied, (1) the maximum beak gape (mm); (2) the head displacement (mm); (3) the time of each mandibulations, and (4) the average head speed (mm s^{-1}) during each mandibulation (calculated by the ratio between displacement and time). Such an initiative is vital in determining the proper feed particle size for each bird's age, leading to appropriate nutrition and performance.

Vocalization acoustic parameters (energy, bandwidth, first and second formant) may offer hints on broiler's condition. By assessing young broilers vocalization, researchers [34] were able to separate the birds according to the genetic strain and gender. The authors also suggest a theoretical model in which implementation may help estimate welfare conditions for broilers. Using the sounds' peak potency [35], they propose a model that predicts the weight of broilers allowing farmers to monitor the growth of broiler chickens during the production cycle. This model is a useful tool to integrate the information provided by the on-site and real-time automatic weighing scale placed in the broiler houses.

Dairy and Beef Production

Several models have been applied to control management actions in ruminant production. The simple and straightforward virtual fence is placed as a visually unseen boundary around individual animals, much like traditional fencing. The separation of individual animals is standard management to prevent fighting while maintained together in the same paddock with a cow in estrous. The general virtual fencing application is keeping animals in or out of an area on the landscape using, for instance, the radio-frequency (RF) signals [36]. Radio-frequency signals have been used in early livestock containment studies that came from ground-based transmitters. It is also being used in the implementation of traceability programs. However, systems in which the RF signals emanate from satellites appear to have the most potential for free-ranging animal control [37]. Because virtual fencing can prompt changes in the landscape rapidly, virtual fencing should only be used in conjunction with adequate soil and plant management practices. Monitoring associated with decision-making involving soils, plants, and animals must be practiced rather than attempting to apply virtual fencing as just a management tool. Such technology may involve less physical labor than conventional animal control methods, but it needs proper use to be profitable [18].

Feeding and milking are the most time-consuming activities on dairy farms. The on-farm work organization varies significantly according to the farm size, staff composition, equipment level, and the farmer's objectives. For a long time, many dairy farmers have used electronic milk meters to precisely measure milk production, pedometers to detect the increased numbers of steps linked to cows in heat, and analyses of milk conductivity to detect health issues such as mastitis [14]. However, in recent years the equipment is available to identify precisely when cows are in heat, calving, infectious and metabolic disorders, and cows' well-being has expanded considerably [18]. Sensors that allow more precise assessment of behavioral factors (standing/lying down, physical activity in several space dimensions, tail movements, time ruminating) are a reality nowadays. Various milk analysis tools to identify milk composition (fat, protein, lactose), hormones (progesterone), quality (leukocytes, blood), and the presence of enzymes (lactate dehydrogenase) are now incorporated into milking robots [38].

Although the automation of tasks inside a dairy farm might decrease the interaction between cow-stockman, technologies can reduce farmers' workloads and reduce the need to handle animals (e.g., a farmer must direct cows into the milking parlor).

New models favor positive interactions, and human–animal contact can friendly occur, such as a dairy farm has a milking robot. A stockman needs to move freely among the herd to check the equipment.

2 Tools for Implementing Decision-Making Applications in Smart Livestock Farming

The implementation of models in SLF actions has a particular process. First, the mathematical modeling evolves to the development of software. Later on, the architecture of sensors and actuators are arranged in a way to promote the connection into receiving and interpreting the animals' signals and applying decision-making strategies [7, 9].

Even though SLF has great potential use, farmers and other stakeholders cannot use these technologies effectively. It is time-consuming to process and analyze the data derived from different sensors in different formats and frequencies. Hence, the adoption of SLF improvements remains a challenge concerning the collection and processing of relevant data and the development of useful responses for the farmer [19].

Regarding the development of automated technologies for helping (or replacing) the stockman's decision, most actions rely on input or output variables recorded by sensors. According to the earlier development of decision-making software for indicating the welfare status of animals [12, 20], the output of the sensors (such as the activity assessment with a video camera or sound measured using a microphone) is related to animal-based welfare and health indicators (such as aggression or respiratory diseases) [4]. When sensor signals start to diverge from their supposed values, alerts are given to the farmer or the stockman. In this way, the farmer can take immediate action before the detected change in animal response negatively affects production. These actions include solving technical problems such as a blocked feeding line, adapting control settings in the climate and feed controller, or still launching a preventive medical treatment in the animals.

Most of the software and hardware employed in these solutions need essential mathematical models that adequately describe the biological signals emitted by the animals. Biological responses do not always present straightforward or linear outputs; on the contrary, the animals' responses most of the time contain inconsistencies and variations, leading the researchers to find solutions in the field of non-classical logic. For instance, Leliveld et al. (2017) [39] found that some pig vocal signals are better suited to provide information than others, indicating a degree of inconsistency that should be considered when pig vocalization is used to identify a stressful condition.

2.1 Paraconsistent Logic Applications

As seen in previous sessions, there is a probability of having uncertain data, lack of data, or conflicting data. As it is known, classical logic and related ones are not suitable for the task. Thus, the use of new logics is needed that support theory that can be automated and implemented, either in the form of software or hardware. In computing, it is known that a good solution depends fundamentally on a good representation, that is, on an appropriate (formal) language. Moreover, there is a need to make sensible inferences from obscure hypotheses frequently. Therefore, more general and formal techniques are necessary for the issues discussed. In what follows, the Paraconsistent Annotated Evidential Logic Eτ is useful for solving problems related to imprecision, incompleteness, and inconsistency [40].

Roughly speaking, logic is called paraconsistent if it can serve as the underlying logic of theories. It can encompass propositions p and $\neg p$ (the negation of p) both true, without the danger of trivialization. A logic is called paracomplete if it serves as underlying logic that theories based on it contain propositions p and $\neg p$ both false, theories based on it can encompass propositions p and $\neg p$ both false, without trivialization [41].

The focus here is on a particular paraconsistent and paracomplete logic, namely the paraconsistent annotated evidential logic Eτ—logic Eτ. The atomic formulas of the logic Eτ is the type $p_{(\mu, \lambda)}$, where $(\mu, \lambda) \in [0, 1]^2$ and [0, 1] is the real unitary interval (p denotes a propositional variable). An order relation is defined on $[0, 1]^2$: $(\mu_1, \lambda_1) \leq (\mu_2, \lambda_2) \Leftrightarrow \mu_1 \leq \mu_2$and $\lambda_2 \leq \lambda_1$ that is constituting a lattice symbolized by τ. A detailed account of annotated logics can be found in [41].

$p(\mu, \lambda)$ can be intuitively read (among others): "It is assumed that p's favorable evidence degree (or belief, probability, etc.) is μ and contrary evidence degree (or disbelief, etc.) is λ." Thus, (1.0, 0.0) naturally indicates the total favorable evidence, (0.0, 1.0) indicates the total unfavorable evidence, (1.0, 1.0) indicates the total inconsistency, and (0.0, 0.0) indicates the total paracompleteness (absence of information). The operator $\sim|\tau| \rightarrow |\tau|$ defined in the lattice $\sim[(\mu, \lambda)] = (\lambda, \mu)$ works as the "meaning" of the logical negation of Eτ.

The consideration of the values of the belief degree and disbelief degree is made, for example, by specialists who use heuristics knowledge, probability, or statistics.

We can consider several important concepts (all considerations are taken with $0 \leq \mu, \lambda \leq 1$):

- Segment *DB*—segment perfectly defined: $\mu + \lambda - 1 = 0$
- Segment *AC*—segment perfectly undefined: $\mu - \lambda = 0$
- Uncertainty Degree: $G_{un}(\mu, \lambda) = \mu + \lambda - 1$; Certainty Degree: $G_{ce}(\mu, \lambda) = \mu - \lambda$;

To fix ideas, with the uncertainty and certainty degrees, we can get the following 12 regions of output: *extreme states* that are, False, True, Inconsistent and Paracomplete, and *non-extreme states*. All the states are represented in the lattice of Fig. 3: the standard Cartesian system can represent such lattice τ (Table 1).

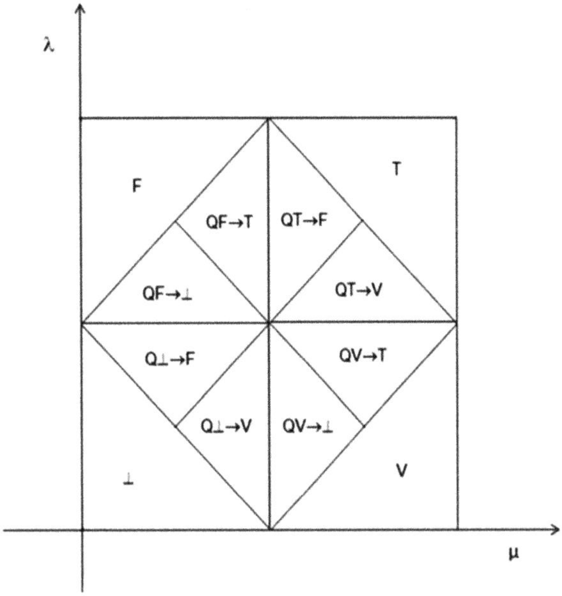

Fig. 3 Representation of the extreme and non-extreme state's regions

Table 1 Extreme and non-extreme states

Extreme states	Symbol	Non-extreme states	Symbol
True	V	Quasi-true tending to Inconsistent	$QV \rightarrow T$
False	F	Quasi-true tending to Paracomplete	$QV \rightarrow \perp$
Inconsistent	T	Quasi-false tending to Inconsistent	$QF \rightarrow T$
Paracomplete	$\perp$	Quasi-false tending to Paracomplete	$QF \rightarrow \perp$
		Quasi-inconsistent tending to True	$QT \rightarrow V$
		Quasi-inconsistent tending to False	$QT \rightarrow F$
		Quasi-paracomplete tending to True	$Q\perp \rightarrow V$
		Quasi-paracomplete tending to False	$Q\perp \rightarrow F$

These states can be described with the values of the certainty and uncertainty degrees using suitable equations. In this text, we have chosen resolution 12 (number of the regions considered according to Fig. 3). However, the resolution is entirely dependent on the precision of the analysis required in the output, and it can be externally adapted according to the applications.

So, such limit values called Control Values are:

$$V_{cic} = \text{maximum value of uncertainty control} = C_3$$

$$V_{cve} = \text{maximum value of uncertainty control} = C_1$$

$$V_{cpa} = \text{maximum value of uncertainty control} = C_4$$

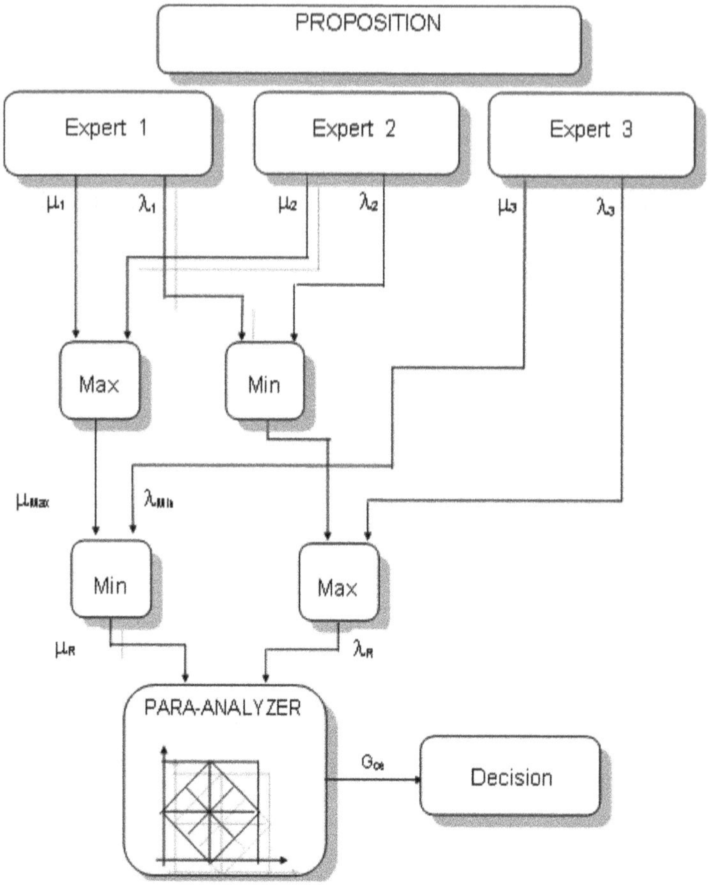

Fig. 4 Scheme of an expert system based on logic Eτ and Para-analyzer algorithm

$$V_{cfa} = \text{maximum value of uncertainty control} = C_2$$

For the discussion in the present text, we used: $C_1 = C_3 = \frac{1}{2}$ and $C_2 = C_4 = -\frac{1}{2}$.

In what follows, a piece of software programmed (based on logic Eτ) using Artificial Intelligence techniques is presented by using databases of expert knowledge to offer advice or make decisions. The expert system is as follows:

(1) Choose a statement regarding the problem considered—for instance, "the pig is stressed."
(2) Regarding the problem, we proceed to consider all parameters needed to analyze the problem.
(3) Once the data is obtained, it is needed to list the experts. In this case, for "expert," it can mean any agent that has expertise on the theme (Fig. 4).

The problem must be analyzed in all necessary aspects for the most complete analysis possible. Thus, for a parameter, it is desirable to have several groups of specialists who will analyze it from different angles. Parameters can have different importance. Each parameter is assigned a weight to map this characteristic. Finally, the idea is for the parameter to be analyzed globally by the various groups of specialists. Another important observation is that each specialist is different from the other. Thus, specialists may also receive different weights.

In what follows, we present how evidence (favorable and contrary) inferred by specialists can be combined. The reader can find the details of each issue in the reference [41].

When we have a complex problem, we express it as a proposition p. All relevant parameters (factors, characteristics, etc.) are considered in the analysis. Also, we find a group of experts capable of giving their opinions about such parameters expressed as favorable evidence and also unfavorable evidence plus external data. Here lies a fundamental difference regarding the other logics. The majority of logical systems (including Fuzzy logic) consider the positive fact of *p*. In this way, the language of the logic Eτ is more faithful in describing the portion of reality in consideration.

Applications

The following examples were selected to present the application of Paraconsistent Annotated Evidential Logic Eτ in processes on smart livestock farming.

Poultry Production

The walking ability of poultry is an essential issue in the assessment of animal welfare. The evaluation is typically done based on visual evaluation of gait scores (GS) from 0 (sound bird) to 5 (lame bird). The extremes are easily detectable; however, the intermediate scores (GS2 and GS3) are not evident. A study developed a paraconsistent algorithm using broiler's speed and acceleration from the gait score of individual broiler's displacement in a walking platform [32]. The velocity and acceleration data were analyzed using paraconsistent logic, an algorithm was developed, and software could forecast the intermediate values of broiler gait. Such an approach required applying a non-classic logic as the data showed a large number of inconsistencies and overlap results of gait prediction. Results indicated that it was possible to detect lameness in broilers under commercial rearing using the developed software since the speed of displacement can be measured and used as the input data by the growers.

2.2 *Pig production*

Infrared skin temperature was used to assess stressful conditions in piglets [42]. A total of 18 propositions were analyzed using the Paraconsistent Annotated Evidential Logic *Eτ* considering the minimum, maximum, and average infrared skin temperature, in the following conditions: pain, fear, hunger, thirst, and normal (baseline), in the phases of nursery and farrowing. Each group is a mathematical proposition to be solved, and each group was separated into two subgroups (A and B) to be compared among themselves. For the farrowing and nursery stages, the recorded data were analyzed considering three groups (male and female separately, and both sexes grouped) to verify whether the skin temperature was affected by the sex. The imaging technique is subject to certain contradictions and uncertainties that require mathematical modeling. The paraconsistent logic was applied to extract and analyze the contradiction from the data. The stress condition that had higher accuracy in the detection was that predicted by the cold (100%) using the infrared minimum skin temperature (ISTmin), and ISTmin plus the piglet sex, and thirst (91%) using infrared maximum skin temperature (ISTmax) and ISTmax plus the piglet sex. The highest estimate of hunger was found using ISTmin (86%). Results suggest a promising assessment of stressful conditions in piglets using infrared skin temperature (Fig. 5).

Another example of applying the Paraconsistent Annotated Evidential Logic Eτ is the use of pig vocalization for identifying the stressful condition. A study by Silva et al. (2019) [43] indicates the feasibility of using vocal signals intensity to predict if the piglet is stressed. The results of the pig stress using the paraconsistent algorithm are analyzed using the sound wave intensity *vs.* the duration of the vocal signal. Effects of cold/heat stress and pain indicate that most of the sounds related to cold/heat stress lasted 0.2–1.6 s with an intensity between 70 and 86 dB. The vocalization

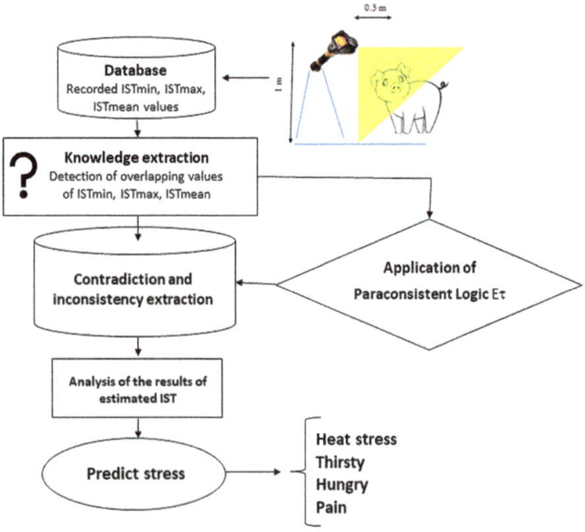

Fig. 5 Schematic view of the decision-making flowchart for predicting stress in piglets using Paraconsistent Annotated Evidential Logic Eτ. Adapted from [42]

related to pain lasted from 0.2 to 1.7 s, and with an intensity that varied mostly from 80 to 93 dB. This knowledge allows the reduction of the contradiction and overlapping in the results making the stress assessment more precise.

2.3 *Use of Machine Learning on Livestock Production*

The challenge to the successful application of smart livestock farming systems relies on an abundance and the quality of data. Sensors and wearable technologies can be implanted on animals to detect their behavior and performance [44], generating a large set of data that might allow researchers to understand a herd or flock or yet the animal response to the environment, a diet, or management. Such large amounts of data with the increased capacity for data storage associated with automated technologies have been quickly generating large-scale data in livestock production and other agricultural activities.

Machine learning is a subfield of artificial intelligence dedicated to the study of algorithms for prediction and inference. The large datasets available in livestock production nowadays provide the prospect to assess beyond classical associations, and data can be used to classify observations into relevant subcategories or forecast outcomes. Learning from data is the essence of machine learning. Data mining takes a comparable venue with machine learning and is often considered in the same perspective. Data mining includes the study of database systems, which becomes crucial in dealing with large datasets. Predictions from decision trees are made using "if-then" rules applied to the dataset, resulting in more understandable predictive models. Ensemble approaches merge multiple models to provide a more accurate predictive model contrasted to single-model methods. Random forests are a popular type of ensemble model that will often provide better accuracy than single trees. In general, multiple trees are built to predict the same target variable using different combinations of predictors and subsets of the data. These multiple models are then associated using various techniques to determine the final projected outcome [45].

Prediction involves distinct analytic methods (such as traditional inferential statistics), machine learning, and data mining to define specific patterns. While predictive analytics generate a model to predict an output, traditional statistics do not describe the relationships among data [46]. The tools used in predictive analytics are most effectively applied to large datasets, and these evaluations may not require the same assumptions regarding homogeneous variances and other traditional statistical assumptions. Comparable to nonparametric statistics, various classification algorithms do not assume the underlying distribution of the covariates.

The predictive analytic process involves selecting a target variable, managing the data, partitioning the data, creating algorithms, improving algorithms, and finally, comparing the created classifiers' accuracy. The model building process is shown in Fig. 6.

Before the model building process, the data need to be randomly partitioned into the data subsets. The amount of data placed in each subset varies with project goals

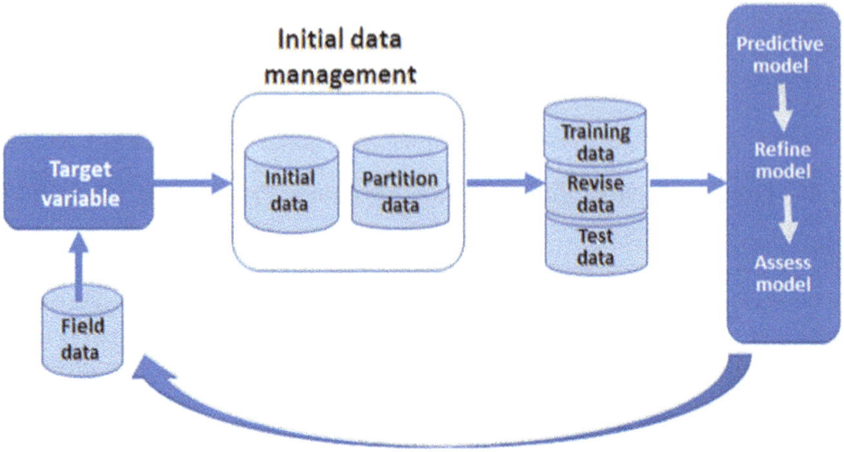

Fig. 6 Schematic diagram representing a systematic framework for predictive workflow containing data management, data partitioning, model building, refining predictive models and accuracy assessment. Adapted from [45]

and frequency of target variable of interest. The nature of the data affects the partitioning data method, and a structured randomized approach may be considered to divide observations among the data partitions.

Applications

We selected some applications of data mining and machine learning in smart livestock farming.

Dairy Production

A data mining model was developed to detect post-calving health problems in robotically milked dairy cows [47]. The model was based on the sensors available in robotic-milking farms assessing individual cows' rumination times, activity, milk yield, body weight, and milking robot visits. The overall model accuracy was up to 92% (sensitivity was 86%, and the specificity was 95% running on the calibration database) down to 84% and 93% on two-party groups. For cows of third or more calving, the accuracy was 72%. According to the authors, the use of additional sensors might improve the model. The presented model seems a useful tool to give the farmer information about the potential health-risk status of a cow, meaning that it identifies the cows that should be separated, and which cows should be left undisturbed.

Poultry Production

A model was developed for predicting broiler breeders' behavior using image analysis and data mining [48]. After analyzing the images from a video where breeders showed several specific behaviors, the authors evaluated each frame's sequence determined for each behavior. Data trends were examined using data mining to develop a classification tree. The model J48 of supervised classification was applied in Software Weka®. Two training algorithms were adopted to validate the tree model (cross-validation and set training). Cross-validation was applied to a dataset divided into ten parts. Nine parts were selected to generate one iteration model, and the remaining data were used for validation. The methodology aimed to identify broiler breeders' behaviors utilizing machine vision techniques presented an accuracy higher than 70% for both validation methods adopted.

In another example, Vale et al. (2008) [49] built a classification model (decision tree) using data mining techniques for predicting broiler mortality caused by heat wave incidence in a tropical country and an open-sided house. The authors used a database from a commercial poultry production and the meteorological characteristics of the region. The root of the found tree was the attribute wind speed (measured daily). Even though the variable wind speed is well known as significant mitigation for heat stress, the class precision found for the flock high mortality was low (0.50). The model infers that when the wind is low, then the mortality is high. When the wind is high, one needs to check the average dry bulb temperature. If the temperature is higher than 26 °C, then there will be high mortality. If the temperature is less than 26 °C, then it is probable that the mortality is low. The predictive model of broiler mortality can be applied to conditions in which the housing environment directly reflects the external environmental conditions, for example, only natural ventilation is used.

Pig Production

Using the sound signals parameters in piglets, Cordeiro et al. (2013) [26] developed a classification data mining model. The decision tree was built using the C4.5 algorithm in the software Weka®, considering the minimum number of objects per leaf equal to nine. The authors were able to classify three types of distress (normal was the baseline) during the farrowing phase (pain, cold, and hunger) with an accuracy rate of 82%. The class with the highest precision rate (a type of accuracy for a specific category of the data) was the pain distress (99%), followed by the normal welfare status (baseline) (0.90), the cold distress (89%), and the hunger distress (69%).

An image-based walk-through system was developed to predict the pig's live weight without restricting the animal [50]. The idea was to predict the weight based on standard measurements of the pig body (Fig. 7). The machine learning approach was used to correlate an array of physical characteristics extracted from the walk-through images to pig live weight in an attempt to enhance the accuracy of the live

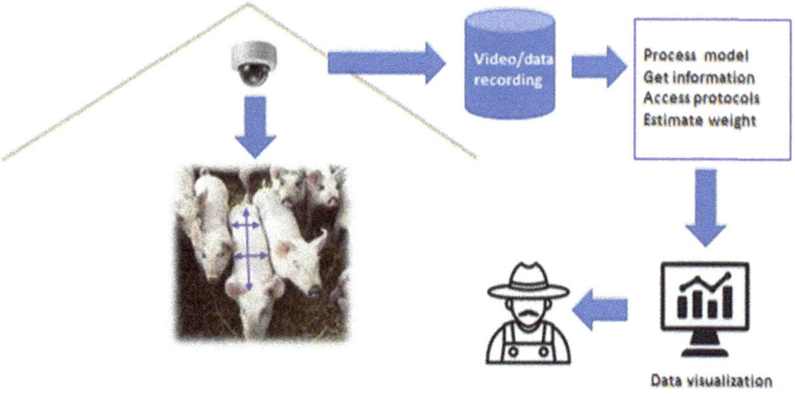

Fig. 7 Schematic flow chart representing the pig weight prediction from the data collection to data processing and the prediction

weight prediction. Results showed that the average relative error of the walk-through weighing system was around 3%. The proposed approach has made it easier for stockmen to forecast the live weight of pigs during production.

2.4 *Technical Challenges*

The use of smart farming is related to the technological development of robust, low-cost sensing systems, data-based models of the important biological processes with significant parameters, and control systems that can manage two or more interacting physical and/or biological processes [17]. The appropriate use of sensors output towards the accuracy and reliability of the collected data are significant issues that directly impact the design of embedded electronics appliances. Data transmission is a complicated task in farm indoor and outdoor environments due to multipath propagation shadowing or even signal attenuation in the field [20]. We believe such a challenge will be solved as new sensors are continuously developing and reaching the market.

The main challenge on data collection and transmission remains in maintaining an optimal trade-off among different aspects such as accuracy, reliability, and transmission together with adequate size and weight of fixed devices carried by animals in harsh environments. Moreover, offering enough power autonomy to assure the correct operation during extended periods is a way to overcome this issue [4].

3 Final Remarks

Developments in modern IT tools have enabled the use of advanced online mathematical identification techniques in smart farming. These modeling methods estimate the unknown model parameters of an abstract mathematical model structure based on an online assessment of process inputs and outputs. The model parameters can be estimated online during the process, resulting in an adaptive model that can associate with the characteristics of most biological processes. However, the model constraints and structure do not support a biological understanding of the causal mechanisms. In the simplest case, a smart farming process may have one system input and one system output; however, in practical applications, there are several process inputs and outputs that interact, resulting in a complex multiple-input, multiple-output systems [17]. Moreover, biological processes are not linear, and the issue is to build tools to establish the biological meaning of a model's structure, order, and parameters. The use of non-classic logic might allow solutions for the biological response models and managerial processes in smart farming since it deals with uncertainties. So far, current literature acknowledges the parameters to be interpreted based on the underlying biological or physical processes. The challenge is to define the biological meaning of a model's structure, order, and constraints to serve as the basis of decision-making tools.

Smart livestock farming offers possibilities to automate the assessment of several procedures on-site. The concept allows the coordinated use of sensors to assess parameters (behavioral, physiological, and production) in livestock and the characteristics of the rearing environment (temperature, relative humidity, and ventilation), and enough accurate information and adequate communication technologies to exchange, store, transform, and restore this information [51]. In doing so, we enable farmers to support decision-making in combination with their observations and expertise. More recently, several commercial systems (Weight-Detect by PLF-Agritech, eYeScan by Fancom BV, Pigwei by Ymaging, OptiScan, GroStat, and WUGGL) have been introduced to the market as Precision livestock farming solutions [52]. Several technologies are available nowadays, and soon updated solutions will be presented to entitle farmers and stockmen the use of automated and straightforward monitoring systems. Such a development is connected to the advancement in sensors capturing behavior, vocalization, and other animal characteristics to act as actuators in decision-making processes. The idea is to provide the license to produce and meet the increasing demand for cheap and healthy food products, as previously proposed by researchers [3, 4, 8].

The players in the food-chain segment will understand the value of the information they produce soon. Not only to themselves, but other stakeholders (feeding companies, veterinarians, breeding companies, consultants, processing plants, retailers), and the consumers. As a result, the desire for data follow-up will increase, resulting in a completely different approach to animal protein production (meat, egg, and milk). This is the imminent smart livestock farming.

References

1. FAO. Food and Agriculture Organization of the United Nations. Retrieved October 15, 2020, from http://www.fao.org/faostat/en/#data/QL.
2. Brueckner, J. K., & Lall, S. V. (2015). Cities in developing countries: Fueled by rural-urban migration, lacking in tenure security, and short of affordable housing. In G. Duranton, J. V. Henderson, & W. Strange (Eds.), *Handbook of regional and urban economics* (pp. 1399–1455). Elsevier.
3. Berckmans, D. (2004). Automatic online monitoring of animals by precision livestock farming. In: *Proceedings of the ISAH conference on animal production in Europe: The way forward in a changing world, Saint-Malo, France*, pp. 27–31.
4. Halachmi, I., & Guarino, M. (2016). Editorial: Precision livestock farming: A 'per animal' approach using advanced monitoring technologies. *Animal, 10*, 1482–1483. https://doi.org/10.1017/S1751731116001142
5. Parsons, D. J., Green, D. M., Schofield, C. P., et al. (2007). Real-time control of pig growth through an integrated management system. *Biosystems Engineering, 96*, 257–266. https://doi.org/10.1016/j.biosystemseng.2006.10.013
6. Frost, A. R. (2001). An overview of integrated management systems for sustainable livestock production. In C. M. Wathes, A. R. Frost, F. Gordon, et al. (Eds.), *Integrated management systems for livestock* (pp. 45–50). Selwyn College.
7. Banhazi, T. M., & Black, J. L. (2009). Precision Livestock Farming: A suite of electronic systems to ensure the application of best practice management on livestock farms. *Australian Journal of Multi-Disciplinary Engineering, 7*(1), 1–14. https://doi.org/10.1080/14488388.2009.11464794
8. Schofield, C. P., Wathes, C. M., & Frost, A. R. (2002). Integrated management systems for pigs—Increasing production efficiency and welfare. In D. K. Revell & D. Taplin (Eds.), *Animal production in Australia* (pp. 197–200). Adelaide.
9. Nääs, I. A. (2002). Application of mechatronics to animal production. *CIGR E-Journal, 4*, 1–14.
10. Moura, D. J., Silva, W. T., Nääs, I. A., et al. (2008). Real time computer stress monitoring of piglets using vocalization analysis. *Computers and Electronics in Agriculture, 64*, 11–18. https://doi.org/10.1016/j.compag.2008.05.008
11. Thompson, P. B. (2015). *From field to fork: Food ethics for everyone*. Oxford University Press.
12. Berckmans, D. (2014). Precision livestock farming technologies for welfare management in intensive livestock systems. *Revue Scientifique et Technique OIE, 33*, 189–196.
13. Werkheiser, I. (2020). Technology and responsibility: A discussion of underexamined risks and concerns in Precision Livestock Farming. *Animal Frontiers, 10*, 51–57. https://doi.org/10.1093/af/vfz056
14. Senger, P. L. (1994). The estrus detection problem: New concepts, technologies, and possibilities. *Journal of Dairy Science, 77*, 2745–2753.
15. Kettlewell, P. J., & Moran, P. A. (1992). Study of heat production and heat loss in crated broiler chicken: A mathematical model for a single bird. *British Poultry Science, 33*, 239–252.
16. Carvalho, V., Nääs, I. A., Mollo, M., et al. (2005). Prediction of the occurrence of lameness in dairy cows using a fuzzy-logic based expert system.—Part I. *CIGR E-Journal, 7*, 1–12.
17. Wathes, C. M., Kristensen, H. H., Aerts, M., et al. (2008). Is precision livestock farming an engineer's daydream or nightmare, an animal's friend or foe, and a farmer's panacea or pitfall? *Computers and Electronics in Agriculture, 64*, 2–10. https://doi.org/10.1016/j.compag.2008.05.005
18. Borchers, M. R., & Bewley, J. M. (2015). An assessment of producer precision dairy farming technology use, pre-purchase considerations, and usefulness. *Journal of Dairy Science, 98*, 4198–4205. https://doi.org/10.3168/jds.2014-8963
19. Van Hertem, T., Rooijakkers, L., Berckmans, D., et al. (2017). Appropriate data and visualization is key to Precision Livestock Farming acceptance. *Computers and Electronics in Agriculture, 138*, 1–10. https://doi.org/10.1016/j.compag.2017.04.003

20. Terrasson, G., Villeneuve, E., Pilnière, V., et al (2017). Precision livestock farming: A multidisciplinary paradigm. In *Proceeding of the SMART 2017, the sixth international conference on smart cities, systems, devices and technologies*. Retrieved April 15, 2020, from https://www.researchgate.net/profile/Eric_Villeneuve2/publication/331373949_Precision_Livestock_Farming_A_Multidisciplinary_Paradigm/links/5c764a2fa6fdcc47159e9873/Precision-Livestock-Farming-A-Multidisciplinary-Paradigm.pdf.
21. Ramirez, B. C., Hoff, S. J., & Harmon, J. D. (2018). Thermal environment sensor array: Part 2 applying the data to assess grow-finish pig housing. *Biosystems Engineering, 174*, 341–351. https://doi.org/10.1016/j.biosystemseng.2018.08.003
22. Li, H., Rong, L., & Zhang, G. (2016). Study on convective heat transfer from pig models by CFD in a virtual wind tunnel. *Computers and Electronics in Agriculture, 123*(Suppl C). https://doi.org/10.1016/j.compag.2016.02.027
23. Schofield, C. P., Marchant, J. A., White, R. P., et al. (1999). Monitoring pig growth using a prototype imaging system. *Biosystems Engineering, 72*, 205–210. https://doi.org/10.1006/jaer.1998.0365
24. Kashiha, M., Bahr, C., Ott, S., et al. (2014). Automatic monitoring of pig locomotion using image analysis. *Livestock Science, 159*, 141–148. https://doi.org/10.1016/j.livsci.2013.11.007
25. Manteuffel, G., Puppe, B., & Schön, P. C. (2004). Vocalization of farm animals as a measure of welfare. *Applied Animal Behaviour Science, 88*, 163–182. https://doi.org/10.1016/j.applanim.2004.02.012
26. Cordeiro, A. F. S., Nääs, I. A., Oliveira, S. R., et al. (2013). Understanding vocalization might help to assess stressful conditions in piglets. *Animals, 3*, 923–934. https://doi.org/10.3390/ani3030923
27. Finger, G., Hemeryck, M., Duran, C. O., et al. (2014). Practical application of the pig cough monitor in a German fattening pig herd with PRDC. In: *Proceedings of the 23rd IPVS congress, Cancun, Mexico*, pp. 207–208.
28. Dawkins, M. S., Lee, H.-J., Waitt, C. D., et al. (2009). Optical flow patterns in broiler chicken flocks as automated measures of behaviour and gait. *Applied Animal Behaviour Science, 119*, 203–209. https://doi.org/10.1016/j.applanim.2009.04.009
29. Dawkins, M. S., Cain, R., & Roberts, S. J. (2012). Optical flow, flock behaviour and chicken welfare. *Animal Behaviour, 84*, 219–223. https://doi.org/10.1016/j.anbehav.2012.04.036
30. Kashiha, M., Pluk, A., Bahr, C., et al. (2013). Development of an early warning system for a broiler house using computer vision. *Biosystems Engineering, 116*, 36–45. https://doi.org/10.1016/j.biosystemseng.2013.06.004
31. Aydin, A. (2017). Development of an early detection system for lameness of broilers using computer vision. *Computers and Electronics in Agriculture, 136*, 140e146. https://doi.org/10.1016/j.compag.2017.02.019
32. Nääs, I. A., Lozano, L. C. M., Mehdizadeh, S. A., et al. (2018). Paraconsistent logic used for estimating the gait score of broiler chickens. *Biosystems Engineering, 173*, 115–123. https://doi.org/10.1016/j.biosystemseng.2017.11.012
33. Mehdizadeh, S. A., Neves, D. P., Tscharke, M., et al. (2015). Image analysis method to evaluate beak and head motion of broiler chickens during feeding. *Computers and Electronics in Agriculture, 114*, 88–95. https://doi.org/10.1016/j.compag.2015.03.017
34. Pereira, E. M., Nääs, I. A., & Garcia, R. G. (2014). Identification of acoustic parameters for broiler welfare estimate. *Engenharia Agrícola, 34*, 413–421. https://doi.org/10.1590/S0100-69162014000300004
35. Fontana, I., Tullo, E., Carpentier, L., et al. (2017). Sound analysis to model weight of broiler chickens. *Poultry Science, 96*, 3938–3943. https://doi.org/10.3382/ps/pex215
36. Llaria, A., Terrasson, G., Arregui, H.. et al. (2015). Geolocation and monitoring platform for extensive farming in mountain pastures. In *Proceedings of the IEEE International Conference on Industrial Technology (ICIT 15)*, pp. 2420-2425.
37. Anderson, D. M. (2007). Virtual fencing-past, present and future. *Rangeland Journal, 26*, 65–78. https://doi.org/10.1071/RJ06036

38. Hostiou, N., Fagon, J., Chauvat, S., et al. (2017). Impact of precision livestock farming on work and human-animal interactions on dairy farms. A review. *Biotechnology, Agronomy, Society and Environment, 21*, 268–275.
39. Leliveld, L. M. C., Düpjan, S., Tuchscherer, A., et al. (2017). Vocal correlates of emotional reactivity within and across contexts in domestic pigs (Sus scrofa). *Physiology & Behavior, 181*, 117–126. https://doi.org/10.1016/j.physbeh.2017.09.010
40. Abe, J. M. (2015). Paraconsistent intelligent based-systems: New trends in the applications of paraconsistency. *Book Series: Intelligent Systems Reference Library, Springer-Verlag, 94*, 306.
41. Abe, J. M., Akama, S., & Nakamatsu, K. (2015). *Introduction to annotated logics—Foundations for paracomplete and paraconsistent reasoning. Series Title Intelligent Systems Reference Library* (Vol. 88, 1st ed., p. 190). Springer International Publishing.
42. Fonseca, F. N., Abe, J. M., Nääs, I. A., et al. (2019). Automatic prediction of stress in piglets (Sus Scrofa) using infrared skin temperature. *Computers and Electronics in Agriculture, 168*, 105–148. https://doi.org/10.1016/j.compag.2019.105148
43. Silva, J. P., Nääs, I. A., Abe, J. M., et al. (2019). Classification of piglet (Sus Scrofa) stress conditions using vocalization pattern and applying paraconsistent logic Eτ. *Computers and Electronics in Agriculture, 166*, 105–020. https://doi.org/10.1016/j.compag.2019.105020
44. Neethirajan, S. (2017). Recent advances in wearable sensors for animal health management. *Sensing and Bio-Sensing Research, 12*, 15–29. https://doi.org/10.1016/j.sbsr.2016.11.004
45. White, B. J., Amrine, D. E., & Larson, R. L. (2018). Big data analytics and precision animal agriculture symposium: Data to decisions. *Journal of Animal Science, 96*, 1531–1539. https://doi.org/10.1093/jas/skx065
46. Morota, G., Ventura, R. V., Silva, F. F., et al. (2018). Big data analytics and precision animal agriculture symposium: Machine learning and data mining advance predictive big data analysis in precision animal agriculture. *Journal of Animal Science, 96*, 1540–1550. https://doi.org/10.1093/jas/sky014
47. Steensels, M., Antler, A., Bahr, C., et al. (2016). A decision-tree model to detect post-calving diseases based on rumination, activity, milk yield, body weight and voluntary visits to the milking robot. *Animal, 10*, 1493–1500. https://doi.org/10.1017/S1751731116000744
48. Pereira, D. F., Miyamoto, B. C. D., Maia, G. D. N., et al. (2013). Machine vision to identify broiler breeder behavior. *Computers and Electronics in Agriculture, 99*, 194–199. https://doi.org/10.1016/j.compag.2013.09.012
49. Vale, M. M., Moura, D. J., Nääs, I. A., et al. (2008). Data mining to estimate broiler mortality when exposed to heatwave. *Science in Agriculture, 65*, 223–229. https://doi.org/10.1590/S0103-90162008000300001
50. Wang, Y., Yang, W., Winter, P., et al. (2008). Walk-through weighing of pigs using machine vision and an artificial neural network. *Biosystems Engineering, 100*, 117–125. https://doi.org/10.1016/j.biosystemseng.2007.08.008
51. Fournel, S., Rousseau, A. N., & Laberge, B. (2017). Rethinking environment control strategy of confined animal housing systems through precision livestock farming. *Biosystems Engineering, 155*, 96e123. https://doi.org/10.1016/j.biosystemseng.2016.12.005
52. Vranken, E., & Berckmans, D. (2017). Precision livestock farming for pigs. *Animal Frontiers, 7*, 32–37. https://doi.org/10.2527/af.2017.0106

Part III
Environment

Programmable Process Structures of Unified Elements for Model-Based Planning and Operation of Complex Agri-environmental Processes

Monika Varga, Gergo Gyalog, Janos Raso, Balazs Kucska, and Bela Csukas

1 Introduction

Application of the latest developments of ICT in agriculture is motivated by the pressing need to utilize the locally accessible land, water, and solar energy to produce appropriate amount of food, as well as renewable materials and energy for the growing population, sustainably. Especially, the reasonably designed, climate-adapted or even "climate-smart" solutions have keynote importance [1]. These systems are inherently embedded in the vulnerable ecosystem, associated with the water and soil resources, while they relate to non-agricultural technologies. Having recognized this highly increasing complexity, process modeling engineers also try to implement their well-established and new methodologies for the agri-environmental systems. The challenge is to develop unified modeling and simulation methodologies for the broad, diversified set of multidisciplinary processes, to enhance cooperation between the human-operated technologies and the environment.

M. Varga (✉) · B. Csukas
Hungarian University of Agriculture and Life Sciences, Kaposvar Campus, Kaposvar, Hungary
e-mail: varga.monika@uni-mate.hu; csukas.bela@uni-mate.hu

G. Gyalog
Hungarian University of Agriculture and Life Sciences, Institute of Aquaculture and Environmental Safety, Research Center of Fisheries and Aquaculture, Szarvas, Hungary
e-mail: gyalog.gergo.sandor@uni-mate.hu

J. Raso
University of Sopron, Forestry Research Institute, Puspokladany, Hungary
e-mail: raso.janos@uni-sopron.hu

B. Kucska
Hungarian University of Agriculture and Life Sciences, Institute of Aquaculture and Environmental Safety, Department of Applied Fish Biology, Kaposvar, Hungary
e-mail: kucska.balazs@uni-mate.hu

D. D. Bochtis et al. (eds.), *Information and Communication Technologies for Agriculture—Theme III: Decision*, Springer Optimization and Its Applications 184, https://doi.org/10.1007/978-3-030-84152-2_11

Recently, computer-assisted decisions about planning and operation of complex agri-environmental systems are supported by new generations of sensors and data acquisition methodologies of Information Technology. However, neither statistical, nor structural analysis of the underlying networks is enough, and the integrated development of agriculture-related complex systems also needs the consideration of the causally determined balances of the underlying (multiscale and nonlinear) interactions. The decisions also require the involvement of the predictive models, considering the conservation law-controlled relationships amongst the coexisting sub-processes.

Considering these challenges, this chapter introduces a generally applicable method for the unified, dynamic modeling of agri-environmental systems. Accordingly, the following part of Sect. 1 gives a general overview about existing functional and structural modeling approaches, while Sect. 2 introduces the non-conventional method of Programmable Process Structures (PPS) and its characteristics beyond the state-of-the-art. Section 3 illustrates the application of PPS through three case studies, highlighting, the most important feature of PPS, namely the general applicability for quite different process systems, that enables the combined consideration of dynamics also for the interacting systems and their environment. Finally, Chapter 4 summarizes the conclusions.

1.1 Functionality Modeling of Complex Process Systems

Many efforts for process modeling and simulation started from chemical engineering. The respective methodologies and tools were developed in Process Systems Engineering (PSE, [2]). At millennium Grossmann and Westerberg [3] highlighted the research challenges, as well as formulated a broader definition for PSE. According to the authors' prognosis, in the next period PSE moved beyond the traditional fields of chemical process engineering with a stronger link to natural and other engineering sciences. In another overall report 3 years later [4], authors emphasized the future directions and challenges in terms of developmental needs, as well as addressed the role of chemical process engineers, as an interface between field-specific engineering and method-specific (applied mathematical and computer) sciences [5].

In the meantime, PSE stepwise became a field that covers the application of engineering approaches and tools for design, planning, control, and operation of complex process systems in various scales, supported by increasing computational capacity. By now, having exceeded the "chemical supply chain," the problem domain extended to directions of both nano-scale and tera-scale, significantly. Enhanced performance with special regard to the sustainability issues became another inevitable direction of developments in the past decade [6].

The expanding problem domains were also demonstrated in a Special Issue from 2013 [7], where editors summarized the experiences of the 22nd event of Computer-Aided Process Engineering. Topics from "product and process design," through "sustainable development," "energy management," until "systems biology and healthcare engineering" are emphasized, again.

Conventional models of PSE are represented in the so-called balance equations that are described by a set of ODEs (Ordinary Differential Equations), PDEs (Partial Differential Equations), or DAEs (Differential Algebraic Equations), etc. The related software tools (e.g., Aspen Plus, ProSim, gPROMS, Aspen Custom Modeler) presume the availability of a comprehensive and coherent database of the necessary physico-chemical data. Although affordable sensors and new data acquisition methods of Info-Communication Technology provide more and more data for decision support also in the agri-environmental processes, in most cases it supports only the local decisions of a given segment, but data are not utilized in higher level, territorial decision-making.

In the past decades, PSE methodologies have moved toward nano-scale biosystems and material sciences, as well as toward tera-scale industrial and agri-environmental process networks. Regarding nano-scale, Systems Biology claimed for the integration and quantification of biological data [8] in the past period. Process system modeling tools can contribute significantly to the analysis, understanding, and design of biological systems (e.g., in analysis of healthy or diseased states in drug discovery) through the quantification and exploration of the complex chemical and physical interactions [9].

Regarding tera-scale, the model-based management of water- [10–12], energy- [13], and trans-sectorial agri-food [14, 15] networks, as well as the solution of complex environmental problems [16–18] became substantially important.

The most challenging area is the management of complex, interacting processes, where the multiple spatial/temporal scales and the multi-objective optimization require new approaches and capabilities from PSE. Nowadays, on the basis of the well-established chemical engineering roots, PSE is ready to further develop the engineering-based problem-solving for the newly emerging fields.

Considering optimization, PSE focused on the extensions of (originally linear programming based) methodologies of operation research, developed Mixed Integer Linear (and Nonlinear) Programming. Recently, there are very comprehensive optimization models and tools, like GAMS [19]. The basic optimization paradigm is to determine an exact and global optimum for a suitably simplified model. On the contrary, engineering way of thinking often prefers the multi-objective sub-optimal solutions, obtained from more detailed models.

Artificial and computational intelligence-based methods also became popular for optimization in various fields (e.g., [20] for waste water treatment, [21] for bioprocess modeling; [22] for scheduling). These meta-heuristic approaches cover various nature-inspired techniques (neural networks, genetic algorithms, particle swarm optimization, bee colony optimization, etc.), which have advantageous characteristics, like simplicity, easy parallelizability, and easy integration with detailed simulation models.

A comprehensive overview about the application of conventional (mainly mathematical programming based) PSE methodologies for Water-Food-Energy Nexus has been published by Garcia and You [17]. Nevertheless, PSE-based solution of complex WFEN networks is still in an early stage of application.

System Dynamics (SD) is another approach, initiated by Forrester [23] to understand the behavior of complex systems. It is to be noted that SD does not clearly distinguish between the semantics of balance-based additive measures and signals. One of the most commonly used SD implementations is Stella that contains a modular interface to an ODE-based simulator [24, 25].

Agent-based approach is also an often-used tool, where the problem-solving is based on the communication of the agents, executing their autonomous local programs. This principle was successfully applied for various fields in agriculture, as well [26, 27].

1.2 Structural Modeling of Complex Systems

Regarding structural modeling, networks are used for the simplified mapping of real systems, where nodes represent the elements (compounds, objects, individuals, actors etc.) while edges describe the relationships (interaction, transport, communication, friendship, etc.) existing between these nodes. Edges can be non-directed, directed, or bi-directionally directed and they can be associated with a weight that differentiates the importance, cost, speed, etc. of the relationships.

The topological characteristics of the networks offer useful insights into the structure of the studied system (e.g., [28–30]). The most often used properties of the networks are degree, assortativity, closeness, betweenness, density of links, average path length, largest minimum path (diameter), and modularity (e.g., [31, 32]). Average values of the properties describe the global characteristics of the network while statistical distributions characterize the set of the single nodes [31].

Importance of structural features appeared in Systems Biology (e.g., [33] Systems Ecology (e.g., [34], and Social Networks (e.g., [35]).

Many authors emphasized that dynamic characteristics of network properties play a crucial role in the functioning of the system under investigation (e.g., [36, 37]). These works tried to combine network analysis with the linear state space models. However, linearity seems to be a quite arbitrary assumption while there might be more complex, multiscale, event-driven, and nonlinear processes with considerable time delays in the complex systems.

From another aspect, the essential structural features of process models can be represented by nets. While networks (i.e., the commonly known simple graphs) can be interpreted as relations within a set of elements, nets can be defined as relations between two sets [38, 39]. In the past decades, extensive efforts have been made for the further development of Petri Net that is a special case of the general net structures

[39]. There were straightforward implementations of quantitative, time, and event-driven functionalities of the structural models in the form of higher order Quantitative Petri Nets [40, 41].

Appropriate integration of structural and functional approaches is still an important issue in the new fields of applications. The introduced methodology of Programmable Process Structures (PPS) in Sect. 2 utilizes a combined structural and functional approach for the dynamic modeling and simulation of various process networks.

2 Methodology

Agriculture, aquaculture and forestry have a determining role in the sustainable utilization of the locally accessible solar energy, land and water to produce appropriate amount of healthy food, as well as renewable materials and energy. The ongoing development results in increasing coupling amongst the involved industrial and agricultural technologies, as well as the underlying ecosystem. This needs a new framework that is capable both for the holistic overview of these complex processes and for the consideration of the nonlinear dynamics of embedded sub-processes. A goal is to afford a possible modeling component for the evolving Digital Agriculture.

On the one hand, the structure and functionalities of the complex, coupled natural and human-built processes are more sophisticated than the networks in sense of network science. On the other hand, the case-specific local functionalities of the elementary building blocks of these structures are less complex than the various specific mathematical methods. As an intermediate solution, we have developed Programmable Process Structures. Main features of the method can be studied in the referred papers, in detailed [42, 43]. However, main characteristics and differences from other available methods can be summarized, as follows. Methodology of Programmable Process Structures (PPS, see the flowchart in Fig. 1) allows the automatic generation, local programming and dynamic simulation of extensible and connectible process models for various natural and technological processes. PPS uses two kinds of meta-elements (i.e. states and transitions) for the process network to be described. As a first step, having started:

- from these two meta-prototypes,
- from a generalized ontology for the description of process network, and
- from the initial data and parameters.

PPS generates the empty structure of the model. Afterwards, the next step is the description of functionalities (in the form of local program codes) in sense of the underlying physical, chemical, and biological processes. The functionalities of the state and transition elements are declared by the local programs of the so-called prototype elements which can be derived from the general meta-prototypes. These prototype elements contain symbolic input, parameter, and output variables, as well as an editable local program (representing equations, expressions, rules, constraints,

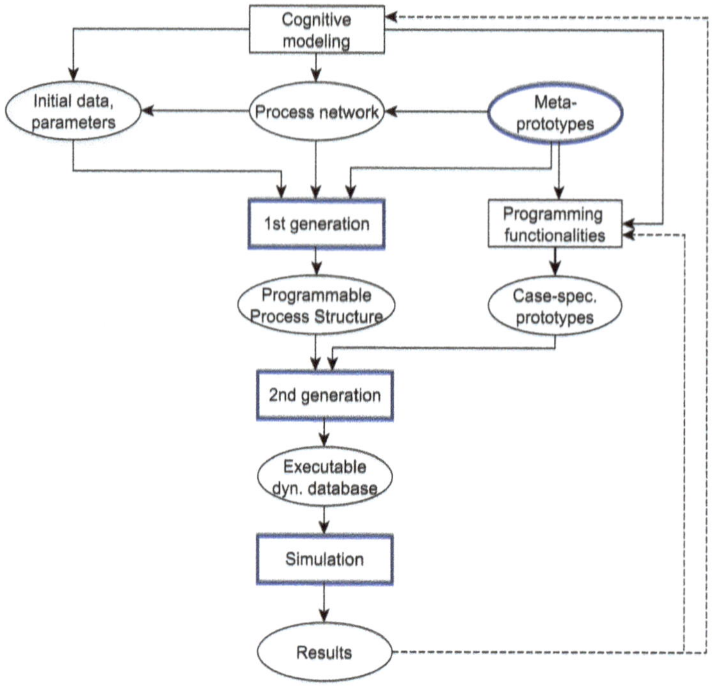

Fig. 1 Generation of Programmable Process Structures

etc.). In the various applications, many state and transition elements usually can be modeled with the same prototype elements (i.e. with the same local programs).

The state and transition elements of the actual model can be parameterized and initialized concerning their case-specific prototypes during the structure generation, automatically, and/or by editing of the generated structure. Next, the executable code of the model (in the form of a declarative dynamic database) can be prepared in the second step of generation, automatically. The actual model elements are executed by their associated prototype elements. The execution, as the connection-based communication amongst the state and transition elements of the Programmed Process Structure is solved by a general-purpose kernel program. During the simulation, the actual elements start with their initial conditions and parameters, and the output values are recalculated stepwise, in the knowledge of input and parameter data, by the associated local program prototypes. The distinguished input and output connections for the conservation law-based mass and energy components and for the signals, make the combined execution of the balance-based and signal-based (e.g. control) functionalities possible.

In the past years, PPS has been successfully applied for modeling and simulation of various process systems from the field of:

- Chemical Process Engineering (e.g., plastic waste pyrolysis, [44]; photo-Fenton process [45]; hydrogen supply chain [46])
- Biotechnology (anaerobic fermentation, [47])
- Systems Biology (multiscale p53/miR34a signaling [48])
- Systems Ecology (lake food web, [42])
- Complex Environmental Processes (watershed hydrology, [12])

3 Results and Discussion

In addition to the previously mentioned fields of applications, in the past years PPS has been applied for studying of various complex agri-environmental processes (Recirculating Aquaculture Systems in [49]; pond aquaculture in [50]; and agroforestry systems in [51]). The chapter overviews these new applications in order of their increasing complexity. Furthermore, it aims to show the role of dynamic modeling and simulation to support planning and decisions at various levels, using the same framework.

3.1 *Recirculation Aquaculture System*

Challenge

Recirculation Aquaculture System (RAS) became a commonly used solution to contribute food supply with the most productive utilization of limited land and water resources, as well as with the lowest possible load on environment [52]. It is an almost closed technological system, including biological, biotechnological and chemical processes in a recycle loop that involves many challenges due to its complexity. There are many interactions amongst the system components, which are difficult to manage by an experimental analysis alone. Accordingly, RAS is a typical example, where the coexisting sub-processes require the application of dynamic simulation models with process engineering way of thinking.

Experimental Unit

The experimental unit (see Fig. 2) contained 9 fish tanks of 655 L, a drum filter, and a Moving Bed Biofilter. Two different groups of Common carp (*Cyprinus carpio*) were stocked (72 pieces of 485 g and 40 pieces of 813 g in two lines (2 $\times$ 3 tank) while one line operated without fish. Fish was fed with Aller Master two times daily. Oxygen supply of tanks (cca. 6 L min^{-1}, with purity 93%) came from an oxygen generator through diffusors. Recycle ratio was set for 90% of tank volumes per hour. The pilot experiments were carried without oxygen enrichment of the recycling water.

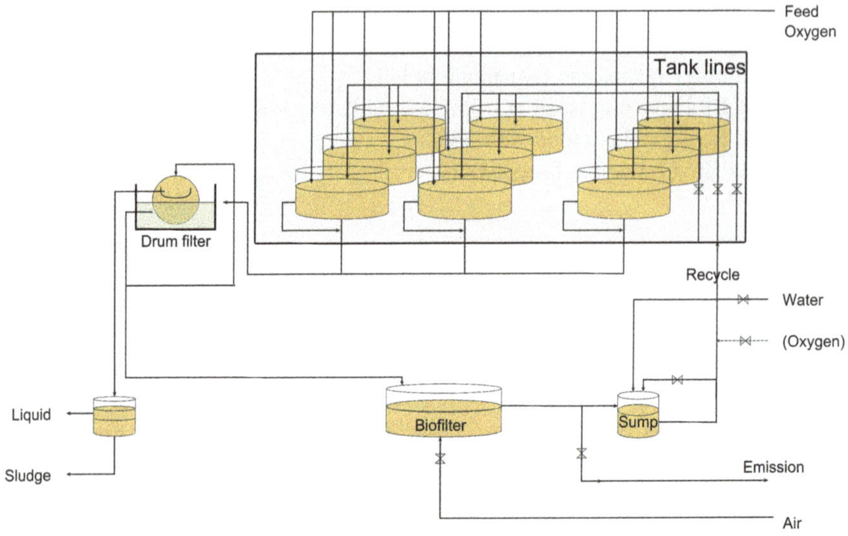

Fig. 2 Architecture of the experimental pilot system (based on [49])

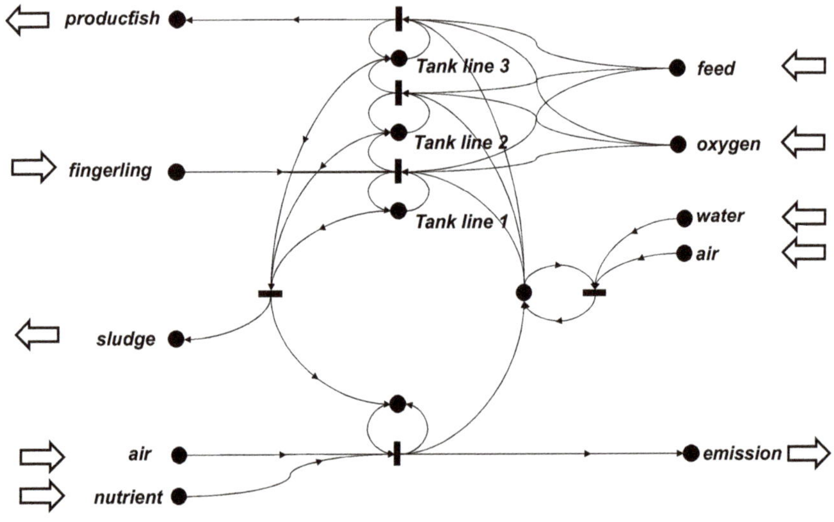

Fig. 3 The flux network of the example RAS

Conceptual Model

The flux network of a three-stage RAS is shown in Fig. 3. It also illustrates the environmental input and output of flows. The grading of fish is solved by appropriately scheduled cyclic change of fish (from Stage 3 to Product, from Stage 2 to Stage 3, from Stage 1 to Stage 2, and a new fingerling in Stage 1).

The essential feature of RAS is that the high recycle rate of purified water makes possible to decrease the fresh water supply, with a considerably less environmental load (in the form of sludge and purified emission). It is associated with the additional cost of biofilter, drum filter, and recycle pump.

PPS Implementation of the Model

Design and operation of RAS need easily generable, modifiable, and reusable dynamic models for simulation of the coupled technological, biological, microbiological, and chemical processes. Especially, the effective design and control of multi-stage fish rearing need an appropriate model.

The PPS model of a three-stage RAS is illustrated in Fig. 4. The model can be generated from two general meta-prototypes, from the textual description of the process network and from the Excel file of the data.

The editable transitions (rectangles) and states (ellipses) have input and output connectors for materials and signals. The actual model (in the left side) and the programmable prototypes (in the right side) can be generated from the general meta-prototypes (upper right).

The technological and model parameters, as well as the simulation controlling data can be edited in an input MS Excel file, while the simulated results appear in an output Excel sheet. The processing of the input and output data is supported by MS VBA macros.

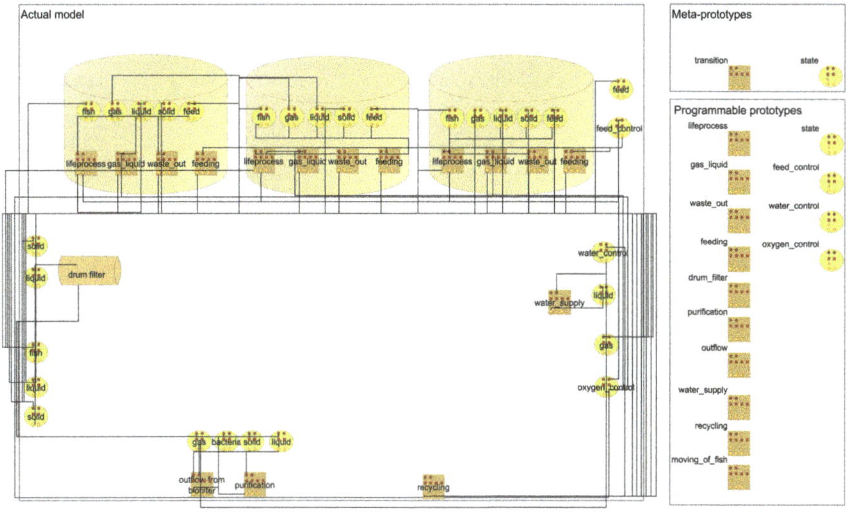

Fig. 4 Programmable Process Structure of the RAS system (based on [49])

Validation of a Pilot Experiment

For the validation of the model, we used the collected data about Common carp (occasional individual weight measurements, biomass measurements), as well as occasional NO_3 measurements.

Normalized root mean square error (NRMSE, %) values (see in Table 1) were calculated to evaluate the model predictions, in line with Eq. (1), as follows:

$$\mathrm{NRMSE} = \frac{\sqrt{\frac{\sum_{i=1}^{n}(m_i - c_i)^2}{n}}}{\overline{m}} * 100 \tag{1}$$

where:

c_i is the ith calculated value
m_i is the ith measured value
n is the number of observations
$\overline{m}$ is the mean of measured values

Study of a Complete Fish Grading Process

Modeling makes possible to simulate a fictitious situation where the calculated volume of a single fish tank continuously follows the prescribed stocking density in the function of changing average fish mass. Figure 5 illustrates the increase of the

Table 1 The calculated NRMSE values

NRMSE values, %	Tank line 1	Tank line 2
Average individual weight	1.16	3.40
Biomass	1.15	1.24
Nitrate	8.01	8.01

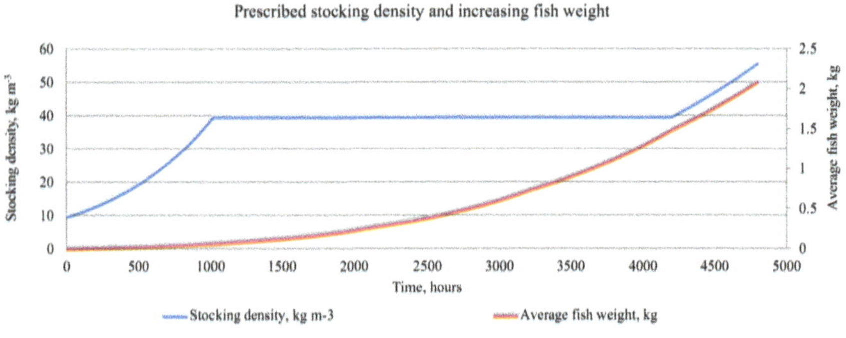

Fig. 5 Increasing average weight of fish and the prescribed change of the stocking density

individual fish in this single fictitious tank while the stocking density follows the prescribed characteristic intervals by the increased volume.

During the simulation of this hypothetical process unit, we can determine how the saturation of recycling water with oxygen (in higher pressure) makes possible to maintain the necessary concentration of dissolved oxygen in the tank along the whole production period. In addition, we can adjust the tolerable ammonium, nitrite and nitrate concentrations by the appropriate recycle rate and water exchange according to the efficiency of Waste Water Treatment system.

Simulation-Based Design

Considering the simulated increasing volume of the single tank model, we can determine the volume of the subsequent stages algorithmically (see Fig. 6) by equidistant partitioning of the necessary rearing time.

Next, the developed model can be tested by the appropriately generated simulation model of the multi-stage system, with the cyclic moving of fish at the prescribed schedule. The change of the population density in the subsequent stages is illustrated in Fig. 7. Further characteristics, such as changing of average fish weight, necessary freshwater input, dynamics of liquid emission and sludge can also be monitored [49].

Having the knowledge of the available tanks, it is easy to design an optional N-stage system. The simulation-based design of multi-stage RAS consists of increasing number of parallel tanks in the subsequent stages with oxygen saturation of the recycling water. The change of dissolved oxygen, ammonium, nitrite, and nitrate concentrations can also be studied in the automatically generated model of the cyclically changing stages.

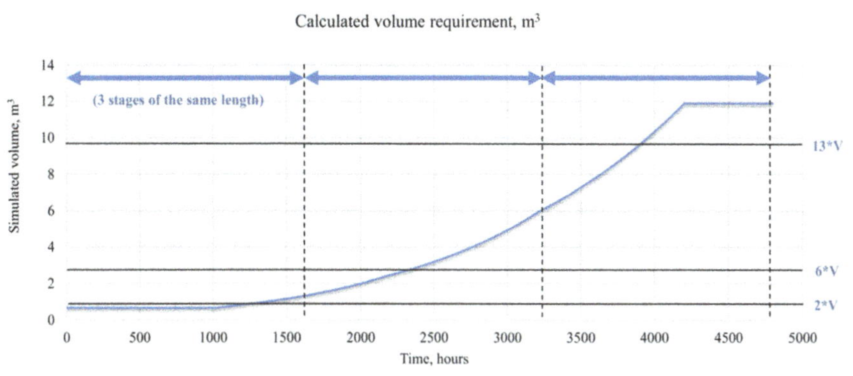

Fig. 6 Integral mean-based partitioning of the increasing volume for three stages

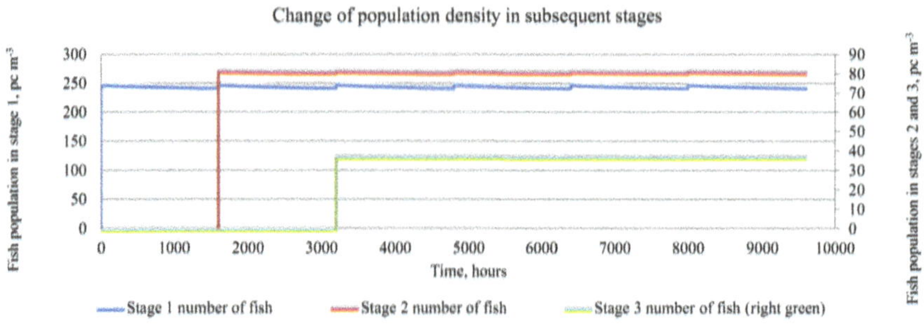

Fig. 7 Population density in the consecutive stages

Experiences About the Applied Methodology

PPS supported the flexible automatic generation and simulation of complex RAS models utilizing the available data and functionalities. The models could be validated by the short time, pilot experiments. Afterwards, the suggested method, based on the hypothetical simulation of a single tank of increasing volume, made possible the simplified analysis of the possible multi-stage systems. This method helps to study the various control strategies, as well as to accomplish a previous optimization, without the combinatorial complexity of grading. In the knowledge of the available tanks, it is easy to design the respective multi-stage system.

3.2 Ecosystem-Involved Fishpond

Challenge

The bulk of fish production comes from pond farms. Taking into account the scarcity of the land and water resources, complex model-based understanding, planning, and management of these systems has an increasing importance, especially if the adaptation to climate change is to be considered [53]. The challenge from methodological side comes from the fact that widely applied and well-developed differential equation solvers sometimes does not fully support the flexible combination of various processes of different spatial and temporal scales in ecosystem containing processes. Spatial multiscale characteristics, origin from the variety of physical, chemical, biological processes, needed to describe the system, while temporal multiscale features appear when environmental effects (including climate change) are considered in a longer time horizon. Challenges related to data availability come from the situation that accurate data are not always available at the start of model building, so flexible methodologies, which allow stepwise development of various model parts, have an increasing importance.

Investigated Production Site

Model development and validation were based on the datasets from the period of 2010–2016 of a production fishpond SzegedFish Ltd., located at the Southern Great Plain in Hungary. The studied fishpond is a 139 ha bi-culture pond with Common carp (*Cyprinus carpio*) and Bighead carp (*Hypophthalmichthys nobilis*). Considering the food web, we involved only those functionally important species (phytoplankton, zooplankton, benthos and detritus) that are usually present in the semi-extensive Central European shallow ponds.

Considering the local weather characteristics, the production season is 214 days from April 1 to October 31 in each year. Water management is in line with the typical levee-type fishpond level control. Production season starts with a level of 1 m, next water supply is from 0.95 m to 1.05 m until August 15 and from 0.6 m to 0.9 m afterwards, while overflow is beyond 1.1 m. After harvesting, the fishpond is drained at the end of the production season.

Conceptual Model

The conceptual model is the representation of the underlying system to be investigated and modeled (covering also the necessarily involved physical, chemical, biological processes). In this case, it refers to the ecosystem-involved fishpond, as well as to the environmental and managerial effects, illustrated in Fig. 8. The

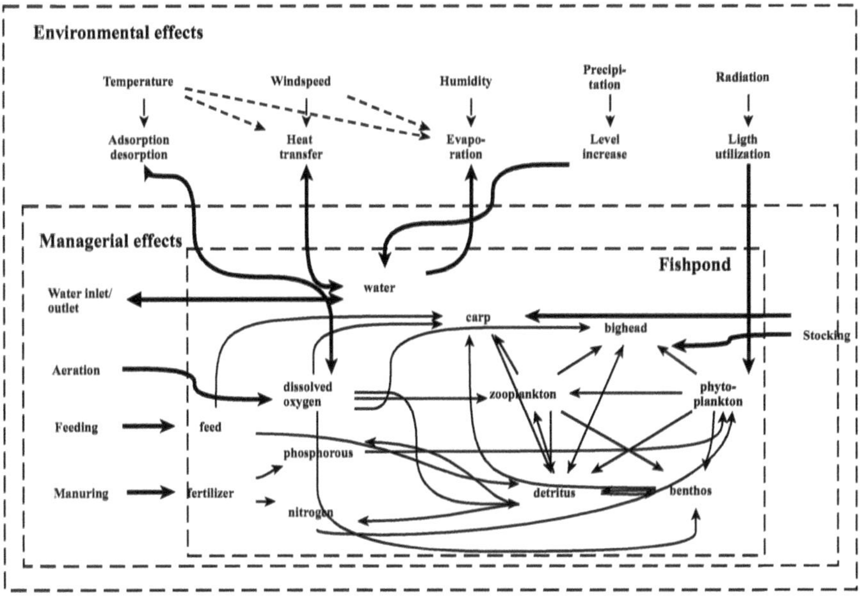

Fig. 8 Conceptual model of the fishpond with the connected environmental and managerial effects

functioning of the coupled natural and human managed process systems is determined and limited by the dynamic processes of the underlying general and component-specific mass and energy conservation.

PPS Implementation of the Model

The generated PPS can be seen in Fig. 9. The state elements refer to those entities, components and signals that have to be considered in the model. The transition elements represent the underlying physical, chemical and biological processes to be taken into consideration. The respective functionalities are described by 6 state and 12 transition prototypes, containing the local program codes for the calculation of the respective class of actual elements. Detailed model description can be studied at [50]. During the execution of the simulation, all of the actual states and transition elements are calculated by the respective prototype program and unifying (binding) the variables with their own actual data and parameters.

The simulator is prepared for various modes of execution, as follows:

- simulation of single or multiple past production seasons with factual meteorological and management data,
- simulation with single and multiple production seasons with factual meteorology but with various hypothetical sets of management scenarios,
- detailed run for climate scenarios (with daily output reporting), and
- simplified run for climate scenarios (with output reporting only in the last day of the production season yearly).

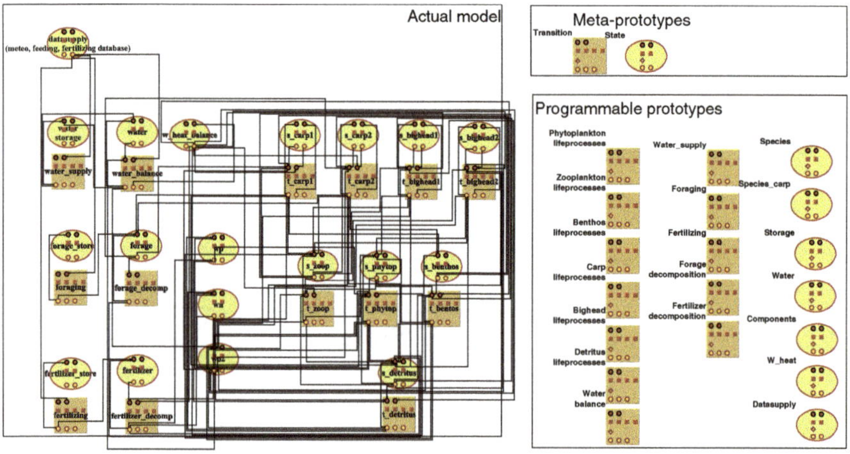

Fig. 9 Generated Programmable Process Structure of food web-involved fishpond model

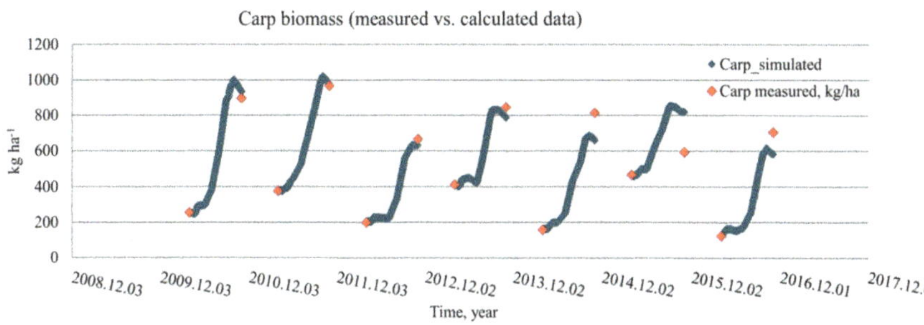

Fig. 10 Validation of carp biomass

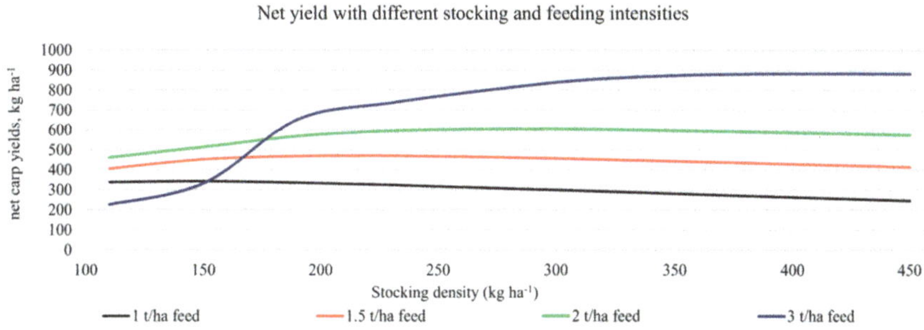

Fig. 11 Possible operations with changing stocking density and feeding rate

Validation of the Model

The validation of the model is illustrated by the comparison of the simulated and measured carp biomass during the reference period (see Fig. 10), as well as by pond level values. The high difference in 2014 in Fig. 10 is the consequence of some non-modeled disturbances. However, NRMSE values of 6.3% for pond level and 15% for carp biomass (calculated by Eq. (1), introduced previously) show an acceptable prediction ability of the model.

Simulation of Various Managerial Strategies

The simulation model supports the managerial decisions about the advantageous operation of fishpond. As an example, we can study the effect of stocking density and feeding intensity on the produced amount of fish. As it is illustrated in Fig. 11, the possible operations cover a large enough space, while the decision may depend on the actual cost coefficients and prices.

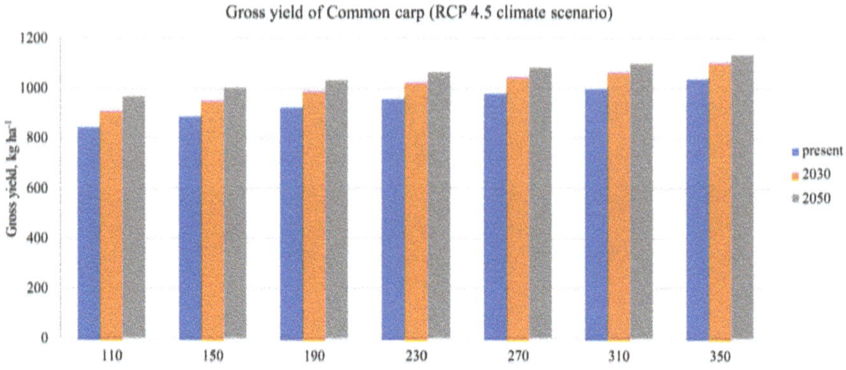

Fig. 12 Simulated average carp weight under various stocking rates (kg ha^{-1}) for NORESM RCP 4.5 climate scenario

Effect of Climate Change on Production of Fishpond

The effect of climate change on the pond aquaculture was studied by 50 years long dynamic simulations utilizing the NORESM RCP 4.5 and RCP 8.5 scenarios. Forecasted data for precipitation and average air temperature were utilized (because according to the previous investigations, air temperature is in good correlation with water temperature for the shallow fishpond during the farming season). Figure 12 shows a prediction for the average individual weight of Common carp. A slowly increasing trend can be observed in the individual weight that may be resulted from the increased anabolic activity in line with the slowly increasing temperature and with the tendentiously higher oxygen production of phytoplankton.

Figure 13 illustrates the increasing demand for water supply in case of the scenarios NORESM RCP 4.5 and NORESM RCP 8.5. The change is non-monotonic because the increasing temperature is associated with more precipitation in the forecasts.

The increasing frequency of extreme weather situation is obvious from the number of days with possible critical dissolved oxygen concentration, as it can be seen in Fig. 14.

Experiences About the Applied Methodology

There is a complex network of influences amongst the interacting natural processes and human activities. To achieve environmentally benign and profitable solutions at the same time, cooperation between the natural and human-controlled processes may be consciously planned and promoted by the validated simulations.

A good example for the possible consensus (useful cooperation) and conflict (harmful cooperation) is the optimal share of food web and crop dosage initiated feeding of fish, also considering the additional manuring to supply nitrogen and phosphorus for phytoplankton.

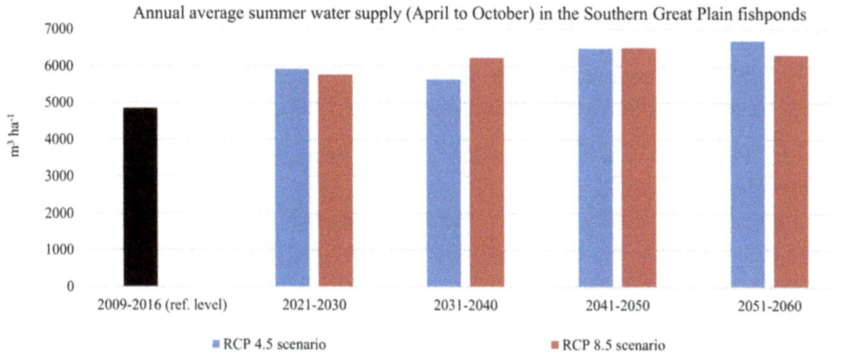

Fig. 13 Effect of climate change on water supply in the two scenarios

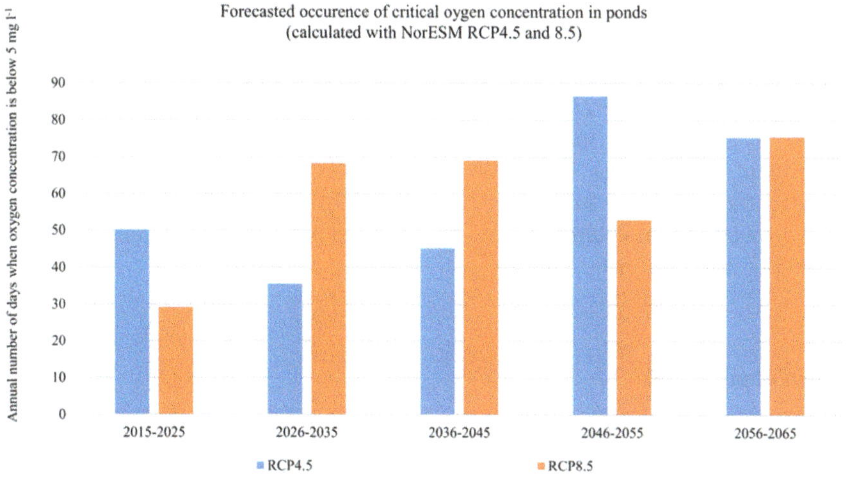

Fig. 14 Effect of climate change on number of days with critical dissolved oxygen concentration

Dynamic simulation-based case studies help to understand and utilize the complex multiple input/multiple output feedback mechanisms in the fishpond. As a good example, consider the dissolved oxygen: partly it comes from the air, partly produced by phytoplankton and it is consumed by many species. Increasing temperature decreases the equilibrium concentrations, but it increases the production of phytoplankton and the anabolic activity of fish (which is a positive side effect).

3.3 Agroforestry Site

Challenge

Planning and operation of complex agroforestry systems require easily modifiable and extensible multiscale dynamic models. The simplifying principles-based cognitive model of agroforestry systems under development utilizes process engineering way of thinking.

The proposed approach helps to synthesize experimental information and experts' knowledge into a framework must support the analysis of:

- trees and crops interacting under various management and meteorological scenarios,
- long-term dynamic balance of components (atoms) within the agroforestry and the environmental and human operated input/output flows,
- specific material balances, like nutrient balance in changing hydrological characteristics, especially comparison of agroforestry with conventional crop production, etc.

Experimental Site

The investigated experimental shelterbelt site (established in 1998) is in Földes, Hungary (47°17′58.3″N 21°23′10.6″E). There are five classes of 26 trees and shrub species (75% *Quercus robur*) in 8 X 3m spacing. Crop rotation between trees consists of dinkel wheat, sweet corn and pumpkin. The same crop of rotation is realized also in a nearby located control area. Left-hand part of Fig. 15 highlights the experimental and the control area, while right-hand part shows the detailed map and scaling of the experimental area.

Conceptual Model

The cognitive model (see Fig. 16) is based on the following process structure and principles:

Fig. 15 Location of experimental agroforestry and control sites

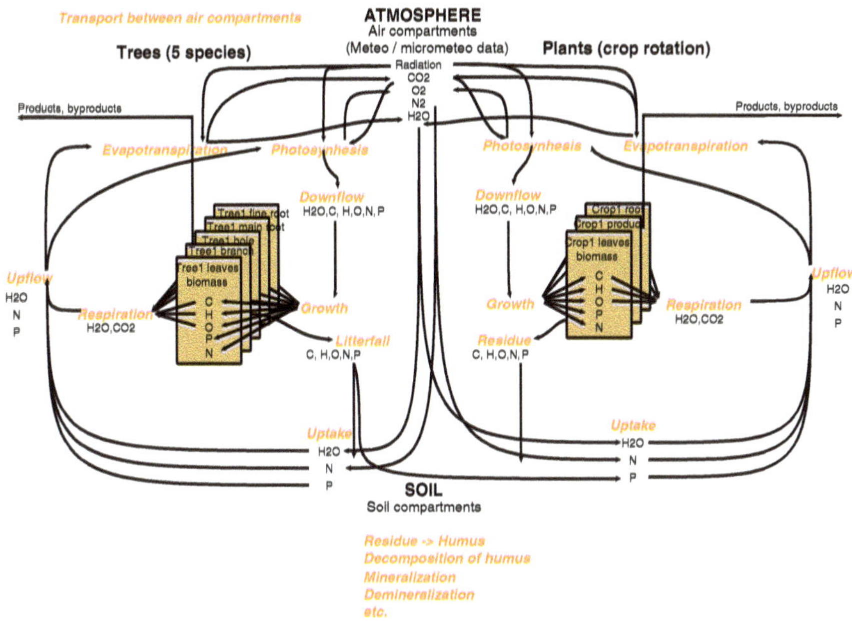

Fig. 16 Cognitive model of the agroforestry system

(a) The model describes the dynamic balance of C, H, O, N, P and optional X atoms, as well as H_20, O_2, CO_2 and N_2 components, represented by the state elements in the compartments of:

- trees (leaves, branch, bole, main root, fine root, up-flow, down-flow),
- plants (leaves/stem, product, root, up-flow, down-flow),
- land parts (residue, humus, solution, inorganic),
- air compartments.

(b) The functionalities (photosynthesis, growth, respiration, evapotranspiration, uptake, seepage flows, air mixing) are calculated by the transition elements, according to the stoichiometric changes of above components within and between the compartments.

(c) The tree- and plant-related main down-flow and up-flow fluxes are basically determined as follows:

- by push logistics of the photosynthesis-driven utilization of CO_2 from air and H_2O from up-flow to produce O_2 into air and [C,H,O,N,P] pool into down-flow,
- by the pull logistics of evapotranspiration-driven emission of CO_2 and H_2O (into air from up-flow).

(d) Trees and plants (as self-controlled living systems) tend to avoid the irrational behavior.

(e) Water and component transports in the soil are primarily driven by concentration difference and seepage.
(f) Component transport in the air is described as a wind-driven mixing.

PPS Implementation of the Model

The generated Programmable Process Structure derived from the general meta-prototypes and from the parameterized description of process network is illustrated in Fig. 17 (one of the right-hand side compartments is shown in more details in the left-hand side of the figure).

Illustration of Simulation-Based Analysis

The above-described model of medium complexity runs daily with two time steps, distinguishing the changing daylight and dark periods, in accordance with the respective meteorological and meteorology-driven hydrological data. An example for temporal change of the photosynthesized biomass in a forest compartment is illustrated in Fig. 18.

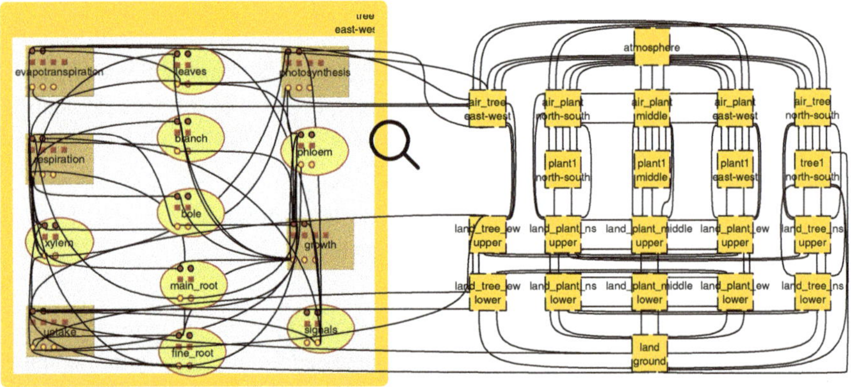

Fig. 17 PPS model of agroforestry system

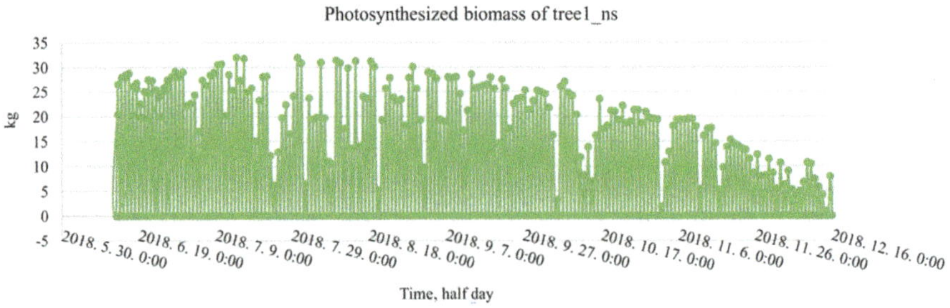

Fig. 18 Photosynthesized biomass in a tree compartment

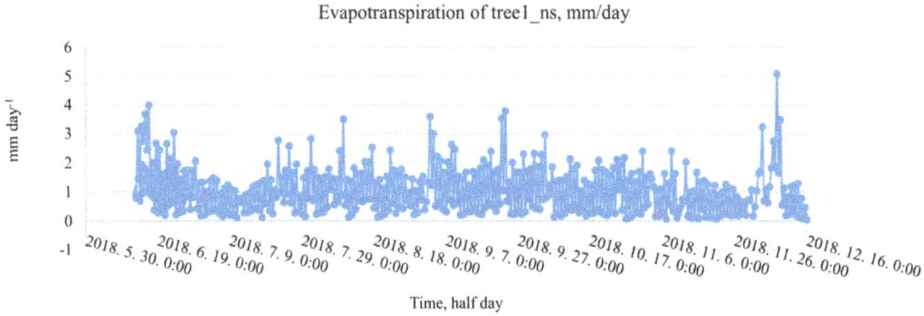

Fig. 19 Evapotranspiration rate in a tree compartment

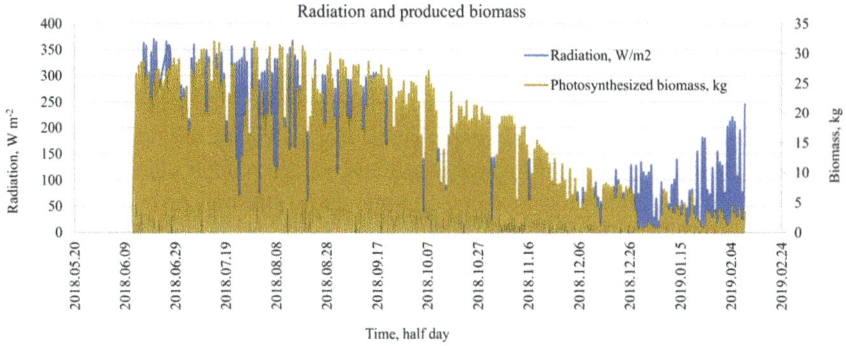

Fig. 20 Radiation and photosynthesized biomass in a tree compartment

Figure 19 shows the calculated evapotranspiration rate in the same tree compartment. The higher peaks are in accordance with the meteorological conditions.

Both Figs. 18 and 19 show how the plants correspond to the seasonal characteristics of meteorology. For example, synthesized amount of biomass is getting to decrease in the fall period. In the model, it is caused by the consideration of decreased radiation in the calculation of photosynthesis, as well as by the time-driven event of litterfall, scheduled in the model, appropriately. Also, frequent changes of evapotranspiration correspond to the daily data of temperature, relative humidity, wind and precipitation.

The radiation and the photosynthesized biomass in a tree compartment are compared in Fig. 20. The figure shows that regardless to the radiation, photosynthesis may be limited by the lack of water in the drought periods of the vegetation. Biomass production decreases at the end of vegetation period.

Experiences About the Applied Methodology

The first trials with the model proved the reasonable application of process engineering inspired simplifying principles. The refinement of functionalities, the parameterization, the testing and the stepwise validation are in progress in the knowledge of additional field measurements.

According to the preliminary results, Programmable Process Structure can be used for the incremental stepwise development of agroforestry model because of the following features:

- It is easy to make any change in the model structure and in the local programs, followed by the automatic re-generation of the whole model.
- The transition-based model implementation supports the stepwise on/off switching and testing of the various functionalities (e.g. following the photosynthesis → growth → respiration → evapotranspiration → uptake cycle, coupled with the transport processes amongst land and air compartments).
- Time- and event-driven rules can be added to the local models.
- Atom balance for the sub-models and for complete model (considering also the environmental and human-decided input and output) may help testing and validation of the model.

A good example for the local cooperation between the functionally connected neighbors is their collaboration in phosphorus and nitrogen supply. Trees can uptake phosphorus from the deeper ground layers, while littering returns part of phosphorus to the upper soil, where plant can utilize it. In contrary, certain plants increase nitrogen fixation that supports also the nitrogen demand of the trees.

4 Concluding Discussion

Nowadays, planning and operation of complex agri-environmental systems are supported by new generations of sensors and data acquisition methodologies of Information Technology. However, the decisions about the integrated development of agriculture-related holistic systems also need the consideration of the causally explicit, dynamic balances of the multiscale and nonlinear interactions. This requires the application of "first principles"-based predictive models, considering the conservation law-controlled relationships of the underlying processes.

Having recognized the existing gap between the increasing amount of useful data and the large-scale long-term decisions, the international mainstream of process model engineering tries to implement the well-established and new modeling frameworks for agri-environmental systems. The challenge is to develop unified modeling and simulation methodologies for the broad, diversified set of multidisciplinary processes. The development of PPS has been motivated by the fact that dynamic simulation of agri-environmental processes requires easily modifiable, extensible and connectable models with unified representation of functional and structural features.

In this chapter, we have illustrated the application of the methodology through the examples of three recently studied complex systems. These examples show how Programmable Process Structure manages the structure and the functionalities of these quite different process systems. The analyzed models can be generated from the same two meta-prototypes and from the standardized description of the process network, resulting in a net of unified state and transition elements, automatically. The prototype elements, representing the functionalities of the model can also be derived from the general meta-prototypes. The case-specific prototype elements contain symbolic input, parameter and output variables as well as a local program code. The state and transition elements of the actual model can be parameterized and initialized concerning their case-specific prototypes. In various applications, many state and transition elements usually can be modeled with the same reusable local programs.

Taking the examples in turn, Recirculating Aquaculture System is a simple case for an almost closed, environmentally benign recycling process. It illustrates the additional complexity, caused by the closed loop. Also, it demonstrates how a causally transparent, dynamic balance model supports the understanding of the coupled cooperative system of fish tanks and Waste Water Treatment. Another lesson is that the approximately validated "first principles"-based dynamic balance model makes possible to generate a fictitious model for the simplified analysis of the design space and control strategies for a multi-stage system, with a reduced complexity. Next, the results of this fictitious model can be tested by the analysis of the appropriately generated complex system. This effective way of process design would not be realized by the formal analysis of the available data alone, obviously.

The ecosystem-based pond aquaculture represents the next level of complexity, caused by the multiple interactions between the environmental and managerial effects. The subsystem, connecting of these two kinds of input/output flows and influences, is a complex nonlinear food web itself. Accordingly, the comprehensive managerial decisions need causally right understanding of the underlying dynamic balances. Nevertheless, this needs more sensors and more intelligent data interpretation, but modeling and simulation may increase the effective utilization of the elaborated data. The causally right holistic balance model is also significant in analyzing effects of climate change in the future operation of fishpond aquaculture. Considering the multiple feedback amongst the interacting components and sub-processes, the statistical and intelligent data mining-based analysis of the available (often disjunct) knowledge elements has to be supplied with the quantitative, conservation law-based interpretation of nonlinear balance models. Moreover, the mutual learning between the data acquisition-based and "first principles"-based models seems to be the best approach.

The most complex agroforestry example illustrates a spatially compartmentalized system with time-driven and event-driven processes, involving daily and seasonally changing life processes of trees and plants. The consciously planned rotation of crops coexists with the long-term life cycle of trees. The actual interactions between the subsystems are affected highly by the (more and more extreme) meteorological and hydrological conditions, while the lifetime of trees is comparable with the time horizon of climate scenarios. The preliminary results of our case study proved that some fundamental principles, coming from the experiences of conventional process systems, help to develop a simplified but rigorous stoichiometric atom balance model. The simplified, transparent transitions, describing the functionalities between the components in the various land, air, tree and plant compartments, can be represented as naturally organized supply/demand chains. In this case, the preliminary model also contributes to the conscious design of the continuous and sampling-based measurements.

Acknowledgments The research was supported by the European Social Fund and Hungarian Government via the EFOP-3.6.2-16-2017-00018 "Let's co-produce with the nature! Agro-forestry, as a new outbreak" project.

References

1. CCAFS and UNFAO. (2014) Questions & Answers: Knowledge on Climate-Smart Agriculture. United Nations Food and Agriculture Organisation (UNFAO), . https://hdl.handle.net/10568/42432
2. Stephanopoulos, G., & Reklaitis, G. V. (2011). Process systems engineering: From Solvay to modern bio- and nanotechnology. A history of development, successes and prospects for the future. *Chemical Engineering Science, 66*, 4272–4306. https://doi.org/10.1016/j.ces.2011.05.049
3. Grossmann, I. E., & Westerberg, A. W. (2000). Research challenges in process systems engineering. *AICHE Journal, 46*, 1700–1703. https://doi.org/10.1002/aic.690460902
4. Breslow, R., et al. (2003). Beyond the molecular frontier: Challenges for chemistry and chemical engineering. In *Committee on challenges for the chemical sciences in the 21st century*. National Research Council.
5. Klatt, K. U., & Marquardt, W. (2009). Perspectives for process systems engineering-Personal views from academia and industry. *Computers and Chemical Engineering, 33*, 536–550. https://doi.org/10.1016/j.compchemeng.2008.09.002
6. Bakshi, B. R. (2014). Methods and tools for sustainable process design. *Current Opinion in Chemical Engineering, 6*, 69–74. https://doi.org/10.1016/j.coche.2014.09.005
7. Bogle, D., & Fairweather, M. (2013). Special issue—Computer Aided Process Engineering (CAPE) tools for a sustainable world. *Chemical Engineering Research and Design, 91*, 1371–1372. https://doi.org/10.1016/j.cherd.2013.07.001
8. Stephanopoulos, G., & Schmitt, W. A. (2001). Systems Biology: An emerging theme in biological research. *Computer Aided Chemical Engineering, 9*, 55–68. https://doi.org/10.1016/S1570-7946(01)80006-7
9. Bogle, I. D. L., Allen, R., & Sumner, T. (2010). The role of Computer Aided Process Engineering in physiology and clinical medicine. *Computers and Chemical Engineering, 34*, 763–769. https://doi.org/10.1016/j.compchemeng.2009.10.021
10. Ahmetović, E., Ibrić, N., Kravanja, Z., & Grossmann, I. E. (2015). Water and energy integration: A comprehensive literature review of non-isothermal water network synthesis. *Computers and Chemical Engineering, 82*, 144–171. https://doi.org/10.1016/j.compchemeng.2015.06.011
11. Estrada, V., Di Maggio, J., & Diaz, M. S. (2011). Water sustainability: A systems engineering approach to restoration of eutrophic Lakes. *Computers and Chemical Engineering, 35*, 1598–1613. https://doi.org/10.1016/j.compchemeng.2011.03.003
12. Varga, M., Balogh, S., & Csukas, B. (2016). An extensible, generic environmental process modeling framework with an example for a watershed of a shallow lake. *Environmental Modelling & Software, 75*, 243–262. https://doi.org/10.1016/j.envsoft.2015.10.022

13. Adams, T. A., II. (2015). Future opportunities and challenges in the design of new energy conversion systems. *Computers and Chemical Engineering, 81*, 94–103. https://doi.org/10.1016/j.compchemeng.2015.04.029
14. Eleftherios, I., Bochtis, D., & Vlachos, D. A. (Eds.). (2016). *Supply chain management for sustainable food networks*. Wiley.
15. Varga, M., Balogh, S., & Csukas, B. (2010). Sector spanning agrifood process transparency with Direct Computer Mapping. *Agricultural Informatics, 1*, 73–83.
16. Copado-Méndez, P. J., Pozo, C., Guillén-Gosálbez, G., & Jiménez, L. (2016). Enhancing the ε-constraint method through the use of objective reduction and random sequences: Application to environmental problems. *Computers and Chemical Engineering, 87*, 36–48. https://doi.org/10.1016/j.compchemeng.2015.12.016
17. Garcia, D. J., & You, F. (2016). The water-energy-food nexus and process systems engineering: A new focus. *Computers and Chemical Engineering*. https://doi.org/10.1016/j.compchemeng.2016.03.003
18. Yue, D., You, F., & Snyder, S. W. (2014). Biomass-to-bioenergy and biofuel supply chain optimization: Overview, key issues and challenges. *Computers and Chemical Engineering, 66*, 36–56. https://doi.org/10.1016/j.compchemeng.2013.11.016
19. Rosenthal, R. (2008). *GAMS—A user's guide*. GAMS Development Corporation.
20. Moral, H., Aksoy, A., & Gokcay, C. F. (2008). Modelling of the activated sludge process by using artificial neural networks with automated architecture screening. *Computers and Chemical Engineering, 32*, 2471–2478. https://doi.org/10.1016/j.compchemeng.2008.01.008
21. Nicoletti, M. C., & Jain, L. C. (Eds.). (2009). *Computational intelligence techniques for bioprocess modelling, supervision and control, studies*. Springer Science & Business Media. https://doi.org/10.1007/978-3-642-01888-6
22. Harjunkoski, I., Maravelias, C. T., Bongers, P., Castro, P. M., Engell, S., Grossmann, I. E., Hooker, J., Méndez, C., Sand, G., & Wassick, J. (2014). Scope for industrial applications of production scheduling models and solution methods. *Computers and Chemical Engineering, 62*, 161–193. https://doi.org/10.1016/j.compchemeng.2013.12.001
23. Forrester, J. W. (1961). *Industrial dynamics*. Pegasus Communications.
24. Espinoza, A., Bautista, S., Narváez, P. C., Alfaro, M., & Camargo, M. (2017). Sustainability assessment to support governmental biodiesel policy in Colombia: A system dynamics model. *Journal of Cleaner Production, 141*, 1145–1163. https://doi.org/10.1016/j.jclepro.2016.09.168
25. Nabavi, E., Daniell, K. A., & Najafi, H. (2017). Boundary matters: The potential of system dynamics to support sustainability? *Journal of Cleaner Production, 140*(1), 312–323.
26. Amadou, M. L., Villamor, G. B., & Kyei-Baffour, N. (2018). Simulating agricultural land-use adaptation decisions to climate change: An empirical agent-based modelling in northern Ghana. *Agricultural Systems, 166*, 196–209.
27. Fernandez-Mena, H., Gaudou, B., Pellerin, S., MacDonald, G. K., & Nesme, T. (2020). Flows in Agro-food Networks (FAN): An agent-based model to simulate local agricultural material flows. *Agricultural Systems*, 102718. https://doi.org/10.1016/j.agsy.2019.102718
28. Newman, M. E. J. (2003). The structure and function of complex networks. *SIAM Review, 45* (2), 167–256.
29. Newman, M., Barabási, A. L., & Watts, D. J. (Eds.). (2006). *The structure and dynamics of networks*. Princeton University Press.
30. Strogatz, S. H. (2001). Exploring complex networks. *Nature, 410*, 268–276.
31. Caldarelli, G. (2007). *Scale-free networks: Complex webs in nature and technology*. Oxford University Press.
32. Steuer, R., & Zamora-López, G. (2008). Global properties of networks. Chapter 3. In B. H. Junker & F. Scrheiber (Eds.), *Analysis of biological networks*. Wiley. https://doi.org/10.1002/9780470253489
33. Alberghina, L., & Westerhoff, H. V. (2005). *Systems biology: Definitions and perspectives*. Springer-Verlag.

34. Jordan, F., Scheuring, I., Vasas, V., & Podani, J. (2006). Architectural classes of aquatic food webs based on link distribution. *Community Ecology, 7*, 81–90.
35. Fellman, P. V., Bar-Yam, Y., & Minai, A. A. (Eds.). (2015). *Conflict and complexity, Countering terrorism, insurgency, ethnic and regional violence*. Springer-Verlag.
36. Csermely, P., Korcsmáros, T., Kiss, H. J., London, G., & Nussinov, R. (2013). Structure and dynamics of molecular networks: A novel paradigm of drug discovery: A comprehensive review. *Pharmacology & Therapeutics, 138*(3), 333–408. https://doi.org/10.1016/j.pharmthera.2013.01.016
37. Liu, Y. Y., Slotine, J. J., & Barabasi, A. L. (2011). Controllability of complex networks. *Nature, 473*, 167–173.
38. Brauer, W. (Ed.). (1980). *Net theory and applications* (Vol. 84). Springer Lecture Notes in Computer Science.
39. Petri, C.A. (1966). *Communication with automata*. Supplement1 to Techn. Report RADC-TR-65-377, Vol. 1 Griffiss AFB, New York. Original in German (Kommunikation mit Automaten, Univ. of Bonn, 1962).
40. Chen, M., & Hofestädt, R. (2003). Quantitative Petri net model of gene regulated metabolic networks in the cell. *Silico Biology, 3*(0029), 347–365.
41. Peleg, M., Rubin, D., & Altman, R. B. (2005). Using petri tools to study properties and dynamics of biological systems. *Journal of the American Medical Informatics Association, 12*(2), 181–199.
42. Varga, M., & Csukas, B. (2017a). Generation of extensible ecosystem models from a network structure and from locally executable programs. *Ecological Modelling, 364*, 25–41. https://doi.org/10.1016/j.ecolmodel.2017.09.014
43. Varga, M., & Csukas, B. (2017b). Programmable Process Structures, generated from a network and from functional meta-prototypes. In *2017 AIChE Annual Meeting, Minneapolis, MN*, Paper #448e.
44. Csukás, B., Varga, M., Miskolczi, N., Balogh, S., Angyal, A., & Bartha, L. (2013). Simplified dynamic simulation model of plastic waste pyrolysis in laboratory and pilot scale tubular reactor. *Fuel Processing Technology, 106*, 186–200. https://doi.org/10.1016/j.fuproc.2012.07.024
45. Audino, F., Varga, M., Pérez-Moya, M., Graells, M., Espuña, A., & Csukás, B. (2018). Experiments and direct computer mapping based model for photo-fenton process. *Computer Aided Chemical Engineering., 43*, 223–228. https://doi.org/10.1016/B978-0-444-64235-6.50040-1
46. Varga, M., & Csukas, B. (2018). Meta-modeling of Hydrogen Supply Chains: A Programmable Structure based representation. In C. Azzaro-Pantel (Ed.), *Hydrogen supply chain. Design, deployment and operation* (1st ed.). Elsevier. https://www.elsevier.com/books/hydrogen-supply-chain/azzaro-pantel/978-0-12-811197-0.
47. Tankovics, A., Takács, D., Szendefy, J., Csukás, B., & Varga, M. (2019). Residence time distribution-based analysis of an industrial-scale biogas fermenter. *Hungarian Journal of Industrial Chemistry, 47*(2), 43–51.
48. Varga, M., Prokop, A., & Csukas, B. (2017). Biosystem models, generated from a complex rule/reaction/influence network and from two functionality prototypes. *Biosystems, 152*, 24–43. https://doi.org/10.1016/j.biosystems.2016.12.005
49. Varga, M., Csukas, B., & Kucska, B. (2020a). Implementation of an easily reconfigurable dynamic simulator for Recirculating Aquaculture Systems. *Aquacultural Engineering, 90*, 102073. https://doi.org/10.1016/j.aquaeng.2020.102073

50. Varga, M., Berzi-Nagy, L., Csukas, B., & Gyalog, G. (2020b). Long-term dynamic simulation of environmental impacts on ecosystem-based pond aquaculture. *Environmental Modelling & Software, 134*, 104755. https://doi.org/10.1016/j.envsoft.2020.104755
51. Varga, M., Raso, J., Keseru, Z. S., & Csukas, B. (2019). Generation of configurable and extensible, multiscale models for dynamic simulation of complex agroforestry systems. In *Poster L24.P.16. 4th World Congress on Agroforestry, Montpellier, France.*
52. Heldbo, J., & Klee, P. (2014). Rethinking aquaculture to boost resource and production efficiency. In *Sea and land-based aquaculture solutions for farming high quality seafood The Rethink Water network and Danish Water Forum White Papers*, Copenhagen. Retrieved from www.rethinkwater.dk.
53. FAO. (2011). *The state of the world's land and water resources for food and agriculture (SOLAW)—Managing systems at risk.* Food and Agriculture Organization of the United Nations, Rome and Earthscan, London. Retrieved January 22, 2020, from http://www.fao.org/3/a-i1688e.pdf

Monitoring and Estimation of Sugarcane Burning in the Middle Paranapanema Basin, Brazil, Using Linear Mixed Models

Jéssica Alves da Silva, Edinéia Aparecida dos Santos Galvanin, and Daniela Fernanda da Silva Fuzzo

1 Introduction

Studies on sugarcane burning demonstrate that the use of fire in agriculture has been condemned for centuries by soil conservation manuals since it increases the temperature and decreases the natural moisture of the soil, leading to greater compaction, loss of porosity, erosion, and consequently soil infertility [1]. Sugarcane harvesting is mechanized in the 25% of the total Brazilian production and in 40% specifically in the state of São Paulo; the remainder is cut manually and undergoes pre-cut burning [2]. In the last 30 years in the scope of modernization, both agriculture and the environment have undergone drastic structural changes, causing numerous environmental and social problems [3].

Since the 1970s, issues regarding environmental protection have been discussed in the state of São Paulo, with an emphasis on the burning activities. The prohibition of burning in the cane fields is the target of several decrees and state laws that regulate this practice. In 2002, State Law 11,241 was passed, which stipulated a schedule for the elimination of burning in the cane fields, starting from that same year with a deadline of 2021 for mechanized areas and 2031 for non-mechanized areas.

J. A. da Silva
Faculty of Agronomic Sciences, Mestranda pela Paulista State University – UNESP, Botucatu, Brazil
e-mail: jessica.a.silva@unesp.br

E. A. dos Santos Galvanin (✉)
São Paulo State University – UNESP, Ourinhos, Brazil
e-mail: edineia.galvanin@unesp.br

D. F. da Silva Fuzzo
State University of Minas Gerais – UEMG, Frutal, Brazil
e-mail: daniela.fuzzo@uemg.br

D. D. Bochtis et al. (eds.), *Information and Communication Technologies for Agriculture—Theme III: Decision*, Springer Optimization and Its Applications 184, https://doi.org/10.1007/978-3-030-84152-2_12

Several studies [2, 4–9] pointed out the negative consequences of burning in sugarcane areas, especially regarding the health of the population and the environment. In order to accelerate the process of reducing the activity, the São Paulo State Secretariat for the Environment (Secretaria de Meio Ambiente—SMA) and the Sugarcane Industry Union (União da Indústria de Cana-de-Açúcar—UNICA) signed, in August 2007, a protocol of intentions in which the practice of burning cane straw must be gradually reduced until its complete elimination by 2017, and by 2014 in mechanized areas.

With regard to sugarcane suppliers, the proposal is that the total elimination of burning occurs by 2021 in areas subject to full mechanization of the harvesting process, that is, where automotive harvesters can operate; and by 2031 in areas in which such machines are unable to operate (areas with a slope greater than 12%). Based on these laws, farmers need to notify the practice of burning to the State Secretariat for the Environment, which is responsible for monitoring the activities and checking compliance with the laws (Assembleia Legislativa do Estado de São Paulo—ALESP, [10]).

The government has thus been looking for alternatives to extinguish the practice [11, 12] and the advent of orbital remote sensing makes it possible to systematically acquire information at a global level [13], making data from orbital sensors the main source for the study of fire occurrences [13, 14]. With the possibility of obtaining information quickly at low to moderate costs, data from remote sensors have become viable for studies of cultures such as sugarcane [15].

The objective of research was to evaluate the spatial and temporal distribution of fire incidences in the period from 2000 to 2018 in the Water Resources Management Unit 17 (Unidade de Gerenciamento de Recursos Hídricos 17—UGRHI 17), located in the state of São Paulo—Brazil, and to carry out the future estimate of this activity through mixed linear models. In this context, the Water Resources Management Unit 17 [16] was chosen as a spatial delimitation to assess the spatial and temporal distribution of fire incidences in sugarcane crops. The area is located in the state of São Paulo, which is, according to [17], the largest producer of sugarcane in the country.

2 Material and Methods

The Middle Paranapanema Basin has a drainage area of 16,749 km^2, with a population of approximately 660,475 inhabitants. The hydrographic division of the state of São Paulo is established by the Water Resources State Plan (Plano Estadual de Recursos Hídricos—State Law 9034/94), which determined the integration into UGRHI-17 of the municipalities whose seats are located in its area, a total of 42 cities. Thirteen (13) of these, possess part of their territory in the unit's area; however, with their seats being outside the area of UGRHI-17, they are classified as contained area municipalities [16]. Figure 1 represents the location map of the basin, which is located in the center-west of the state of São Paulo, between the coordinates

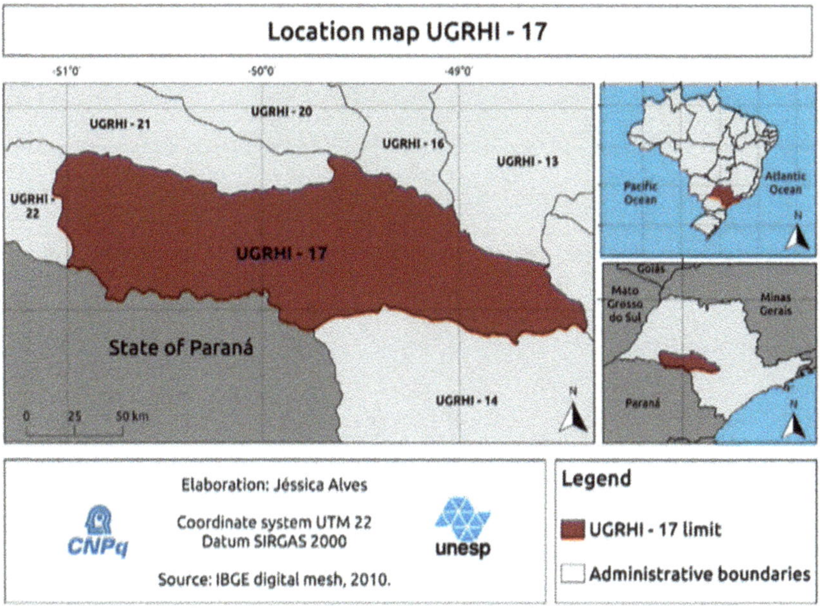

Fig. 1 Location map of the Water Resources Management Unit of the Middle Paranapanema 17 (UGRHI-17)

23°30′00″ and 22°0′0″ to the south and 48°30′30″ and 51°0′0″ to the west, on the border with the state of Paraná.

The area has 1354 km^2 of remaining natural vegetation, which occupies approximately 8% of the area of the Water Resources Management Unit (UGRHI 17). The most frequent categories are Seasonal Semideciduous Forest and Savannah. It has seven Conservation Units (Unidades de Conservação—UC): the Environmental Protection Area Corumbataí-Botucatu-Tejupá, Catetus Ecological Station (ES), ES Assis, ES Santa Bárbara, State Forest (SF) of Avaré, SF Águas de Santa Bárbara, and SF Assis [16].

The geology and aquifer units of UGRHI-17 present different types of soils, and the geological units identified in the area are formed by sedimentary and igneous rocks of the Paraná basin, and recent sedimentary deposits, from the Cenozoic age. Just over 60% of the dimension concerns the sandstones of the Bauru Group and almost 40% concerns the basaltic igneous rocks of the Serra Geral Formation. These two units constitute the two main accessible aquifers in the region: Bauru and Serra Geral (Fig. 2) [16].

Regarding the economic vocation of UGRHI-17 in the urban sphere, the services and commerce sectors stand out, with some industrialization around the largest urban centers. In rural areas, agriculture and livestock are the most expressive activities, the steady expansion of sugarcane crops, and the sugar and alcohol industry being of high importance (Fig. 3) [16].

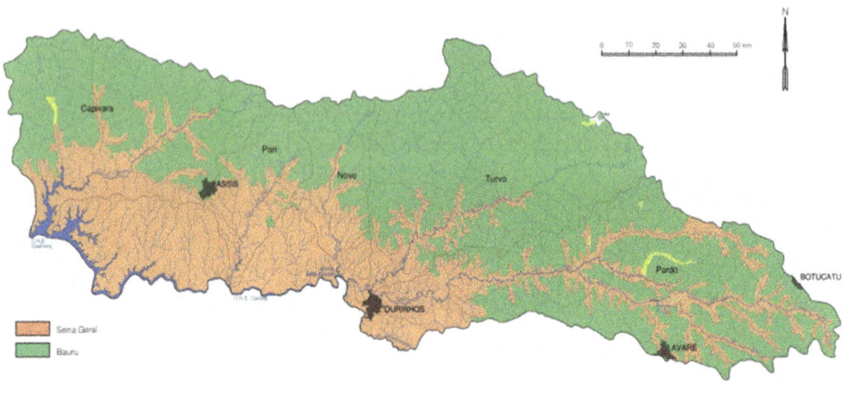

Fig. 2 Main aquifer units present in UGRHI-17, Bauru and Serra Geral. Source: Sistema Integrado de Gerenciamento de Recursos Hídricos do Estado de São Paulo [16]

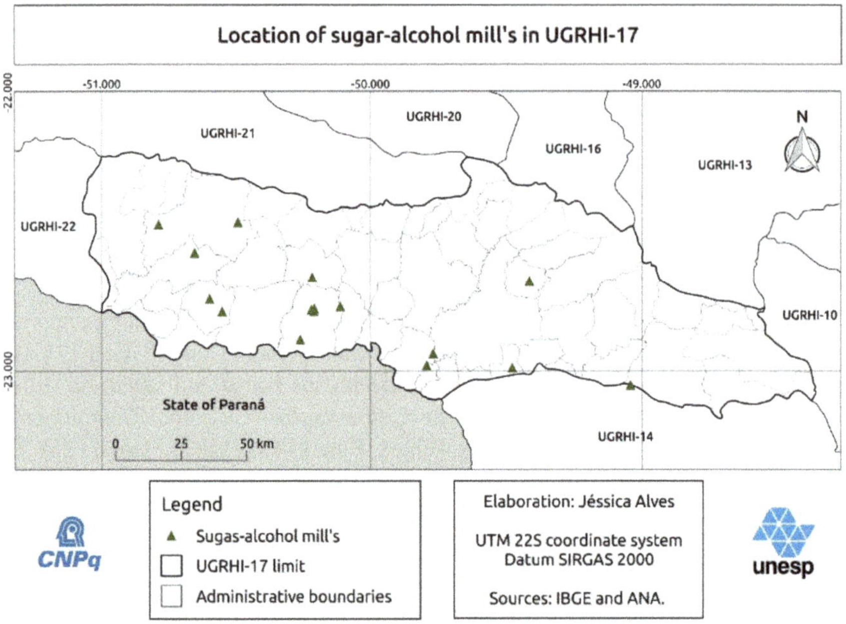

Fig. 3 Location map of sugar and alcohol industries in the UGRHI-17

Figure 3 shows the location of the sugar and alcohol plants in UGRHI-17. Sugarcane areas must be cultivated close to sugar and/or ethanol processing plants in order to reduce transportation costs and minimize the fast post-harvest deterioration. In view of this, sugarcane is planted only in the municipalities that have a processing unit nearby.

2.1 *Topographic Survey*

For the spatiotemporal analysis of the area in study, data regarding fire incidences were collected for free from the catalog of the National Institute for Space Research (Instituto Nacional de Pesquisas Espaciais—INPE) and images from the Landsat 5, 7, and 8 satellites, both obtained from the INPE and the United States Geological Survey (USGS) databases. The Thematic Mapper (TM) sensor was also used, with a spatial resolution of 30 m, aboard the Landsat 5 (bands 3, 4, and 5), available for free in the INPE catalog. The data was captured in the period of April to September, which comprises the dry season of the year; the images refer to the years 2000 and 2006. The methodology flowchart is presented in Fig. 4.

The acquisition of the fire incidences database was possible through the Platform of Monitoring and Warning of Forest Fires in the Cerrado Project (Pró-Cerrado INPE), an initiative made in cooperation between the governments of Brazil and of the United Kingdom, with support from the World Bank.

The ArcGis 10.4 software [18] was used to produce, edit. and quantify thematic maps. All data acquired were superimposed on the altimetric results and on those regarding the shapes referring to the fires, the territorial limit of the Management

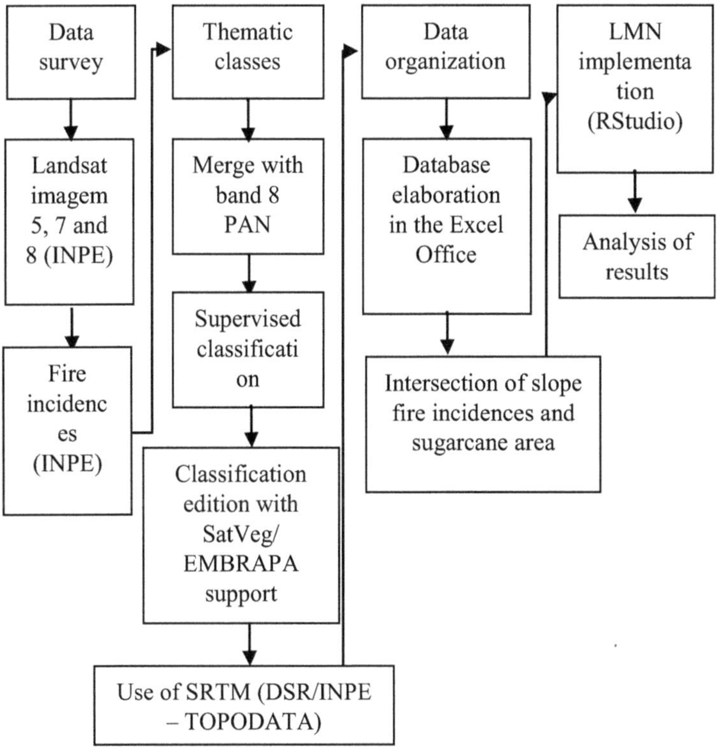

Fig. 4 Methodology flowchart

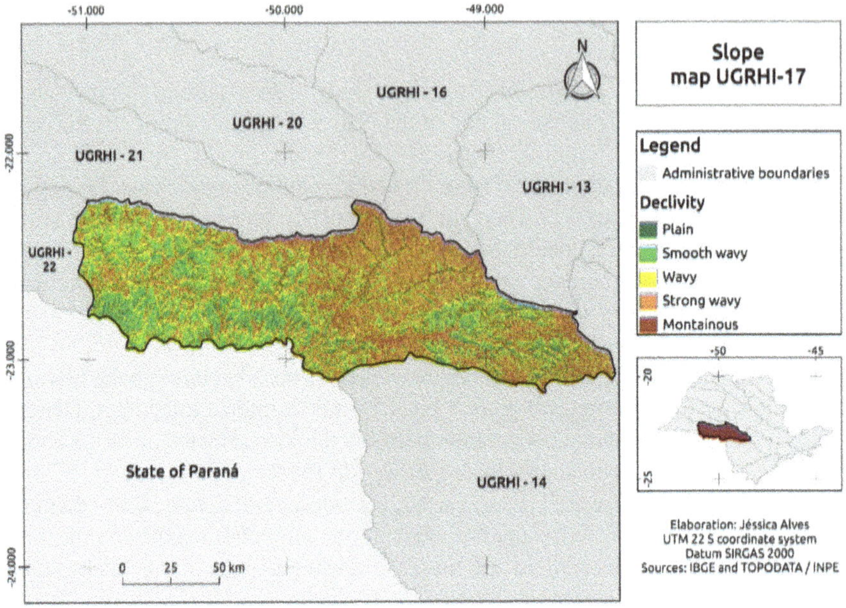

Fig. 5 Slope map for the region of UGRHI-17

Unit and the sugar and alcohol plants present in UGRHI-17, as well as the result of the supervised classification in the years evaluated.

The spatial area of UGRHI-17 occupies a total of 16,763km^2, encompassing 42 municipalities; in light of this, a fusion with band 8 of the OLI instrument was made. Band 8 is panchromatic and has a spatial resolution of 15 m, for Landsat 7 and 8, allowing for better resolution. In order to differentiate sugarcane crops from other cultures, supervised classification was carried out, extracting information from the images to recognize homogeneous regions. In this case, the recognition was made for the topic of interest, that is, areas which comprise sugarcane crops.

In order to filter the information, the results of the supervised classification were intersected with the fire data provided by the Wildfire Program (INPE), concomitant with the slope map generated from the altimetry data, according to the EMBRAPA. The data was obtained from the SRTM topography, originated by the Geomorphometric Database of Brazil—DSR/INPE (TOPODATA).

The TOPODATA project offered the Digital Elevation Model (Modelo Digital de Elevação—MDE), based on the SRTM data provided by USGS, which culminated in the elaboration of the slope classes. From the SRTM data, the slope map was generated using the percentage slope tool available in ArcToolbox in the ArcGis software, following EMBRAPA's relief standards. The geomorphology of the studied area corresponds to the western plateau of São Paulo, characterized by hills with flattened tops. The altimetry of the region (Fig. 5) ranges from 350 to 600 m in altitude and its dominant slope is <6% [16].

2.2 Statistical Modeling

Mixed linear models (MLMs) are a short reference to mixed effect models, in which the terms of fixed and random effect, together with elements of variance and covariance, are considered. These models can be expressed in different, but equivalent forms, appropriate to represent grouped data: that is, data which are collected over time or which consists of clusters of related statistical units. In order to model the burning activity over time, the model is presented according to Eq. (1):

$$\begin{gathered} y_i = X_i\beta + Z_ib_i + \varepsilon_i \\ b_i\tilde{N}_q(0, \psi) \\ \varepsilon_i\tilde{N}_{ni}\left(0, \sigma^2 I_{ni}\right) \end{gathered} \quad (1)$$

where N_q and N_{ni} are multivariate normative distributions of the dimension q and n_i, respectively, and y_i is the $n_i \times 1$ vector of responses representing measurements of areas in the geographical relief category. X_i is the matrix of the model $n \times p$ of fixed effect regressors. β is $p \times 1$, referring to the vector of the fixed effect coefficients, which are considered invariant for all relief classes. Z_i is the $n_i \times p$ regression matrix for the random effects of observations in the relief categories. b_i is the $q \times 1$ vector of random effect coefficients for the slope classes type i, $2i$ is the $n_i \times 1$ error vector. ψ is the $q \times q$ covariance matrix for the random effect and I_{ni} corresponds to the identity matrix of the n_i dimension [19].

The analyzed geocoded data consisted of polygons, each of them identified by a pair of coordinates based on number of fire incidences; polygon sugarcane area (in hectare); and slope (four relief classes).

The database was built in (*txt) format in the column structure, showing the variables, Identifier (ID), sugarcane area in hectares, amounts of fire incidences in sugarcane areas and the slope that represented the place of manual burning. In the Rstudio software [20], two models were implemented, model 1 (M1) represented in equation:

$$\text{area}_{ij} = \beta + b_{ij} \times \text{year}_{ij} + E_{ij} \quad (2)$$

where area_{ij} represents the area measurement for polygon j and i represents the sugarcane category, β corresponds to a fixed value, $b_{ij} \times \text{year}_{ij}$ corresponds to the random slope of the incidences according to the year and E_{ij}. E_{ij} is the random error, i is the type of use—in this case, sugarcane. Following this, the slope was included as a fixed effect in the model (M2), with Eq. (3), which is illustrated by the following:

$$\text{area}_{ij} = \beta + \text{incidences}_{ij} + b_{ij} \times \text{year}_{ij} + E_{ij} \quad (3)$$

where area_{ij} represents the area measurement for the polygon j and i represents the category of sugarcane; β corresponds to a fixed value; incidences_{ij} represents the random intercept by fire incidence; $b_{ij} \times \text{year}_{ij}$ corresponds to the random slope of the fires according to year; E_{ij} is the random error; and i is the type of use, in this case sugarcane.

Before the model underwent adjustments, an exploratory analysis of the data was performed to obtain information on spatiotemporal data. Bar graphs were produced to display groups of area measurements in order to compare grouped values graphically. This process is useful for analyzing data matching and determining the presence of outlying values.

3 Results and Discussion

Among the years analyzed, the year 2000 presented the lowest number of incidences directly linked to the sugarcane activity. Eighty-eight (88) incidences were identified in locations close to the 16 processing plants (Fig. 6). The area in hectares corresponded to 233,973.84, with greater concentration in the western region of the basin (Fig. 6).

The direct change in land use caused by the recent expansion of sugarcane can be evaluated from remote sensing images, previously recorded for sugarcane plantations, allowing to locate and quantify areas that were destined to other uses (pasture,

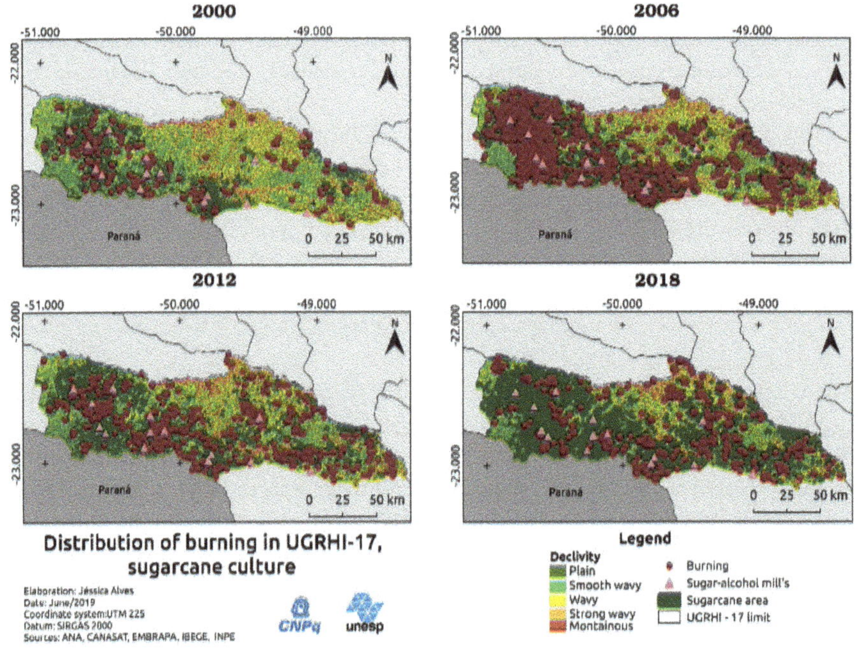

Fig. 6 Distribution of fire incidences in UGRHI-17 in sugarcane cultures during the year 2000

forestry, other monocultures, among others) and were converted into sugarcane plantations, providing relevant information that would be difficult to obtain without the support of this tool.

In the analyzed period, there was a gradual growth in the area dedicated to sugarcane, which can be explained by the high demand for ethanol resulting from the increase in sales of bi-fuel vehicles and the growth in exports of this fuel. Bi-fuel vehicles were introduced to the market in 2003, which boosted production in order to meet the market's demand [21].

With the advent of dual fuel cars, the ability to monitor the expansion of sugarcane in Brazil, which is the largest producer and exporter of sugar in the world and the second largest producer of ethanol, has become relevant due to the culture's agricultural, economic, strategic, and environmental importance. In the last decade, the total area dedicated to sugarcane has doubled in the south-central part of the country.

In 2006, the area in hectares corresponded to 278,378.38; in that period bi-fuel vehicles reached 80% of sales of automobiles and light commercial vehicles (excluding vehicles with diesel engines), which boosted sugarcane production [21].

In that year, 464 fire incidences related to the cultivation of sugarcane were identified. It is important to highlight the western portion of UGRHI-17, where there is a greater concentration of fires, with a focus on the region of the municipalities of Paraguaçu Paulista and Tarumã. In these there are four potentials for processing sugar and alcohol, with the presence of two plants in each municipality. The municipality of Tarumã stood out for hosting the company Raízen, the largest sugar exporter in the international market, with annual production of around 4.1 million tons.

In the same year, the state of São Paulo produced 60% of the national production, about 3.4 million hectares, of which 2.5 million hectares went through the pre-harvest burning process, and 900 thousand hectares were harvested through mechanized means. Regarding the Brazilian production, only 25% of the harvest is done mechanically and the remainder undergoes burning before manual harvesting [22]. The burning period coincides with the season of low rainfall, worsening the conditions and the quality of the air.

Within the state of São Paulo, Law No. 11,241 of September 2002, which regulates the gradual elimination of the burning of sugarcane straw, provides other measures and regulates this activity. The practice of burning was expected to decrease after the Law was implemented; there was, however, an increase of 376 fires between 2000 and 2006.

One of the deadlines stipulated for reducing burning activities, as provided by law, refers to the year 2006, and defines the elimination of 30% in the non-mechanized area. Re-evaluations were stipulated to take place, according to the technological development that makes new machinery feasible in order to enable mechanized harvesting without neglecting socioeconomic aspects [16]. According to the research, there was a greater representation of fires in that year (Fig. 6).

In order to meet the growing domestic demand for ethanol, new areas of sugarcane were cultivated on a large scale. In the 2006 and 2007 harvests alone, cultivation in new areas reached 1.55 million hectares (ha) in the Center-South region of Brazil [23]. As a result of this expansion, the cultivation of sugarcane has replaced the areas of pasture and agriculture in similar proportions and, despite the considerable increase in mechanical harvesting, manual harvesting, which requires burning straw, is still practiced.

The federal government conducted a study to establish agroecological zoning for sugarcane, which considers environmental, economic, and social aspects to guide the adequate expansion of sugarcane and investments in the biofuel sector. The Secretary of Environment of the State of São Paulo also signed an agri-environmental protocol in 2007, which establishes the gradual elimination of the practice of burning until 2014 for mechanized areas (slope $\leq$ 12%) and until 2017 for non-mechanized areas (slope > 12%), anticipating the deadline stipulated in laws [16].

It is important to highlight that in the state of São Paulo, mechanized harvesting started in 1973, but became more widespread with the implantation of Proálcool in 1975, with financial incentives. It was only with the evolution of the machinery' capacity and, above all, with the greater restrictions imposed by the Agro-Environmental Protocol of 2007 between the São Paulo plants and the government of São Paulo, that the sector started to fully engage in the harvest of raw cane.

In 2012, a reflection of the growth in the demand for sugarcane was seen, as well as the decrease in the number of fires. In that year, 426 fires related to the culture were found, in an area of 413,105.18 thousand hectares. This corresponds to the sharp growth of 48.40% of the sugarcane expansion (Fig. 6). It is noteworthy that the incidences dispersed throughout the basin in that year, lowering the concentration in the municipalities of Paraguaçu Paulista and Tarumã (Fig. 6).

The elimination of burning creates dichotomies: while its banning can contribute to improving air quality and contribute positively to the environment and disease prevention, it can extinguish thousands of jobs, creating social and spatial unsustainability. According to [24], cane fields are an important source of jobs for a portion of the population with low education levels. This represents the highest demand for agricultural labor in the State of São Paulo with numbers of 250,907 men per year in 2002, equivalent to 35% of all agricultural labor force.

Unlike most crops, sugarcane has an extensive harvest schedule, and its monitoring requires the acquisition of images captured during the harvest period, which begins in April and extends to December. In this study, the acquisition of images took place from June to September. The identification of the type of sugarcane harvest in the remote sensing images is performed based on the visual differences between the areas in which sugarcane harvest is carried out with or without the practice of burning.

Through the slope map, generated from SRTM images (Shuttle Radar Topography Mission), it was possible to infer the mechanized areas of sugarcane harvesting ($\leq$12% slope) and non-mechanized areas (>12% slope). The intersection between the maps and the slope allowed to suggest the type of harvest by relief class; that is, in areas with the greatest slope the harvesting process occurs most frequently with the practice of burning, as these are areas of difficult access for machinery. On the other hand, on lands with lower slope the use of harvesters is more feasible.

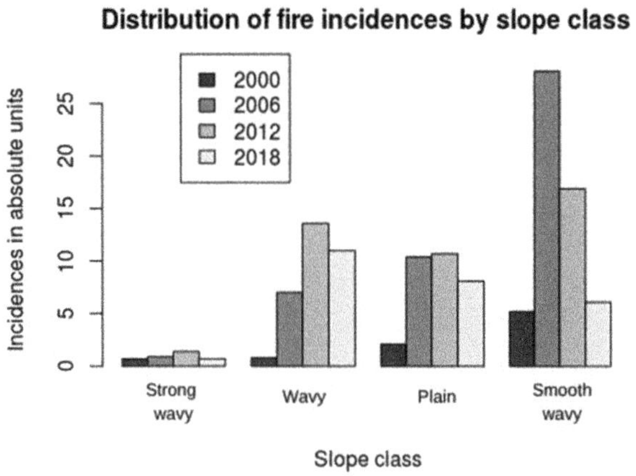

Fig. 7 Distribution of fire incidences by slope class

In 2018, around 259 fire incidences were identified in the sugarcane plantation area (Fig. 6), with a decrease referring to the last 2 years analyzed (Fig. 6). This is a consequence of investments in modernization, as well as the implementation of regulatory instruments and the monitoring of this activity.

The results obtained reveal that the spatial concentration of the areas in which harvesting is carried out through burning has been decreasing in UGRHI-17. This is probably due to the recent changes and investments in mechanization in areas of expansion.

With the application of statistical modeling, it was possible to carry out an exploratory analysis of space-time information, allowing for more accurate results. To characterize the data sets, the bar plot function was used for the construction of a graph with the distribution of the fires according to the slope class (Fig. 7).

In 2006, in plain and smooth undulating terrain, a greater number of incidences were identified. In contrast, in 2012 there was a decrease in fires; however, the areas that carried out the burning are those with undulating and strong undulating relief. The year 2018 shows the same scenario as 2012, further intensifying fires in more rugged terrain.

Figure 8 shows the 95% confidence intervals for random intercepts, represented by b_{ij} in Eq. (3). It is observed that the plain and smooth undulating terrains show a statistically significant negative declivity over time, while the strong undulating slope class shows a positive acclivity.

In Fig. 8, with the values of the intercept and random slope according to the terrain's slope class, demonstrating the increase for the strong wavy relief class and decrease for the other classes. The smooth wavy topography category expresses a negative value over the years: that is, it represents a decrease in fires, corresponding to a decrease of 99.9% per year. In the next 6 years, it would mean a decrease of 599.99%, while for lands with a strong wavy slope it shows a positive ascent, which results in an increase of 2.07% per year. In the next 6 years, it would imply a growth of 12.45%. In the year 2021, there will be an increase of 6.21% and in 2031 the growth will be 26.91%, if all the causes and effects remain fixed.

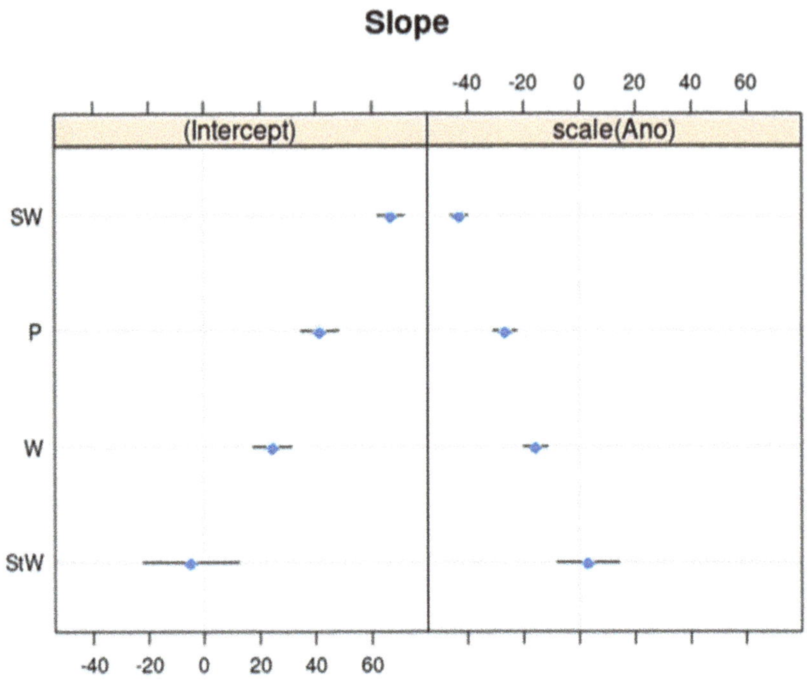

Fig. 8 95% confidence interval for the intercept and random slope (*StW*—strong wavy, *W*—wavy, *P*—plain, *SW*—smooth wavy)

4 Conclusions

The methodologies adopted to analyze the incidences of fires in the study area produced information about the dimensions and distribution of the activity at UGRHI-17 in an efficient and quick manner. The tools employed were of aid regarding the decision-making that contributes to the understanding of future processes of the practice of burning.

The exploratory analysis of the data showed that there was a decrease in fires of 99.9% per year; however, the occurrence in areas with strong undulating relief increased 2.07% per year. This result is shown to be in the opposite direction of State Law 11,241. Statistical analysis indicates that fire incidences in smooth undulating terrain tend to decrease. On steeper terrain, where machines are unable to operate, there is indication of an increase in the practice of burning, if all causes and effects remain fixed.

Geotechnology can support other analyses, such as land use and occupation, and the impact of anthropogenic actions on climate changes. These analyses will provide important information for establishing public policies, as well as for sustainable planning.

References

1. Souza, C. (2007). A influência do ritmo climático na morbidade respiratória em ambientes urbanos. 199f. Dissertação (Mestrado em Geografia). Universidade Estadual Paulista, Faculdade de Ciências e Tecnologia Campus de Presidente Prudente.
2. Ribeiro, H. (2008). Queimadas de cana-de-açúcar no Brasil: Efeitos à saúde respiratória. *Revista de Saúde Pública, São Paulo, 42*(2), 370–376.
3. Kohlhepp, G. (2010). Análise da situação da produção de etanol e biodiesel no Brasil. *Estudos avançados, 24*(68), 223–253.
4. Zancul, A. (1998). O efeito da queimada de canade-açúcar na qualidade do ar na região de Araraquara-SP. Dissertação (Mestrado em Engenharia)—Escola de Engenharia de São Carlos - USP, São Carlos.
5. Ribeiro, H., & Assunção, J. V. (2002). *Efeitos das queimadas na saúde humana* (Vol. 44, pp. 125–148). Estudos Avançados USP.
6. Arbex, M. A. (2001). Avaliação dos efeitos do material particulado proveniente da queima da plantação de cana-de-açúcar sobre a morbidade respiratória na população de Araraquara-SP. Tese (Doutorado em Ciência)—Faculdade de Medicina da Universidade de São Paulo, São Paulo.
7. Lopes, F. S., & Ribeiro, H. (2006). Mapeamento de internações hospitalares por problemas respiratórios e possíveis associações à exposição humana aos produtos da queima de palha de canade-açúcar (Saccharum sp) no Estado de São Paulo. *Revista Brasileira de Epidemiologia, São Paulo, 9*(2), 215–225.
8. Arbex, M. A., et al. (2004). Queima de biomassa e efeitos sobre a saúde. *Jornal Brasileiro de Pneumologia, Brasília, DF, 30*(2), 158–175.
9. Cançado, J. E. D., et al. (2006). Repercussões clínicas da exposição à poluição atmosférica. Jornal Brasileiro de Pneumologia, Brasília. *DF, 32*, S5–S11.
10. ALESP, Assembléia Legislativa do Estado de São Paulo. (2002). Dispões sobre a eliminação gradativa da queima da palha da cana-de-açúcar e dá providências correlatas. Retrieved from https://www.al.sp.gov.br/repositorio/legislacao/lei/2002/lei-11241-19.09.2002.html.
11. Lara, L. L., Artaxo, P., Martinelli, L. A., Camargo, P. B., Victoria, R. L., & Ferraz, E. S. B. (2005). Properties of aerosols from sugar-cane burning emissions in Southeastern Brazil. *Atmospheric Environment, 39*, 4627–4637.
12. Aguiar, D. A., Rudorff, B. F. T., Silva, W. F., Adami, M., & Mello, M. P. (2011). Remote sensing images in support of environmental protocol: Monitoring the sugarcane harvest in São Paulo State. *Brazil. Remote Sens., 3*, 2682–2703.
13. Ichoku, C., Kahn, R., & Chin, M. (2012). Satellite contributions to the quantitative characterization of biomass burning for climate modeling. *Atmospheric Research, 111*, 1–28.
14. Hantson, S. (2013). Pontos fortes e fracos dos pontos ativos MODIS para caracterizar a ocorrência global de incêndio. *Sensoriamento Remoto do Ambiente, 131*, 152–159.
15. Magalhães, I. A. L., Xavier, A. C., & Santos, A. R. (2012). Caracterização da dinâmica espectro-temporal-florestal e da cana-de-açúcar no município de Itapemirim, ES. GEOSUL, v. 28, n. 56, p. 183-210, 2013 (IGNÁCIO, Sergio Aparecido. Importância da estatística para o processo de conhecimento e tomada de decisão). *Revista Paranaense de Desenvolvimento-RPD, 118*, 175–192.
16. São Paulo. Sistema Integrado de Gerenciamento de Recursos Hídricos do Estado de São Paulo. (2019). Comitê da Bacia Hidrográfica do Médio Paranapanema. Retrieved from http://www.sigrh.sp.gov.br/cbhmp/apresentacao.
17. Ronquim, C. C. (2018). Avanço das áreas de cana-de-açúcar e alterações em áreas de agropecuária no interior paulista/Carlos Cesar Ronquim, Marcelo Fernando Fonseca. Campinas: Embrapa Territorial.
18. Esri. (2007). *Arcgis Desktop: Release 9.2*. Environmental Systems Research Institute.

19. Galvanin, E. A. S., et al. (2018). *Mixed-effects modeling for analyzing land use change in the Brazilian Pantanal subregion of Cáceres*. Remote Sensing Applications: Society and Environment.
20. RSTUDIO TEAM. (2015). RStudio: Integrated development for R, RStudio, Inc., Boston, MA. Retrieved from http://www.rstudio.com/.
21. Rodrigues, A. P., & Rodrigues, L. (2008). A nova agenda do setor sucroenergético: O etanol e os desafios do mercado interno. *AgroAnalysis.*, *28*(8), 19–36.
22. Ribeiro, H., & Pesquero, C. (2010). Queimadas de cana-de-açúcar: Avaliação de efeitos na qualidade do ar e na saúde respiratória de crianças. *Estudos avançados, 24*(68), 255–271.
23. CANASAT. (2019) .Monitoramento de cana-de-açúcar via imagens de satélite. Retrieved from http://www.dsr.inpe.br/laf/canasat/.
24. Braunbeck, O. A., & Magalhães, P. S. G. (2004). Colheita sustentável com aproveitamento integral da cana. *Visão Agrícola, Piracicaba, 1*(1), 72–79.

A Decision Support System for Green Crop Fertilization Planning

Efthymios Rodias, Eleftherios Evangelou, Maria Lampridi, and Dionysis Bochtis

1 Introduction

The intensification of agriculture to meet the ever-increasing need for adequate amounts of production has triggered significant progress in agricultural production processes at various levels in the last decades [1]. A portion of the advances regards planning, simulation, and/or operational scheduling in terms of agricultural machinery management, aiming at increasing the efficiency of production [2]. The purpose is to go beyond the formal observance of agronomic protocols and pass to the holistic management of the production process through operations management [3]. This type of standardization of agricultural activities paves the way towards their further automation through the integration of Information and Communication Technologies (ICT) and their eventual robotization [4].

At this light, and towards the optimum management of the various field operations, a series of applications, tools, and simulation methodologies have been performed by various authors [5–7]. Such analyses involve system optimization on operational level focused on minimization of financial cost, energy use, or CO_2 emissions [8, 9]. Each specific in-field or logistic operation includes a large amount

E. Rodias (✉) · M. Lampridi · D. Bochtis
Institute for Bio-Economy and Agri-Technology (iBO), Centre for Research and Technology Hellas (CERTH), Thessaloniki, Greece
e-mail: e.rodias@certh.gr

E. Evangelou
Institute of Industrial and Forage Crops, Hellenic Agricultural Organization "Demeter", Larisa, Greece

D. D. Bochtis et al. (eds.), *Information and Communication Technologies for Agriculture—Theme III: Decision*, Springer Optimization and Its Applications 184, https://doi.org/10.1007/978-3-030-84152-2_13

of inputs associated with it, for the cost or energy consumption evaluation. Several studies exist that assess the field operations related to a specific crop or multiple crop production systems [10–12]. However, each crop production system may differ from others due to its very specific features, according to the type of crop (i.e., food or non-food crop), the agricultural practices, or the agronomic protocols that are followed. For that purpose, other works focus on the contribution of the individual operations and the machinery used in the overall energy cost balance [6, 13] in order to provide with assessment information irrespective to the type of cultivation.

On-farm energy efficiency is becoming increasingly important in the context of concerns for rising energy costs and greenhouse gas emissions in each field operation due to field machinery use, agrochemicals, fertilizers, etc. [14]. Today's agricultural production relies heavily on the consumption of non-renewable fossil fuels [15]. Consumption of fossil energy results in direct negative environmental effects through the release of CO_2 and other combustion gases [16]. Nonetheless, the dependency of agricultural operations to non-renewable energy sources is also related to a number of economic and societal impacts that jeopardize energy security and autonomy [17]. Thus, taking into account the need to address the negative impacts of agricultural activities, it is important to move towards practices that abide to sustainable agriculture [18]. Towards that direction, the monitoring and assessment of individual agricultural operations is imperative.

Among the highest energy-consuming field operations is fertilization. Various forms of fertilizers are applied in agricultural crop production (such as organic fertilizers and chemical fertilizers). Chemical fertilizers come from mineral extraction, and they are enriched with nutrients in a concentrated form that make them more easily available to plants. On the other hand, organic fertilizers are made from natural materials composition such as animal or plant materials and mined minerals, with almost no processing. In this case, nutrients are released slowly compared to chemical fertilizers by natural biological processes, and they have relatively low concentrations. For this reason, fertilizers application is a field operation that should comply with strict scheduling constraints in order to achieve maximum performance of the fertilizers and low environmental impact [19]. Fertilization has been evaluated among other field operations by a number of authors regarding the energy input evaluation [20, 21].

In this chapter, chemical fertilization is evaluated under two different case studies, an agri-food annual crop and a perennial energy crop in order to indicate the variance of fertilization energy cost between these two crops (industrial tomato and *Arundo donax*) on annual basis.

Tomato is on a global scale among the foods that are produced in very large quantities. It is estimated that annually about 126 million tons of tomato are produced globally from which about 30 million tones are processed [22]. Due to the economic value of tomato product, farmers' interest for improving the conditions of tomato cultivation is increasing, adopting progressive production systems such as the integrated management system (IMS). Furthermore, the farmers and their cooperatives understand the necessity of these management systems and are adopting them following the international trends and standards. In Southern European countries, the tomato production industry is highly mechanized and heavily reliant on fossil fuels (electricity and diesel). In such highly mechanized farming systems, machinery inputs are significant and can represent 40–50% of the farm input costs, while the rest of the inputs is mainly related to agrochemicals and fertilizers. Given the major dependence on direct energy inputs and rising energy costs, energy use efficiency is an emerging issue for the tomato industry.

On the other hand, *Arundo donax* is a perennial energy crop that is cultivated not for agri-food but for biomass energy production [23]. Due to its natural features, it is a more resistant crop in relation to its enemies, and it has no significant need for agrochemicals' application. As a biomass crop, it requires high energy consumption through harvesting and transportation operations but during, mainly, its primary establishment years, it requires nutrients contribution by fertilization.

In this study, the comparison of energy consumption of these two crops fertilization operations is performed, based on a decision support system in order to extract the optimal crop for a given set of fields. The rest of the chapter is organized as follows: Section 2 presents the system under examination and the methods for the estimation of energy consumption. Moving further, in Sect. 3 the application of the proposed methodology is presented, describing the crops under examination along with the fertilization scenarios and the input parameters used. Sections 4 and 5 present the results of the assessment as well as the relevant discussion while the chapter concludes with Sect. 6.

2 System Description

The described system (Fig. 1) includes the in-field and the logistics operations related to the farm-field-farm cycle for fertilizers spreading. In this cycle, the field machinery and the fertilizers that are going to be applied in the field are included. As indirect inputs, the embodied energy of tractors and implements, the fertilizers, and the fuels contribute to the system.

For the estimation of energy consumption throughout a given crop production period, a series of input parameters should be included. These are, namely the production inputs (such as the field features and the crop features), the machinery inputs (i.e., the tractors and implement characteristics), the operation-related input parameters (i.e., the parameters related to in-field and transport operations), and the material-related inputs [11]. The overall structure of the proposed system is presented in Fig. 2.

For the in-field fuel consumption estimation, the specific volumetric fuel consumption equation given in the American Society of Agricultural and Biological Engineers (ASABE) standards [24] was used:

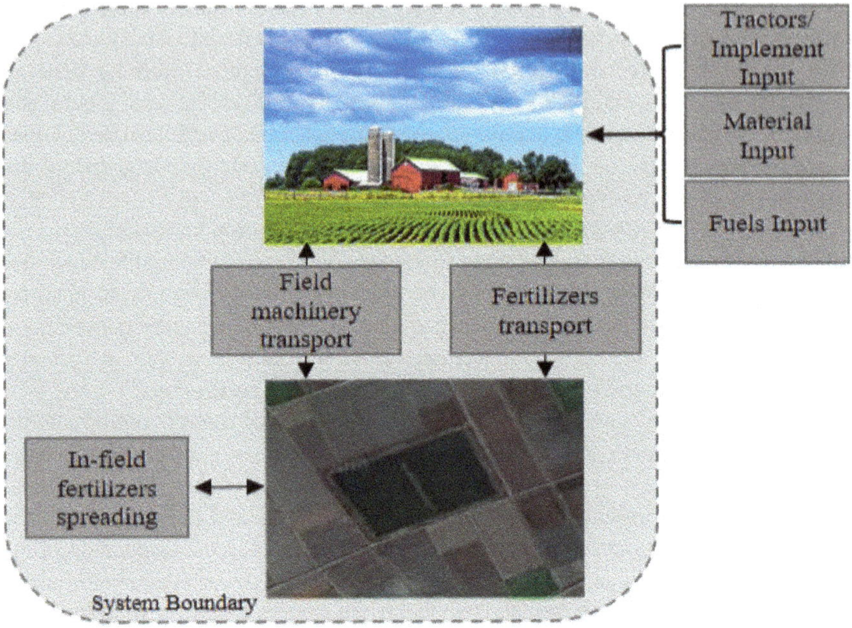

Fig. 1 The described system boundary

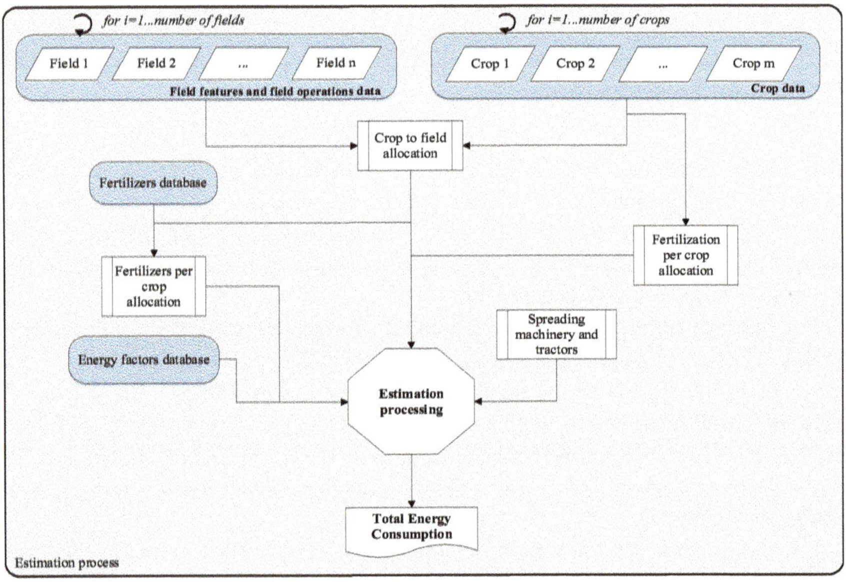

Fig. 2 The overall structure of the estimation processing

$$Q = \left(2.64X + 3.91 - 0.203\sqrt{738X + 173}\right) * X * P_{pto}$$

where Q is the diesel fuel consumption at partial load (in L h^{-1}), $X = P/P_{\text{rated}}$ is the ratio of equivalent PTO (power-take-off) power (P) in the specific operation to the rated PTO power (P_{rated}) that is normally considered as the 83% of the gross flywheel [25].

As a further step, the fuel energy consumption was calculated as follows:

$$\text{FE} = \text{FEC} * Q * \left(\frac{1}{C}\right) * A$$

where FE is the fuel energy (in MJ), FEC is the fuel energy coefficient (in MJ L^{-1}), Q is the fuel consumption (in L h^{-1}), C is the field operational capacity (in m^2 h^{-1}), and A is the field area (in m^2).

For the estimation of in-field machinery embodied energy, it could be presented mathematically as follows:

$$\text{MEE} = \text{EEC} * w * \left(\frac{\left(\frac{A}{C}\right)}{\text{ML}}\right)$$

where MEE is the machinery embodied energy (in MJ), EEC represents the coefficient of both tractor and machinery, w is the associated weight (in kg), and ML is the tractor and machinery estimated lifetime. It is also worth noting that the embodied energy of the machinery is reduced to the duration of each operation.

This system regards fertilizers application; a field operation that includes material flow inputs. The energy consumption is calculated based on the in-field fertilizing energy input and the farm-field transportation energy input. On top of this, the material flow input should be added, mainly related to the fertilizers' embodied energy. As for the fertilizers-related transportation energy cost, there might be variation for each case according to the number of trips that should be carried out to fulfill the fertilization needs of each field/crop.

3 Case Study Demonstration

3.1 The Demonstrated Crops

For the demonstration of the decision support system, a case study that includes two completely different crops (an energy crop—*Arundo donax* and a food crop—industrial tomato) was selected and evaluated only in terms of their fertilization energy cost.

Arundo donax (also known as giant reed) is a perennial crop with several features enhancing its potential as an energy crop. It grows in a wide range of climatic conditions with recorded yields up to 40 t ha^{-1} of dry matter in certain wild populations [26, 27]. Giant reed is a quite aggressive plant and can grow on almost any soil type [28]. Angelini et al. also reported that the average annual energy consumption was 12.5 GJ ha^{-1} for a 12-year production period [28]. In another energy analysis of the giant reed, the estimated annual energy cost varied from approximately 3.9–17.9 GJ ha^{-1} for unfertilized and fertilized crops, respectively, for a 10-year production cycle [23]. Finally, in a 5-year giant reed production cycle, the average annual energy cost was 16.7 GJ ha^{-1} [29].

On the other hand, tomato is a quite common vegetable crop with high potential to be consumed not only in its primary form but as a processed product (juice, puree, paste, sauce, or whole canned), as well. It is a major agricultural product that accounts for over 30% of global production [30]. Various scenarios in terms of various tomato production systems (open-field, greenhouse), and different fertilization applications have been assessed under the criterion of less environmental impact compared to yield production [31, 32]. The energy consumption evaluated in a stake-tomato production system was found to be about 97 GJ ha^{-1} [33]. Another study presented energy requirements for open-field and greenhouse tomato production systems up to 47.6 and 21 GJ ha^{-1}, respectively [34]). Similarly, a case study from Iran presented that the total energy consumption in a tomato production system was 65.2 GJ ha^{-1} of which the 50% came from fertilization processes [35]. Overall, based on the current knowledge of authors, industrial tomato crop production processes have not been under study in terms of in-field operations.

3.2 *Fertilization Scenario*

The production case studies of these two crops are based on the prevailing production practices regarding only the fertilizers application followed by farmers. The production-related parameters (in-field application of fertilizers, implemented machinery, applied dosages, etc.) were selected in all cases after a peer review of the related bibliography and/or real farmers data and according to the real commercial data in order to be as close to the real production procedure as possible.

In this study, the first crop, *Arundo donax*, is a perennial crop and, for this reason, is demonstrated in a 10-year production period targeting to reduce the total fertilization input to an annual basis. Fertilization takes place each year but in variable dosages (Table 1). More specifically, annual applications of nitrogen are recommended at a level up to 100 kg ha^{-1}, according to soil fertility. Moreover, it is essential before the establishment to incorporate sufficient phosphorus into the soil by incorporating a minimum quantity of 200 kg ha^{-1}, especially in phosphorus-poor fields. Potassium fertilization should be applied only where it is required [28]. In the present case study, for the first year, the application of 80 kg N ha^{-1}, 200 kg P_2O_5 ha^{-1}, and 100 kg K_2O ha^{-1} were considered. For each following even year (2nd, 4th, 6th, 8th, and 10th year) 80 kg N ha^{-1} and 50 kg P_2O_5 ha^{-1} were considered. While, for each odd year, excluding the first one (i.e., 3rd, 5th, 7th, and

Table 1 Ten-year fertilization plan for *Arundo donax*

Year	Nitrogen (N) ($kg\ ha^{-1}$)	Phosphorus (P_2O_5) ($kg\ ha^{-1}$)	Potassium (K_2O) ($kg\ ha^{-1}$)
1	80	200	100
2-4-6-8-10	80	50	–
3-5-7-9	80	50	100

Table 2 Annual fertilization plan for industrial tomato (average values)

Nitrogen (N) ($kg\ ha^{-1}$)	Phosphorus (P_2O_5) ($kg\ ha^{-1}$)	Potassium (K_2O) ($kg\ ha^{-11}$)
174	86	106

9th), 80 kg N ha^{-1}, 50 kg P_2O_5 ha^{-1}, and 100 kg K_2O ha^{-1} were considered to be applied.

On the other hand, the industrial tomato crop is an annual crop, and its corresponding fertilizing inputs based on real farmers' data. Fertilization management consists of pre-planting applications of a portion of the total N at the same time with the total applied P and K, and in the season gradually applications of the remaining N. The total average applied quantity of nitrogen was 174 kg ha^{-1} (ranged from 104 up to 235 kg N ha^{-1}), while the application of phosphorus was on average 86 kg P_2O_5 ha^{-1} (ranged from 24 up to 184 kg ha^{-1}), and the average applied quantity of potassium was 106 kg K_2O ha^{-1} (ranged from 26 up to 200 kg ha^{-1}),) as presented in Table 2.

3.3 *Input Parameters*

The operational and field machinery-related inputs that included in both presented cases are detailed in Table 3. It is assumed that same type of machinery (boom-type sprayer) was used in both cases. In the previous chapter, the annual applications of fertilizers have already been presented for each of the crops. In terms of the corresponding energy factors that are included in fertilization are 78.1 MJ kg^{-1} for nitrogen, 17.4 MJ kg^{-1} for phosphorus, and 13.7 MJ kg^{-1} for potassium [37].

4 Results

Both crops presented in this study were assessed for a given set of nine fields with a range of field areas from 1.65 to 6.83 ha. More specifically, the detailed field areas along with their distances from the farm location are presented in Table 4. Given all the energy inputs and other parameters, the energy consumption for fertilizers application was calculated for each one of the crops. The energy consumption for

Table 3 Operational and field machinery inputs

Inputs	
Operating width[a] (m)	18
Operating speed[b] (km h^{-1})	11
Field efficiency[b]	0.7
Tractor embodied energy[c] (MJ kg^{-1})	138
Implement embodied energy[c] (MJ kg^{-1})	129
Tractor weight[d] (kg)	2930
Implement weight[a] (kg)	3350
Tractor estimated life[b] (h)	16,000
Implement estimated life[b] (h)	1200
Fuel energy coefficient[d,e] (MJ L^{-1})	41.2
Tractor power (kW)	90
Lubricants energy coefficient[f] (MJ L^{-1})	46
Nitrogen fertilizer energy content[c] (MJ kg^{-1})	78.1
Phosphorus fertilizer energy content[c] (MJ kg^{-1})	17.4
Potassium fertilizer energy content[c] (MJ · kg^{-1})	13.7

[a]Commercial values
[b]American Society of Agricultural and Biological Engineers [36]
[c]Kitani [37]
[d]Wells [38]
[e]Barber [39]
[f]Saunders et al. [40]

Table 4 Fields features

A/N	Field area (ha)	Distance farm-field (km)
1	6.83	4
2	3.24	4
3	3.00	4.4
4	2.48	6
5	1.65	4.5
6	2.80	2
7	2.72	5.3
8	3.20	6.1
9	1.80	1.5

fertilization of these two crops is depicted in Table 5 in terms of the main energy-consuming categories (i.e., fuels, machinery embodied energy, and fertilizers embodied energy).

The total annual energy consumption of the presented crops is up to 227.50 GJ and 468.71 GJ for *Arundo donax* and industrial tomato, respectively. The detailed energy input per field is presented in Fig. 3. Regarding the average consumed energy per field area unit (i.e., one hectare), for *Arundo donax* was up to 8.21 GJ ha^{-1}, and 16.9 GJ ha^{-1} for industrial tomato. More specifically, in the case of the energy input ranged from 8.18 to 8.25 GJ ha^{-1}, while for the industrial tomato case the energy input was more dispersed among the fields, as presented in Fig. 4.

Table 5 Energy consumption per field and crop

Fields	Crops	Energy consumption (MJ)		
		Fuels	Machinery embodied	Fertilizers
1	Arundo donax	669	122	7,580
	Tomato	1792	901	109,887
2	Arundo donax	335	112	3,596
	Tomato	857	449	52,103
3	Arundo donax	316	121	3,329
	Tomato	817	489	59,275
4	Arundo donax	281	161	2,752
	Tomato	688	442	51,775
5	Arundo donax	191	120	1,831
	Tomato	469	330	26,667
6	Arundo donax	277	59	3,107
	Tomato	667	164	45,508
7	Arundo donax	297	143	3,019
	Tomato	725	395	42,525
8	Arundo donax	349	165	3,551
	Tomato	920	670	42,444
9	Arundo donax	180	44	1,998
	Tomato	417	68	27,190

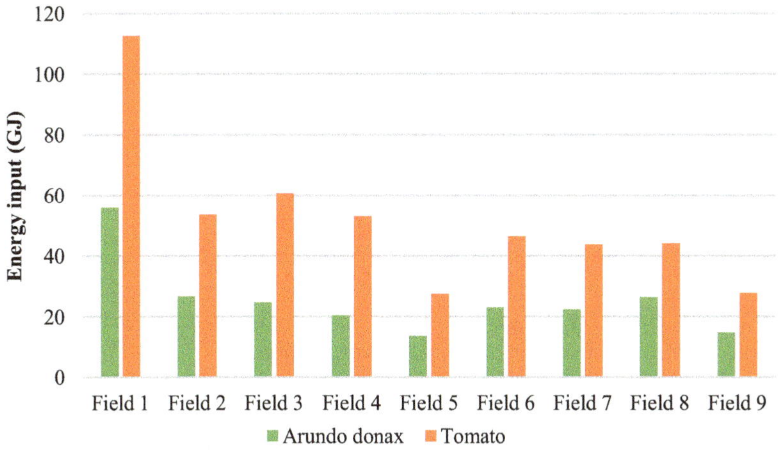

Fig. 3 Annual energy consumption (in GJ) for each crop and field

In terms of *Arundo donax*, the annual energy consumption is extracted as the average annual of the total energy consumption for a 10-year period. As a result, the total energy input for the lifetime of this energy crop would be ten times higher regarding the application of the fertilizers. Of course, it should be underlined the fact that the energy input may differ from year to year due to different fertilizers dosages and/or other factors such as variability in machinery use.

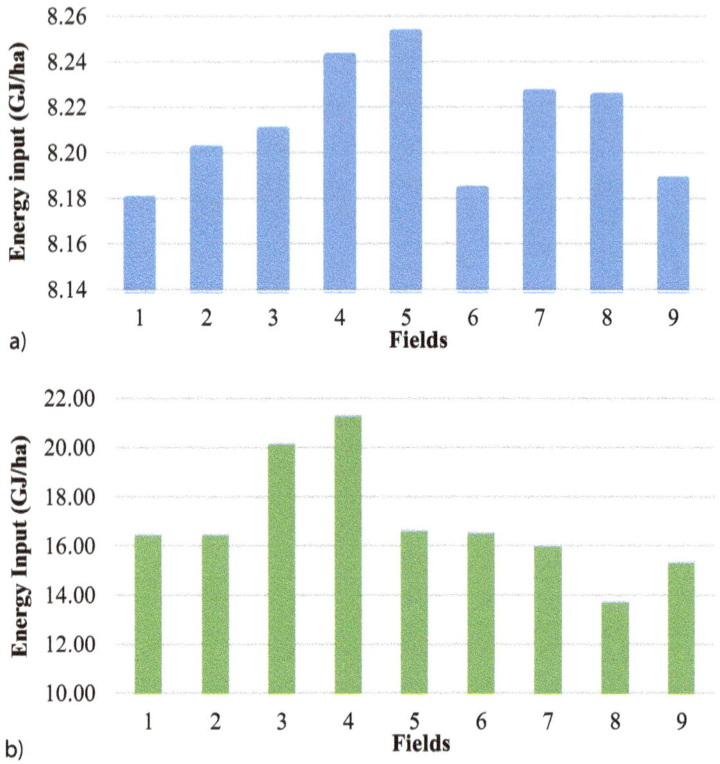

Fig. 4 Energy input per unit area (**a**) *Arundo donax* (Annual values resulting from 10-year average), (**b**) Tomato

5 Discussion

Fertilization is a very energy-consuming field operation that affects the total energy input in a given crop production system taking into account the features of the system. In this chapter, two significantly different crops were assessed by the use of a decision support tool in terms of fertilizers application.

Agri-food crops are usually engaged to high needs in nutrients supplement to achieve a high quality of the final product and maximum yields. On the other hand, energy crops have lower needs compared to food crops although their fertilization needs are not negligible especially when high biomass yield production is required. This has been depicted using the decision support system for the specific field areas.

More specifically, the embodied energy in tractors and field machinery had the lower energy contribution in the system for both cases, while the fuels energy consumption for both in-field and logistic (farm-to-field) use during fertilizing operation was about 2–5 times higher for *Arundo donax* case and about two to six times higher for industrial tomato fields. However, the highest energy consumption for both crops related to the fertilizers embodied energy. In the case of *Arundo*

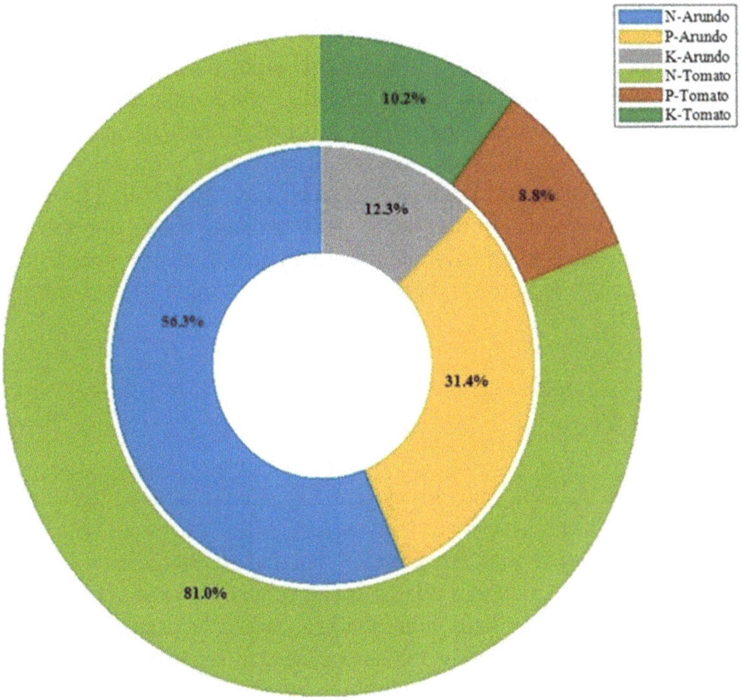

Fig. 5 Energy input allocation (%) for each of the main chemical fertilizers

donax, the associated dosages of the main chemical elements were based on average prices regarding this crop. On the other hand, the fertilizers quantities for industrial tomato were derived from real farmers' data. The embodied energy per kg of the fertilizers together with the applied fertilizers dosages is the main factor that affects the embodied energy of fertilizers per field area unit.

As has been mentioned above, the embodied energy for chemical fertilizers is the energy consumed throughout the fertilizer production cycle. For this reason, the embodied energy per chemical fertilizer varies. The energy input allocation for the main chemical elements of both crops is presented in Fig. 5. This figure underlines the high use of nitrogen fertilizers in both industrial tomato and *Arundo donax* fields. Phosphorus fertilizers energy input contributes more (about 30%) in *Arundo donax* crop than in tomato, while potassium energy input is more significant for tomato.

The significant point that the two case studies differ is their crop production cycle. *Arundo donax* was evaluated for a decade, whereas industrial tomato for a year. To have comparable results, the fertilization energy cost of *Arundo donax* was downsized on annual basis by extracting the average annual energy cost. This fact is responsible for the more intense fluctuation of the industrial tomato line in the previously presented Fig. 4. The variance in the consumption of energy among different crops indicates the need for specialized assessments with the use of operations management tools.

Operations management tools can be integrated within Farming Management Information Systems offering holistic services to farmers. Such systems utilize ICT technologies in the implementation of agronomic protocols resulting in the increase of operation efficiency and the improvement of the sustainability performance through the considerate use of inputs. This is achieved through optimization, with the application of algorithms, as for example it is applied in Precision Agriculture (PA) and Variable Rate Application (VRA) techniques. However, a thorough documentation of conventional processes is a determining factor for the eventual application of the above techniques and the assessment of any proposed alternative process as for example the use of robotics in precision farming, organic farming, etc.

6 Conclusions

In this chapter, a decision support system has been presented for the energy assessment of fertilization operation in various types of crops, both food and non-food. Such a system may be a useful tool for agronomists, policy makers, cooperatives, or other agriculture-related bodies under the objective of minimum energy crop production planning, both strategic and/or tactical depending on the crop cycle. Low-energy crop production is as crucial as low-cost crop production in modern agriculture and for this reason, further research should be conducted regarding innovative tools and applications towards this scope. As a further step, this decision support system may be modified and improved in a way to be more user-friendly and transformed into a mobile online application. This would give the opportunity to any farmer interested to assess his crop fertilization protocol and redesign it if needed. In addition, a more accurate energy fertilization assessment is under authors' planning by including also micro-nutrients or other types of fertilizers such as organic fertilizers.

References

1. Moysiadis, V., Tsolakis, N., Katikaridis, D., et al. (2020). Mobile robotics in agricultural operations: A narrative review on planning aspects. *Applied Sciences, 10*(10), 3453. https://doi.org/10.3390/app10103453
2. Bochtis, D. D., Sørensen, C. G. C., & Busato, P. (2014). Advances in agricultural machinery management: A review. *Biosystems Engineering, 126*, 69–81.
3. Sørensen, C. G., Fountas, S., Nash, E., et al. (2010). Conceptual model of a future farm management information system. *Computers and Electronics in Agriculture, 72*, 37–47. https://doi.org/10.1016/j.compag.2010.02.003
4. Marinoudi, V., Sørensen, C. G., Pearson, S., & Bochtis, D. (2019). Robotics and labour in agriculture. A context consideration. *Biosystems Engineering, 184*, 111–121. https://doi.org/10.1016/j.biosystemseng.2019.06.013

5. Berruto, R., Busato, P., Bochtis, D. D., & Sørensen, C. G. (2013). Comparison of distribution systems for biogas plant residual. *Biomass and Bioenergy, 52*, 139–150. https://doi.org/10.1016/j.biombioe.2013.02.030
6. Lampridi, M., Kateris, D., Sørensen, C. G., & Bochtis, D. (2020). Energy footprint of mechanized agricultural operations. *Energies, 13*, 1–15. https://doi.org/10.3390/en13030769
7. Rodias, E., Berruto, R., Busato, P., et al. (2017b). Energy savings from optimised in-field route planning for agricultural machinery. *Sustainabiity, 9*(11), 1956. https://doi.org/10.3390/su9111956
8. Bochtis, D. D., Sørensen, C. G., Busato, P., & Berruto, R. (2013). Benefits from optimal route planning based on B-patterns. *Biosystems Engineering, 115*, 389–395. https://doi.org/10.1016/j.biosystemseng.2013.04.006
9. Rodias, E. C., Lampridi, M., Sopegno, A., et al. (2019a). Optimal energy performance on allocating energy crops. *Biosystems Engineering, 181*, 11–27. https://doi.org/10.1016/j.biosystemseng.2019.02.007
10. Busato, P., Sopegno, A., Berruto, R., et al. (2017). AWeb-based tool for energy balance estimation in multiple-crops production systems. *Sustain, 9*, 1–18. https://doi.org/10.3390/su9050789
11. Rodias, E., Berruto, R., Bochtis, D., et al. (2017a). A computational tool for comparative energy cost analysis of multiple-crop production systems. *Energies, 10*(7), 831. https://doi.org/10.3390/en10070831
12. Sopegno, A., Rodias, E., Bochtis, D., et al. (2016). Model for energy analysis of Miscanthus production and transportation. *Energies, 9*, 1–16. https://doi.org/10.3390/en9060392
13. Mantoam, E. J., Angnes, G., Mekonnen, M. M., & Romanelli, T. L. (2020). Energy, carbon and water footprints on agricultural machinery. *Biosystems Engineering, 198*, 304–322. https://doi.org/10.1016/j.biosystemseng.2020.08.019
14. Nassi o Di Nasso, N., Bosco, S., Di Bene, C., et al. (2011). Energy efficiency in long-term Mediterranean cropping systems with different management intensities. *Energy, 36*, 1924–1930. https://doi.org/10.1016/j.energy.2010.06.026
15. Dyer, J. A., Kulshreshtha, S. N., McConkey, B. G., & Desjardins, R. L. (2010). An assessment of fossil fuel energy use and CO2 emissions from farm field operations using a regional level crop and land use database for Canada. *Energy, 35*, 2261–2269. https://doi.org/10.1016/j.energy.2010.02.013
16. Gündoğmuş, E. (2006). Energy use on organic farming: A comparative analysis on organic versus conventional apricot production on small holdings in Turkey. *Energy Conversion and Management, 47*, 3351–3359. https://doi.org/10.1016/j.enconman.2006.01.001
17. Arizpe, N., Giampietro, M., & Ramos-Martin, J. (2011). Food security and fossil energy dependence: An international comparison of the use of fossil energy in agriculture (1991-2003). *CRC Critical Reviews in Plant Sciences., 30*, 45–63. https://doi.org/10.1080/07352689.2011.554352
18. Lampridi, M., & Kateris, D. (2020). Primary Production Sustainability. In D. Bochtis, C. Achillas, G. Banias, & M. Lampridi (Eds.), *Bio-economy and agri-production: Concepts and evidence* (1st ed., p. 348). Academic Press, Elsevier.
19. Rodias, E. C., Sopegno, A., Berruto, R., et al. (2019b). A combined simulation and linear programming method for scheduling organic fertiliser application. *Biosystems Engineering, 178*, 233–243. https://doi.org/10.1016/j.biosystemseng.2018.11.002
20. Samavatean, N., Rafiee, S., Mobli, H., & Mohammadi, A. (2011). An analysis of energy use and relation between energy inputs and yield, costs and income of garlic production in Iran. *Renewable Energy, 36*, 1808–1813. https://doi.org/10.1016/j.renene.2010.11.020
21. Stolarski, M. J., Krzyżaniak, M., Warmiński, K., et al. (2017). Energy efficiency of perennial herbaceous crops production depending on the type of digestate and mineral fertilizers. *Energy, 134*, 50–60. https://doi.org/10.1016/j.energy.2017.05.195

22. Karakaya, A., & Özilgen, M. (2011). Energy utilization and carbon dioxide emission in the fresh, paste, whole-peeled, diced, and juiced tomato production processes. *Energy, 36*, 5101–5110. https://doi.org/10.1016/j.energy.2011.06.007
23. Angelini, L. G., Ceccarini, L., & Bonari, E. (2005). Biomass yield and energy balance of giant reed (Arundo donax L.) cropped in central Italy as related to different management practices. *European Journal of Agronomy, 22*, 375–389. https://doi.org/10.1016/j.eja.2004.05.004
24. ASABE. (2015). *D497.6: Agricultural machinery management data.*
25. Bochtis, D., Sorensen, C. G., & Kateris, D. (2018). *Operations management in agriculture* (1st ed.). Elsevier.
26. Lewandowski, I., Scurlock, J. M. O., Lindvall, E., & Christou, M. (2003). The development and current status of perennial rhizomatous grasses as energy crops in the US and Europe. *Biomass and Bioenergy, 25*, 335–361.
27. Vermerris, W. (2008). *Genetic improvement of bioenergy crops*. Springer.
28. Angelini, L. G., Ceccarini, L., Nassi o Di Nasso, N., & Bonari, E. (2009). Comparison of Arundo donax L. and Miscanthus x giganteus in a long-term field experiment in Central Italy: Analysis of productive characteristics and energy balance. *Biomass and Bioenergy, 33*, 635–643. https://doi.org/10.1016/j.biombioe.2008.10.005
29. Mantineo, M., D'Agosta, G. M., Copani, V., et al. (2009). Biomass yield and energy balance of three perennial crops for energy use in the semi-arid Mediterranean environment. *Field Crops Research, 9*, 204–213. https://doi.org/10.1016/j.fcr.2009.07.020
30. Vidyarthi, S. K., & Evans, M. E. (2019). Chapter 7: Development of new tomato products in a very consolidated market. In *Food chemistry, function and analysis*.
31. Martínez-Blanco, J., Muñoz, P., Antón, A., & Rieradevall, J. (2009). Life cycle assessment of the use of compost from municipal organic waste for fertilization of tomato crops. *Resources, Conservation and Recycling, 53*, 340–351. https://doi.org/10.1016/j.resconrec.2009.02.003
32. Muñoz, P., Antón, A., Paranjpe, A., et al. (2008). High decrease in nitrate leaching by lower N input without reducing greenhouse tomato yield. *Agronomy for Sustainable Development, 28*, 489–495. https://doi.org/10.1051/agro:2008024
33. Esengun, K., Erdal, G., Gündüz, O., & Erdal, H. (2007). An economic analysis and energy use in stake-tomato production in Tokat province of Turkey. *Renewable Energy, 32*, 1873–1881. https://doi.org/10.1016/j.renene.2006.07.005
34. Moghaddam, P. R., Feizi, H., & Mondani, F. (2011). Evaluation of tomato production systems in terms of energy use efficiency and economical analysis in Iran. *Notulae Botanicae Horti Agrobotanici Cluj-Napoca, 3*, 58–65.
35. Jadidi, M. R., Sabuni, M. S., Homayounifar, M., & Mohammadi, A. (2012). Assessment of energy use pattern for tomato production in Iran: A case study from the Marand region. *Research in Agricultural Engineering, 58*, 50–56. https://doi.org/10.17221/32/2010-rae
36. American Society of Agricultural and Biological Engineers. (2011). *ASAE D497.7 MAR2011 agricultural machinery management data.* ASABE. https://doi.org/10.1126/science.85.2210.446-a.
37. Kitani, O. (1999). *CIGR handbook of agricultural engineering*. Volume V, CIGR–The I, ASAE Publication: St. Joseph, MI, USA.
38. Wells, C. (2001). *Total energy indicators of agricultural sustainability: Dairy farming case study* (Technical Paper 2001/3).
39. Barber, A. (2004). Seven *case study farms: Total energy & carbon indicators for New Zealand Arable & Outdoor Vegetable Production*. p. 46.
40. Saunders, C., Barber, A., & Taylor, G. (2006). *Food miles—Comparative energy/emissions performance of New Zealand's agriculture industry*. Agribusiness & Economics Research Unit, Lincoln University.

Knowledge Elicitation and Modeling of Agroecological Management Strategies

Roger Martin-Clouaire

1 Introduction

Agricultural producers operate in a dynamic and complex environment in which incremental innovation is a standard practice. However, environmental problems (e.g., pollution, reduced access to water, soil degradation), a changing climate, and shifting social demands are large contemporary challenges that may mean this "business as usual" approach is no longer sufficient. Specifically, agricultural producers may need to adapt to new contexts and, more likely, embrace a more profound or transformative change if they are to secure a future that is desirable, viable, and sustainable. The agroecological movement occurs in this transformation to more sustainable agriculture [1–3]. Basically, agroecology is a farming approach that aims at making the most beneficial use of ecological processes within the agroecosystem to produce food and fiber in a sustainable and ethical manner [4].

For many farmers, however, a rapid shift to agroecology is neither possible nor practical. Farming that relies primarily on the natural features of the agroecosystem requires new agricultural practices and management attitude [5]. Farmers attracted by agroecology must be involved in active learning, (re)inventing technologies, and adapting their farming systems and habits. The transition from conventional agriculture to agroecology requires developing tools that can help analyze and design biodiversity-friendly production systems [6, 7] and practices that intentionally use ecosystem services [8–10]. Adopting, discovering, or implementing service-centered agroecological principles requires fundamentally different ways of designing, monitoring, and managing agroecosystems to consider specific features, such as

R. Martin-Clouaire (✉)
INRAE, UR875 Mathématiques et Informatique Appliquées Toulouse, Castanet-Tolosan, France
e-mail: rogermc@club-internet.fr

D. D. Bochtis et al. (eds.), *Information and Communication Technologies for Agriculture—Theme III: Decision*, Springer Optimization and Its Applications 184, https://doi.org/10.1007/978-3-030-84152-2_14

a wide range of partially known and interacting ecological processes, several spatio-temporal scales, and the importance of farmers' cognition, as well as multiple social drivers [6]. Farmers engaged in the transition process can, through professional awareness, experience, and intuition, distinguish what is unsustainable from what is at least more sustainable. Nonetheless, farmers need to understand agroecosystems in more detail by increasing their (i) biophysical knowledge about ecosystem functions [11] and the functions' local expression, and (ii) management skills [12], which help them make multiple decisions in interaction with agroecosystems to achieve desired production goals that are consistent with agroecological principles. Moreover, agroecological innovations are collective and integrative; they are typically based on the co-creation of knowledge, combining science with the traditional, practical, and local knowledge of producers [6].

Agroecology implementation is hampered by knowledge barriers relative to constitutive elements of practices. Knowledge may be missing regarding for instance the likely consequences of an action, its appropriateness regarding short- and long-term objectives, its compatibility with other actions, or its edge on alternative actions depending on the situation. Filling these knowledge gaps can emerge through a slow learning process of farmers based on evidence-based reasoning and critical thinking about their concrete personal experience with agroecological principles. The use of information and communications technologies (ICT) can greatly enhance and speed the learning process through more rigorous handling of various components and processes of agroecosystem, agroecology-focused and design-oriented discussions in groups of farmers, and computer evaluation of management alternatives in a dynamic simulation environment. Integrated simulation models contribute important insights to the analysis of farming systems. Simulation helps unraveling the complex and dynamic interactions and feedbacks among biophysical components across scales. It provides a framework for integrative contributions by functioning as learning platforms in participatory processes.

The objective of this chapter is to capture, in a modeling framework, the management strategies of farmers at a level of detail that enables the operational management process, (i.e., decision-making about and implementation of production-management actions) to be simulated. The approach adopted is inspired by the Belief-Desire-Intention (BDI) theory that provides a conceptual framework structuring the sequential decision-making behavior of rational agents. The theory was initiated by philosophical work of Bratman [13] on practical reasoning (reasoning directed towards action) and further formalized and implemented by artificial intelligence and multi-agent researchers [14–16]. The BDI model states how agents decide, moment by moment, which actions to perform to pursue their goals consistently in time given their beliefs and values.

Although there are several simulation approaches developed to investigate agricultural systems (especially biophysical aspects) few are working up to farm scale [17, 18] as in the present project. The objective is to enable deep exploration of whole-farm management aspects including planning of production activities, goal-based adaptation to circumstances and dynamic allocation of limited resources in operational context of a production strategy implementation. The effectiveness of management strategies for governing agroecological systems depends on a thorough understanding and intelligible representation of management strategies and farmers' decision-making behavior [19–22]. The modeling framework discussed in this chapter must represent the biophysical system as perceived and understood by farmers, the elements of decision-making that underlie the management process, and the required resources that can be used or consumed. The decision-making models built within this framework are only abstractions; they cannot replicate all aspects of human knowledge or reasoning. However, they should make it possible to assess the physical feasibility and risk of intended plans realistically. The model of a specific farm (real or imagined) should help groups explore the planning of new agroecological management strategies in the face of knowledge gaps about the results of these strategies (especially highly innovative ones) and uncertainties about uncontrollable factors (e.g., weather, market, regulations). Learning in this process of adaptive management occurs through the informative practice of management itself, with the current management strategy being altered as understanding improves. In other words, the present approach is driven by knowledge-management objectives to develop a consistent set of processes to represent, organize, and disseminate agroecological knowledge. It uses dynamic simulation models and relies on a knowledge-creating and knowledge-sharing culture embodied in participatory-design approaches [23]. The growth of practical knowledge depends largely on accumulating and organizing information produced by experimental research. We recommend using virtual experiments (e.g., computer simulation) with an "integrated approach," in which progress is made by combining existing generic and site-specific knowledge from multiple domains (e.g., agronomy, ecology, farm-management science). Exploring and disseminating agroecological principles and achievements require uncovering and sharing knowledge among actors in a specific context. Knowledge expressed in the context of problem solving must be explicit for subsequent expansion and use. Some of this knowledge is tacit, which is generally difficult to articulate. Tacit knowledge must be converted into explicit knowledge (e.g., document or model) before it can be discussed, improved, and ultimately used.

This chapter is a preliminary attempt to develop guidelines for eliciting and modeling farmers' knowledge about management of their agroecological systems, which is necessary to test and disseminate the knowledge. Practical, empirical, and tacit knowledge in a partially known domain is best leveraged through social interactions, which justifies our focus on designing management practices through collective efforts in simulation-based participatory workshops.

2 Agroecological Farm Management

A production system such as a farm requires strategic and operational management. Strategic management consists of setting the overall direction of the system: choosing the type of production and configuring the overall long-term course [24]. In comparison, operational management consists of (1) committing to a plan of action to obtain desired and feasible results, (2) making contextual changes to the management strategy (especially contingency plans) to keep the system moving in the chosen direction, and (3) performing short-term actions that comply with the plan and circumstances. By committing to a plan, the farmer can pursue long-term goals and articulate its actions with intermediate sub-goals. Management behavior is also influenced by events that trigger adaptation of the plan when unexpected situations occur.

Farming systems require operational management that conforms to business objectives (e.g., maintain mid-term economic viability, satisfy contracts). Agroecology emphasizes sustainable production strongly by exploiting ecosystem services, maintaining long-term production, making farm work healthier and enjoyable, and contributing to improvement of environmental issues. Sustainable agriculture is more than a set of practices; it requires skillful management that promotes potential beneficial ecological processes while exploiting synergies between them and finding trade-offs when necessary. Farmers' management behaviors must be designed for the purpose and local context of the farm and be adaptable to changing conditions.

There are as many farming systems as there are land-based configurations and types and combinations of management practices. When deciding to implement a farming system, farmers consider interactions among system components and the many spatial and temporal characteristics of possible actions. For instance, establishing a particular crop rotation may change the amounts and types of nutrients to apply and the appropriate pest management. No-till farming generally reduces soil loss and conserves soil moisture but may also change weed management. Irrigation generally increases yields but also the amounts of nutrients that need to be applied. Management decisions that span seasons (e.g., crop rotations) interact dynamically with weather events and determine production constraints (e.g., soil moisture, pest populations).

Cropping practices can be divided into categories of agricultural activities that farmers perform to produce food, fiber, or energy. Major activities include [25]:

- Soil and crop management: determining which varieties, crops, and crop mixtures to grow; their spatial distribution and temporal rotation; and the type of soil tillage (if any), cultivation, and conservation to perform to increase soil quality (e.g., reduce compaction, reduce erosion, and increase organic matter) and conserve soil moisture.
- Pest management: determining weed, insect, disease, and other threats to crop growth and quality, and preventive actions (e.g., crop spacing and intercropping) or remedial actions (e.g., natural pesticides, biological pest control, planting directly in a cover crop or crop residues, and trap cropping) to perform.
- Nutrient management: determining the additional nutrients the soil needs for crop growth and the kind and number of applications (e.g., manure and compost) required to improve soil biological activity, increase crop yields, and reduce the risk of contaminating ground and surface water.

- Water management: determining the amount of water required for crop growth and applying it efficiently (scheduling irrigation) given limitations on water availability, using practices such as a cover crop and mulch to reduce evaporation and decrease soil compaction.

When agroecological principles are applied in a suitable management strategy, several ecological functions (e.g., biodiversity conservation, soil and water conservation, biological pest control) can be generated and used, and yield can be increased and stabilized [25]. Supporting farmers orient the complex dynamics of the processes involved means helping them develop the logic required to make rational management decisions which result from combining their understanding of biophysical functioning with the effects of actions that are expected given farm characteristics.

3 Developing a Farm-Management Model

Exploring the behavior of a new system first involves modeling and simulating the entire existing farming system and then modeling and simulating changes to it. To be relevant to farming-system design, research requires better understanding of the internal human representation of management practices (i.e., the management strategy and its resulting actions in the farm's context). A farming-system management strategy is embodied in how a farmer organizes, coordinates, and prioritizes production activities in time, and how these activities are adapted as production progresses. We assume that a farmer willing to engage in agroecological transition follows goal-oriented management behavior, which raises several issues, including:

- identifying the goals
- understanding how goals and actions are articulated in plans
- deciding on final trade-offs in order to initiate actions from the plan

Thus, a major focus is given on determining the role and pattern of mental models that farmers use to make judgments about pursuing, interrupting, or abandoning intended activities and scheduling executable actions depending on the current situation. Farmers' decision-making behavior, which is based strongly on their management strategies, involves physical and observation activities in the biophysical system, and interactions with the external context (e.g., weather, market, regulations). Most decisions depend on the circumstances but remain consistent with a pre-existing management strategy.

Developing a farm-management model requires eliciting and collecting details about many activities, goals, and available resources:

1. Assess the world surrounding the farm (market opportunities and limitations, funding availability, potential to collaborate with peers, availability of information and technical support, threat from regulations).

2. Assess the resources available, such as physical features of the land (e.g., field size, soil properties, layout, topography, livestock production), machinery, equipment, and amount of labor (farmer, family, employees) available during the year.
3. Define the farm's mission and goals.
4. Describe the "primitive" activities (the most basic unit of activity (i.e., not broken down further)) that describe all possible actions by the farmer, including observation activities (e.g., of pest populations).
5. Examine dependencies between activities (e.g., demand for the same resource; timing, sequencing, and synchronization requirements) and develop a plan of action that can achieve the farm's goals by aggregating the primitive activities (using temporal and procedural operators) applied to spatial entities and the required resources.
6. Establish preferences and priorities to use when selecting activities for immediate execution and allocating resources among competing activities.
7. Identify events that can trigger revision of goals and the plan and incorporate the corresponding adjustment into the previous plan.

In particular, the model designer must identify nominal management behavior (i.e., how activities and resources are coordinated under ideal circumstances). It can be modeled as a collection of primitive activities that are organized in a plan involving different actors and tools [26]. Activities in the plan can be scheduled to occur simultaneously or sequentially, only during certain temporal windows, or with a risk of conflict (e.g., multitasking) or constraints (e.g., infeasible to perform) that must be resolved eventually. Once the nominal behavior is determined, the model designer should consider what could go wrong and then introduce flexibility into the plan by making optional the activities that are not mandatory, by creating one or more branching points, or by enabling changes to the plan in response to events [27]. Since resources are often limited and may be unavailable when needed, deadlines may not be met, especially if the context has changed in an unexpected manner or an external event makes the plan obsolete. In these cases, other management activities may be required beyond the nominal behavior.

4 Decision-Relevant Concepts

This section is devoted to a closer look at the fundamental conceptual ingredients of the modeling framework with which decision-making behavior in farm management can be represented, examined, and discussed as an object of study.

4.1 Activities, Operations, and Resources

Activities are the most important construct in our conceptual model of farm-management behavior [26, 28, 29]. There are two different activity types: primitive and composite. Each primitive activity represents an action to be performed (e.g., grass cutting) to an entity (e.g., a field) by one or more actors. In other words, it denotes a technical operation (something to be done) to be applied to a particular biophysical object or location (if feasibility conditions are satisfied) by executing agents and possibly other resources (e.g., machinery, inputs). A primitive activity has local opening and closing conditions, which are defined by temporal windows and/or predicates (Boolean functions) that refer to biophysical states or indicators. The incremental change to the biophysical system as the operation is performed is a functional attribute of the operation. Operations can be instantaneous or occur over time, in which case they are executed at a context-dependent speed and might be interrupted. A primitive activity can be viewed as an abstract description of an action. Questions to ask about a primitive activity include the following:

- What operation must be performed?
- Where will it be performed?
- How early or in what situation can it be performed?
- When must it be completed?
- How will it be performed?
- How much time, energy, and resources will be required to perform it?

The ability to benefit from organizational and timing flexibility depends on effective execution, which is determined by the resources involved (e.g., operation resources, actors). Carefully representing the resources [27] and their availability is essential to understand the situation and the potential for improvement. In short, a resource is an entity that supports or enables the execution of activities. Typically, resources include those who perform activities, required machinery and facilities, and inputs (e.g., seeds, fertilizer, water, fuel, funding). Resources are, by definition, finite and have a strong influence on when and how activities can be performed. A resource's availability is restricted by constraints that specify conditions under which it can be used or consumed. Constraints are temporal (statically or dynamically determined temporal windows of availability), capacity-related (the amount available), or state-related. Temporal constraints on the availability of actors may be flexible; for instance, working time per day can vary slightly if needed, but total working time per year must comply with strict laws.

Observation activities differ in that they do not change the biophysical system directly; nonetheless, they play an essential role in agroecology. Technology for sustainable farming must emphasize measurement and observation equipment or services (e.g., soil analysis, manure analysis, pest identification) that help farmers assess their situations. It also must focus on larger system scales. The predators and parasitoids that control pests often require habitats larger than those found on small farms. They require instruments and indicators that illustrate ecological states and relationships on and among farms at the landscape scale.

4.2 Goals and Plans

An important principle of most decision-making processes is that the motives for performing management behaviors range from explicit to implicit. A variety of values and goals motivate farmers' actions. Values are more permanent characteristics of farmers, such as traits they want to embody (e.g., integrity, safety, recognition, sustainability). Some goals simply promote these values, while others relate directly to the farmer's production vision or purpose of the farm. Some goals are ends in themselves (e.g., increase free time, improve soil fertility), while others are only means to an end (e.g., introduce a legume cover crop). The goals specify milestones and end points towards or away from which relevant actions should lead. Examples include:

1. state-based goals such as reducing vulnerability to pests, reducing carbon emissions, maintaining soil health, increasing input efficiency, reducing weed pressure, and avoiding severe water stress
2. more abstract goals such as achieving economic viability of the farm, obtaining social approval, maintaining a stable income, increasing the standard of living, creating good conditions in which to pass the farm to the next generation, and maintaining or improving environmental quality.

The following list of goal-related questions helps model designers analyze a goal thoroughly:

- What are the reasons for pursuing this goal, and why do these reasons matter?
- What specific results are intended?
- What benefits are expected from achieving this goal, and what are potential consequences or costs of not doing so?
- How can achievement of this goal be determined?
- How does this goal align with support, or advance the farmer's overall mission, values, and principles?
- When will the goal be achieved? Do milestones need to be met along the way?
- What obstacles could arise while pursuing this goal? What can be done now to prevent them or address them if they do arise?

Goals, once carefully considered, require a response by the farmer in terms of intention to do. Thus, in a planning process, the farmer's goals must be converted into a plan that will achieve them. Decisions regarding the strategic direction of the farm need to be cascaded into operational decisions in a plan that describes what actions are required or suitable and how to coordinate them on the farm in the context in which they will be performed. A plan specifies how to achieve, over time, a series of connected goals in which the achievement of each satisfies an immediate need and provides a steppingstone to a more ultimate goal. Identifying an action that is useful for achieving a goal is not always simple because several actions may be necessary and certain actions may conflict with other goals (e.g., short-term profitability vs. long-term viability). Little is known about the cognitive process that farmers use to relate desired situations to a consistent plan of action. Individual farmers' goals and the goals' perceived relationship with plans are hidden within conscious and unconscious cognitions and emotions. Farmers use some kind of library of recipes rather than plan from first principles. Thus, we do not attempt to model this planning process, especially because:

1. agroecology remains far from a normalized procedure, which would have made it possible to identify elements of the mental model necessary for planning
2. collective design is more complex than design by an individual

Instead, we focus on making characteristics of management behavior (especially goals and plans) explicit to make them objects that can be examined, discussed, and communicated.

Certain characteristics of a goal are useful when developing a consistent and feasible plan of action, such as importance (i.e., attractiveness, relevance, priority), difficulty of achievement, temporal range (i.e., proximal or distal), and specificity (i.e., abstract or concrete). The goals also provide temporally bound and measurable results to be achieved. The timescale may be the seasonal horizon of crop production, or longer.

Activities can be further constrained by expressing temporal relationships among them and the potential for iteration, logic associations, and relaxation [30]. Composite activities can be constructed using programming constructs (e.g., *before*, *co-start*, *iterate*, *or/and optional)* that specify temporal sequence, iteration, alternatives, aggregation, and optional execution of activities (primitive or composite). These constructs connect all activities except the overall plan, which is the highest level composite activity. Committing to a plan transforms it into an intention that flexibly describes the when, where, and how of what is appropriate to achieve the corresponding goals. With this flexibility, plans play a directional role, providing general guidelines, focus, and flexibility in execution. As an ongoing process, plans can also be adjusted in reaction to events. Plans can change, but nonetheless persist once farmers commit to them; they change only when important events occur and when conditions warrant. Farmers' actions are governed by a continuous process of pursuing the goal and adapting it to the changing environment and its contextual constraints. When farmers engage in joint activities, their actions are also governed by the actions and intentions of the other co-actors.

Temporal specifications in plans reveal much about the timing of decisions and actions. Most intra-seasonal decisions correspond to stages of crop growth (e.g., land preparation, sowing, and harvest) and intermediate operations (e.g., fertilization, pest management, and irrigation). However, the timing of decisions and actions are also determined strongly by events, such as completion of an activity or a weather-related event. Often, several sub-plans (conjunction of composite activities) are used to express similar interventions to be performed on different spatial entities (e.g., fields and rows). The order in which these entities are treated may also need to be specified through priorities.

4.3 Preferences and Priorities

Preferences are personal criteria that help to differentiate what is useful, good, beneficial, or important. Preferences influence the choices that individuals make and are a key characteristic of farmers' management behavior. Practical reasoning involves deciding what to do or justifying what one has done. Farmers can influence how the world changes and have preferences for the resulting states. Since their ability to act is limited, they may need to choose among several beneficial actions that, from their viewpoint, would improve the state. Their preferences are essential discriminating criteria. Actions have expected effects but can also have negative consequences; thus, preference-based decision-making is required to decide whether an action's potential trade-offs are acceptable or whether other actions have better trade-offs. Conscious preferences or priorities must be expressed. They are also necessary when choosing the resources to allocate to primitive activities, which restrict which primitive activities can be executed. Doing one thing means that other things cannot be done.

More generally, preferences and priorities convey internal criteria for evaluating alternative goals and primitive activities and for allocating resources. They help to choose among alternatives and resolve conflicts. They are also used to make decisions in response to different kinds of pressure (e.g., finishing an activity, achieving a state, or freeing up resources by a given date), opportunistic or personal attitude (e.g., continuity and harmonization of activities, and cost/benefit considerations), and commitment to the future.

In decision-making, preferences are informally connected to values (i.e., aspirational traits). In action-determination situations, motivational attributes are used to make certain alternatives more preferable or less preferable than others. Prevention (i.e., avoidance) goals may involve sensitivity to loss when punishment is feared, while target goals may involve sensitivity to gain. This attitude must be expressed through preferences and priorities.

4.4 Events and Reactions

Due to uncontrollable factors (especially weather and pest infestations), farmers cannot rely on routine, calendar-based activities. Their actions must be based on observation and prediction. A plan may encounter certain events (e.g., a long drought) that lie outside its bounds. Specifying changes to a nominal plan as a function of circumstances is called "conditional adjustment." Identifying alternative actions that can be performed if the original plan is inadequate due to changing circumstances is crucial for a robust management strategy. Anticipating a suitable change when things do not go as expected is a cautious and safe attitude.

The element that triggers conditional adjustment is either a calendar condition that becomes true on a specific date, or a state-related condition that becomes true when circumstances meet the condition. The adjustment can be any change to the current plan, such as removing or adding activities, or changing the resources used in certain activities. In this way, management can respond rapidly to address unexpected (but still possible) fluctuations in the external environment and other contingencies. Not all (and usually few) elements of a plan need to be stated in full detail; some can emerge spontaneously through conditional adjustment, and interactions between the situations and constraints that are considered each time decisions are made about which actions to perform.

5 Example of an Agroecological Management Strategy

As an example, a virtual farm in the agrarian region west of Toulouse, France, is considered. From his personal experience, the farmer is aware of consequences of conventional agriculture on the environment and food system, as well as of changes in societal demand. The farmer has also experienced increasing threats to cash crop income due to an overall decrease in productivity and climate change. The farmer begins redesigning the farming system by deeply reconsidering its production and management and breaking them down into goals. This reflection prompts the farmer to look into the potential of organic agriculture. Being relatively unfamiliar with it, the farmer adopts a cautious approach to address production risks and decides to convert only one section of the farm, with the initial agronomic objective of improving soil fertility to ensure acceptable yields over long periods without agrochemical inputs. Each field of both the conventional and organic sections then requires an annual or multi-year cropping plan. The planning process is knowledge-intensive and especially challenging for the organic section, which requires new combinations of agroecological practices, techniques, and equipment. For simplicity, the sample plan includes only sowing, weeding, and harvest (Fig. 1), thus excluding storage, distribution, and observation activities (e.g., for weeds, insects, and diseases).

The plan is based on combining the nominal sequence of activities on each field. For example, the sequence for "field1" starts with alfalfa sowing (Fig. 1). Repeatedly alfalfa is then mechanically weeded, if necessary, and then harvested. The plan continues with the iteration of direct drilling of high-stalk wheat into the alfalfa for 2 consecutive years. Primitive activities occur during temporal windows and require resources (Table 1), while iteration activities have specifications (Table 2).

The plan of activities, which is developed or reconsidered once per year, is necessarily tentative. It may need to be changed due to overly wet critical periods, unforeseen lack of availability of required equipment or labor, pest-related events, or important contingencies. Thus, the future behavior described by the plan is accompanied by adjustment rules for making last-minute changes. For instance, if the soil is too wet or too dry in October, alfalfa sowing will need be postponed to [15 Mar–15 Apr]. More generally, if the crop fails to establish, it can be abandoned, replanted, or replaced with a cover crop or another cash crop. The adjustment may also influence

```
And(
    Before(
        Sowing-alfalfa,
        Iterate (
            Before (
                Optional (Mechanical-weeding-alfalfa),
                Harvest-alfalfa)),
        Iterate (
            Before (
                Direct-drilling-wheat-into-alfalfa,
                Harvest-wheat))
        ),
    .
    .
    .
    )
```

Field1

Fig. 1 Simplified 3-year plan for one field. Dots indicate placeholders for other fields

Table 1 Sample temporal windows and material resources involved in primitive activities

Primitive activity	Time window	Resources
Sowing-alfalfa	[15 Sep–30 Sep]	Rotary harrow, alfalfa seed drill, roller
Mechanical-weeding-alfalfa	[1 Mar–15 Mar]	Weeding harrow
Harvest-alfalfa		Mower, drying system, and labor provided by an external company
Direct-drilling-wheat-into-alfalfa	[1 Oct–20 Oct]	Rotary harrow, wheat seed drill, roller
Harvest-wheat	[1 Jul–10 Jul]	Combine

Table 2 Sample characteristics of the two iteration activities

Iteration	Specifications
Alfalfa harvest	First harvest: at bud stage Subsequent harvest: at flowering stage
Wheat sequence	Two times

the dates or conditions of an activity and even the priorities ultimately invoked to determine which actions of the plan to select and perform in the current situation. For instance, in a dry year, priorities among crops and thus among activities may change because some crops carry more risk than others.

The basic concept is to rely on ecosystem services and preventive measures in nature to maintain or improve fertility and regulate pests and diseases in crops. The main issue is to determine the practical actions that can maintain or enhance these ecosystem services to correspond to the farm's current situation. Reduced tillage, intensive use of cover crops, intercropping, living mulches, and rotations containing crops with large amounts of post-harvest residue, are options that need to be tested for each field to assess their effectiveness and feasibility at the farm scale under a variety of weather scenarios. Thus, for each field of the organic section, the farmer should first determine what types of economic-oriented crops, green-manure crops, and cover crops should be planted; when and how they should be planted; and when they should be harvested, killed, or incorporated into the soil. The farmer has a few cash crops that can generate high income. Crops are first determined for fields based on market, agronomic, and logistical considerations. Key crops are assigned to the most suitable fields as long as they do not compromise the fields' soil quality and long-term productivity. The logistics of labor- and equipment-intensive activities (e.g., sowing, harvesting) must be considered when developing an annual plan. In addition to meeting the current year's production goals, the farmer designs crop sequences to prepare for future key crops.

6 Discussion and Conclusion

The modeling framework outlined in this chapter enables farmers to describe their operational decision-making practices that determine crop-production management at the farm scale clearly and non-ambiguously. It also considers characteristics of the farm's land as well as the labor and material required. Farmers need be encouraged to experiment, improve, share, and spread their own perception and knowledge of farm-production processes and how to influence them. The modeling provides important benefits, including the increased range and complexity of management behaviors that can be examined, the additional insights that can be gained for a given management strategy, and a greater ability for expert-based model validation. Explicit representation helps to reveal broad gaps in current understanding that would be overlooked easily in solely verbal assessments.

The cognitive architecture presented can be supplemented with biophysical process models and embedded in a computing environment that simulates event-based dynamics [28]. Using these farm-production models in simulation experiments can help in understanding complexities of interactions of the processes involved. Simulation provides a basis for experiential learning in realistic farming environments. The models can incorporate farmers' estimates for uncertain or incomplete data and farmers' opinions about difficult-to-measure parameters based on personal knowledge and experience. Most studies focus on only one or a few elements of a specific farming system and thus fail to consider trade-offs and conflicts between interacting components. Only a simulation model that represents the entire range of factors that influence farm households can address the behavior and performance of a farm as a whole and how it could evolve if uncontrollable factors change unexpectedly.

A simulation model can help screen out management options that would be inapplicable in the farm context or unacceptably risky given uncertainty about the response to certain actions. Simulation can reduce uncertainty and reveal relationships within the agroecosystem and between the biophysical context and human actions. Simulating the management strategies presented in this chapter can help assess the suitability of their timing, the pertinence of observation activities, and the ability of resource availability to meet demand. Simulation can easily identify a management strategy that relies on overly tight deadlines or overly small temporal windows or that demands too much labor during certain periods. Seasonality tends to place high premiums on timely performance of critical agricultural tasks (e.g., sowing, harvesting). Although the available labor pool may seem large enough to provide the required amount of labor during the year for all crops, significant labor bottlenecks may occur when tasks must be performed quickly at specific times. Labor is frequently the limiting factor, but other resources (e.g., fertilizer, seeds, irrigation water) must also be available at specific times. Simulation is also essential to test the robustness of management strategies in the face of uncertainty in external factors, especially climate change. Since weather is uncertain, farmers choose varieties, crops, or crop mixtures that tolerate weather variations better; these options can be explored deeply. Equally important, simulation helps determine appropriate reactions to weather variations by changing the plan to sow or harvest a second crop in an extremely dry season. Such conditional adjustments require a mental model of context-dependent possibilities in the future.

Efforts to develop effective tools to facilitate participatory elaboration of goal- and plan-based decision-making strategies are increasing as an important step to facilitate cooperation, mutual understanding, and capacity for generating innovative situation-specific solutions [23]. In the context of agroecology, which lacks a universal model that can consider variations identified in practice, it is important to use all knowledge and experience available to build an empirical understanding of what can or must not be done, depending on the context. To this end, farm models play an essential role in focusing discussion and reasoning when used in participatory workshops involving 3–5 farmers and 1–2 facilitators. Dialog is a central element in developing critical thinking and challenging norms and assumptions. The entry point of a workshop is often the farm of one of the participants used as a case study that the group examines by developing and testing ways to transition to an agroecological system. The process involves collective problem solving, with the basic philosophy of learning primarily from each other in the group rather than being taught from the outside. Farmers can increase their understanding by combining existing knowledge from different domains (e.g., agronomy, entomology, ecology, and management) and different perspectives of the workshop participants.

Thus, modeling supports collective experiential learning in complex, realistic farming environments. It contributes to collaborative development of new management strategies due to well-focused peer-to-peer discussions supported by agronomic, environmental, and economic results of simulation experiments. Models are helpful in simulation experiments that map inputs from a specific case study onto outputs that can reveal essential and understandable insights about the resulting system behavior. The model must facilitate exploration of diverse cases and support collective development of sound argumentation that sheds light on the design problem and enables new ideas to emerge. Groups can often be more intelligent than individuals, combining information from a variety of sources, and overcoming individuals' biases, errors, and limitations. Modeling and simulation foster the diversity of cognitive skills in the group and improve participants' ability to understand the situation, learn, and adapt.

There is extensive recognition of the importance of participatory processes to collecting available knowledge, making it explicit, establishing common understanding of system dynamics, and finding agroecological solutions through innovative management actions. This participatory process combines farmer knowledge and experience with scientific knowledge of ecological dynamics and constraints, which can help participants build a shared view of causal relationships that explain the nature of ecosystem-service trade-offs. Capturing differences in knowledge and understanding of biophysical functioning of agroecosystems is a critical advantage of our modeling approach, which combines the potential to represent and communicate management strategies, making it possible to analyze them critically, with arguments for how to improve them. Salient aspects of the target agroecological system can be inspected, criticized, and modified in its corresponding model until the model agrees with the mental models shared by workshop participants. For a framework to function well as an evaluation tool in a participatory context, it must provide a valid and useful representation of system dynamics and be simple enough for non-specialists from different backgrounds to understand and use. This framework provides a powerful way to use both empirical data and farmers' experience to build a mental model collectively to identify important information gaps and develop tentative solutions. Although hypothetical and superficially realistic, a model can organize knowledge in a consistent manner and provide inexpensive and reliable trial-and-error learning. For the model in the approach presented, validation consists of identifying consistency, or rather lack of inconsistency, from the viewpoint and aim of workshop participants.

This modeling approach has some limitations. As mentioned, it cannot replicate innate human knowledge-management skills such as planning. Information management is another essential skill that the model represents superficially. We did not attempt to model automatic learning from experience (i.e., the information that should be stored, how can it be summarized by decision-relevant indicators) although it would be useful in a new domain such as agroecology. In addition, it does not address farm management at the scale of multiple farms, which raises issues of shared goals and collaborative decision-making involving several farmers. While individuals' farm-management decisions are critical to their business, their decisions about coordination with other farmers (e.g., whether to coordinate, with whom, and how) are equally important in agroecology. Although farm-scale coordination decisions are motivated by individual farmers' objectives of increasing profit and/or decreasing risk, they can have far-reaching and complex implications for the overall provision of ecosystem services. This in turn affects the results and long-term sustainability of the farmers' agroecosystems. To accomplish common goals that cannot be achieved by an individual farmer, two or more actors must coordinate their plans and actions.

The inherent strengths and weaknesses of the farm-management models addressed in this chapter have crucial implications for application to agroecological farming-system design. Their recommended use in participatory workshops emphasizes their primary function of supporting elicitation of farmers' agroecological and production-management knowledge. This use promotes dialog among participants when building a model of a specific farm, analyzing simulation results, and revising the model. The dialog during modeling results in shared understanding based on clarifying assumptions and implicit beliefs of the participants' mental models. The discussion after a simulation experiment focuses on collective verification of the consistency of the results and generates intuition for improving aspects of the simulated farm, including the management strategy. Agroecology requires a shift in the way we generate, learn from, and act on evidence. The use of virtual worlds and simulations to enhance the evidence extracted from real-world experiments is clearly promising although more research is needed for large-scale application of the approach.

References

1. Bennett, E. M., Carpenter, S. R., Gordon, L. J., Ramankutty, N., Balvanera, P., Camp-bell, B., Cramer, W., Foley, J., Folke, C., Karlberg, L., Liu, J., Lotze-Campen, H., Mueller, N. D., Peterson, G. D., Polasky, S., Rockström, J., Scholes, R. J., & Spirenburg, M. (2014). Toward a more resilient agriculture. *Solutions, 5*, 65–75.
2. Caporali, F. (2015). History and development of agroecology and theory of agroecosystems. In M. Monteduro, P. Buongiorno, S. Di Benedetto, & A. Isoni (Eds.), *Law and agroecology* (pp. 3–29). Springer-Verlag.
3. Gliessman, S. (2015). *Agroecology: The ecology of sustainable food systems*. CRC Press.
4. Altieri, M. A., Nicholls, C. I., Henao, A., & Lana, M. A. (2015). Agroecology and the design of climate change-resilient farming systems. *Agronomy for Sustainable Development, 35*, 869–890.
5. Sandhu, H., Wratten, S. D., Porter, J. R., Costanza, R., Pretty, J., & Reganold, J. P. (2016). Mainstreaming ecosystem services into future farming. *Solutions, 7*(2), 40–47.
6. Duru, M., Therond, O., Martin, G., Martin-Clouaire, R., Magne, M.-A., Justes, E., Journet, E.-P., Aubertot, J.-N., Savary, S., Bergez, J.-E., & Sarthou, J.-P. (2015). How to implement biodiversity-based agriculture to enhance ecosystem services: A review. *Agronomy for Sustainable Development, 35*, 1259–1281.
7. Garbach, K., Milder, J. C., Montenegro, M., Karp, D. S., & DeKlerck, F. A. (2014). Biodiversity and ecosystem services in agroecosystems. *Encyclopedia of Agriculture and Food Systems, 2*, 21–40.
8. Dendoncker, N., Boeraeve, F., Crouzat, E., Dufrêne, M., König, A., & Barnaud, C. (2018). How can integrated valuation of ecosystem services help understanding and steering agroecological transitions? *Ecology and Society, 23*(1).
9. Haines-Young, R., & Potschin, M. B. (2018). *Common International Classification of Ecosystem Services (CICES) V5.1 and guidance on the application of the revised structure*. Retrieved from www.cices.eu
10. Martin-Clouaire, R. (2018). Ontological foundation of ecosystem services and the human dimension of agroecosystems. *Agricultural Sciences, 9*, 525–545.
11. Zhang, W., Ricketts, T. H., Kremen, C., Carney, K., & Swinton, S. M. (2007). Ecosystem services and dis-services to agriculture. *Ecological Economics, 64*, 253–260.

12. Power, A. G. (2010). Ecosystem services and agriculture: Tradeoffs and synergies. *Philosophical Transactions of the Royal Society B, 365*, 2959–2971.
13. Bratman, M. (1987). *Intentions, plans and practical reason*. Harvard University Press.
14. Bratman, M., Israel, D., & Pollack, M. (1988). Plans and resource-bounded practical reasoning. *Computational Intelligence, 4*, 349–355.
15. Pollock, J. L. (2006). *Thinking about acting: Logical foundations for rational decision making*. Oxford University Press.
16. Wooldridge, M. (2009). *An introduction to multi-agent systems* (2nd ed.). Wiley.
17. Power, B., Rodriguez, D., deVoil, P., Harris, H., & Payero, J. (2011). A multi-field bio-economic model of irrigated grain-cotton farming systems. *Field Crops Research, 124*(2), 171–179.
18. Rotz, C. A., Corson, M. S., Chianese, D. S., Montes, F., Hafner, S. D., & Coiner, C. U. (2013). *The integrated farm system model reference manual, version 4.0*. Retrieved from https://www.ars.usda.gov/SP2UserFiles/Place/19020500/Reference%20Manual.pdf.
19. Edwards-Jones, G. (2006). Modelling farmer decision-making: Concepts, progress and challenges. *Animal Science, 82*, 783–790.
20. Feola, G., & Binder, C. (2010). Towards an improved understanding of farmers' behaviour: The integrative agent-centred (IAC) framework. *Ecological Economics, 69*, 2323–2333.
21. McCown, R. (2012). A cognitive framework to inform delivery of analytic support for farmers' intuitive management under seasonal climatic variability. *Agricultural Systems, 105*, 7–20.
22. Öhlmer, B., Olson, K., & Brehmer, B. (1998). Understanding farmer's decision making processes and improving managerial assistance. *Agricultural Economics, 18*, 173–290.
23. Sautier, M., Piquet, M., Duru, R., & Martin-Clouaire, R. (2017). Exploring adaptations to climate change with stakeholders: A participatory method to design grassland-based farming systems. *Journal of Environmental Management, 193*, 541–550.
24. Martin-Clouaire, R. (2017). Modelling operational decision-making in agriculture. *Agricultural Sciences, 8*, 527–544.
25. Wezel, A. (Ed.). (2017). *Agroecological practices for sustainable agriculture: Principles, applications, and making the transition*. World Scientific.
26. Martin-Clouaire, R., & Rellier, J.-P. (2009). Modelling and simulating work practices in agriculture. *International Journal of Metadata, Semantics and Ontologies, 4*, 42–53.
27. Martin-Clouaire, R., & Rellier, J.-P. (2011). Dynamic resource allocation in farm management simulation. In *19th international congress on modelling and simulation, Perth*, 12–16 December 2011, pp. 808–814.
28. Martin, G., Martin-Clouaire, R., Rellier, J.-P., & Duru, M. (2011). A simulation framework for the design of grassland-based beef-cattle farms. *Environmental Modelling & Software, 26*, 371–385.
29. Martin-Clouaire, R., Rellier, J.-P., Paré, M., Voltz, M., & Biarnès, A. (2016). Modelling management practices in viticulture while considering resource limitations: The Dhivine model. *PLoS One, 11*(3).
30. Martin-Clouaire, R., & Rellier, J.-P. (2012). *Fondements Ontologiques des systèmes pilotés*. Internal Report. Retrieved from http://carlit.toulouse.inra.fr/diese/docs/ri_ontologie.pdf.

www.ingramcontent.com/pod-product-compliance
Ingram Content Group UK Ltd.
Pitfield, Milton Keynes, MK11 3LW, UK
UKHW021304100626
13040UKWH00003B/25
* 9 7 8 3 0 3 0 8 4 1 5 1 5 *